Applied Mathematics:
A Multidisciplinary Approach

Applied Mathematics: A Multidisciplinary Approach

Edited by
Lucas Lincoln

Willford Press
www.willfordpress.com

Published by Willford Press,
118-35 Queens Blvd., Suite 400,
Forest Hills, NY 11375, USA

ISBN: 978-1-68285-428-0

Cataloging-in-Publication Data

Applied mathematics : a multidisciplinary approach / edited by Lucas Lincoln.
p. cm.
Includes bibliographical references and index.
ISBN 978-1-68285-428-0
1. Mathematics. I. Lincoln, Lucas.
QA36 .A67 2018
510--dc23

For information on all Willford Press publications
visit our website at www.willfordpress.com

Willford Press

Contents

Preface

Applied mathematics is the application of mathematical techniques in various fields such as engineering, computer science, etc. Computational modeling techniques and actuarial science rely entirely on applied mathematics to analyze data and information. The field of applied mathematics investigates various branches such as artificial intelligence, electromagnetics and meteorology through a multidisciplinary approach. Through this book, the readers would gain knowledge that would broaden their perspective about applied mathematics. It is a complete source of knowledge on the present status of this important field.

This book has been the outcome of endless efforts put in by authors and researchers on various issues and topics within the field. The book is a comprehensive collection of significant researches that are addressed in a variety of chapters. It will surely enhance the knowledge of the field among readers across the globe.

It gives us an immense pleasure to thank our researchers and authors for their efforts to submit their piece of writing before the deadlines. Finally in the end, I would like to thank my family and colleagues who have been a great source of inspiration and support.

Editor

1

A class of completely monotonic functions involving the gamma and polygamma functions

Bai-Ni Guo[1,*] and Feng Qi[2,3]

*Corresponding author: Bai-Ni Guo, School of Mathematics and Informatics, Henan Polytechnic University, Jiaozuo City, Henan Province, 454010, China
E-mail: bai.ni.guo@gmail.com;
bai.ni.guo@hotmail.com

Reviewing editor: Shaoyong Lai, Southwestern University of Finance & Economics, China

Abstract: In the paper the authors show that neither the function

$$\ln\Gamma(x)-\left(x-\frac{1}{2}\right)\ln x+x-\frac{1}{2}\ln(2\pi)-a_i\left[(-1)^{i+1}\psi^{(i)}(x+b_i)\right]$$

nor its negative is completely monotonic on $(0,\infty)$, where $a_i>0$ and $b_i\ge 0$ are real numbers, $i\ge 2$, $\Gamma(x)$ is the classical Euler's gamma function, and $\psi^{(i)}(x)=\left[\frac{\Gamma'(x)}{\Gamma(x)}\right]^{(i)}$ are polygamma functions. Moreover, some other results are given in the form of remarks.

Subjects: Advanced Mathematics, Analysis - Mathematics, Integral Transforms & Equations, Mathematical Analysis, Mathematics & Statistics, Real Functions, Science, Special Functions

Keywords: complete monotonicity, gamma function, polygamma function

AMS Subject Classifications: Primary 26A48, 33B15, Secondary 44A10

1. Introduction

A function $f(x)$ is said to be completely monotonic on an interval I if it has derivatives of all orders on I and satisfies

$$(-1)^n f^{(n)}(x)\ge 0 \tag{1.1}$$

ABOUT THE AUTHORS

Bai-Ni Guo is a full Professor in Mathematics at Henan Polytechnic University and received the Bachelor degree in Mathematics for Education at Henan Normal University in China. Her research interests are in the areas of mathematical inequalities, mathematical means, special functions, combinatorics, number theory, and the like. She has published over 150 research articles in reputed international journals of mathematics. She is referees and editors of several mathematical journals.

Feng Qi is a full Professor in Mathematics at Tianjin Polytechnic University in China. He was the Founder and the former Head of the School of Mathematics and Informatics at Henan Polytechnic University. He was ever a Visiting Professor at Victoria University in Australia and the University of Hong Kong. He received his PhD degree of Science in Mathematics from the University of Science and Technology of China. He is Editors of several international journals. He has published over 500 research articles in reputed international journals. For more information, please see his home page at http://qifeng618.wordpress.com.

PUBLIC INTEREST STATEMENT

In the paper, the authors deny the complete monotonicity of a class of functions involving the logarithm for the classical Euler gamma function and polygamma functions. A nonnegative and infinitely differentiable funciton is called a completely monotonic function if itself and all of its derivatives change their signs alternatively. A function is completely monotonic on the positive semi-axis if and only if it is a Laplace tranform.

for $x \in I$ and $n \geq 0$. For more information on this class of functions, please refer to Mitrinović, Pečarić, and Fink (1993, Chapter XIII) and Widder (1946, Chapter IV).

It is common knowledge that the classical Euler's gamma function is defined by

$$\Gamma(x) = \int_0^\infty t^{x-1} e^{-t} dt \tag{1.2}$$

for $x > 0$, that the logarithmic derivative of $\Gamma(x)$ is called psi or digamma function and denoted by

$$\psi(x) = \frac{d}{dx} \ln \Gamma(x) = \frac{\Gamma'(x)}{\Gamma(x)} \tag{1.3}$$

for $x > 0$, and that the derivatives $\psi^{(i)}(x)$ for $i \in \mathbb{N}$ and $x > 0$ are called polygamma functions.

In Alzer and Batir (2007, p. 779), Alzer and Batir obtained that

(1) the function

$$G_a(x) = \ln \Gamma(x) - x \ln x + x - \frac{1}{2} \ln(2\pi) + \frac{1}{2} \psi(x+a) \tag{1.4}$$

for $a \geq 0$ is completely monotonic on $(0, \infty)$ if and only if $a \geq \frac{1}{3}$,

(2) so is the function $-G_a(x)$ if and only if $a = 0$.

We remark that the function $G_a(x)$ may be reformulated as

$$G_a(x) = \ln \Gamma(x) - \left(x - \frac{1}{2}\right) \ln x + x - \ln \sqrt{2\pi} + \frac{1}{2}[\psi(x+a) - \ln x] \tag{1.5}$$

In Merkle (1998, Theorem 1), Merkle proved that the function $F_0(x)$ is strictly concave and the function $F_a(x)$ for $a \geq \frac{1}{2}$ is strictly convex on $(0, \infty)$, where

$$F_a(x) = \ln \Gamma(x) - \left(x - \frac{1}{2}\right) \ln x - \frac{1}{12} \psi'(x+a) \tag{1.6}$$

for $a \geq 0$. See also Qi (2010, p. 46, Section 4.3.3).

Stimulated by Merkle (1998, Theorem 1), the authors considered in Qi (2015), its preprint Qi (2013), and Şevli and Batir (2011) the function

$$f_a(x) = \frac{1}{2} \ln(2\pi) - x + \left(x - \frac{1}{2}\right) \ln x - \ln \Gamma(x) + \frac{1}{12} \psi'(x+a) \tag{1.7}$$

for $a \geq 0$ and established that

(1) the function $f_a(x)$ is completely monotonic on $(0, \infty)$ if and only if $a = 0$,
(2) so is the function $-f_a(x)$ if and only if $a \geq \frac{1}{2}$.

After analysing the functions (1.5) and (1.7), we naturally introduce the function

$$f_{a_i, b_i}(x) = \ln \Gamma(x) - \left(x - \frac{1}{2}\right) \ln x + x - \frac{1}{2} \ln(2\pi) - a_i \left[(-1)^{i+1} \psi^{(i)}(x + b_i)\right] \tag{1.8}$$

on $(0, \infty)$, where $a_i > 0$ and $b_i \geq 0$ are real numbers and $i \geq 2$ are integers, and instinctively ask for complete monotonicity of it.

The main result of this paper is that neither $f_{a_i,b_i}(x)$ nor $-f_{a_i,b_i}(x)$ is completely monotonic on $(0,\infty)$, which can be formulated as the following theorem.

THEOREM 1.1 *For all real numbers $a_i>0$ and $b_i\ge 0$ and all integers $i\ge 2$, neither the function $f_{a_i,b_i}(x)$ defined by (1.8) nor its negative is completely monotonic on $(0,\infty)$.*

In the final section of this paper, we also give several remarks about complete monotonicity of $f_{a_i,b_i}(x)$ and other functions related to the gamma function.

2. Proof of Theorem 1.1

The idea of this proof comes from the second proof of Qi (2013, Theorem 3.1) and its formally published version in Qi (2015, Theorem 2).

The famous Binet's first formula of $\ln\Gamma(x)$ for $x>0$ is given by

$$\ln\Gamma(x)=\left(x-\frac{1}{2}\right)\ln x-x+\frac{1}{2}\ln(2\pi)+\theta(x) \tag{2.1}$$

where

$$\theta(x)=\int_0^\infty\left(\frac{1}{e^t-1}-\frac{1}{t}+\frac{1}{2}\right)\frac{e^{-xt}}{t}\,dt \tag{2.2}$$

for $x>0$ is called the remainder of Binet's first formula for the logarithm of the gamma function $\Gamma(x)$. See Magnus (1966, p. 11) or Qi and Guo (2010, p. 462). Combining this with the integral representation

$$\psi^{(k)}(x)=(-1)^{k+1}\int_0^\infty\frac{t^k}{1-e^{-t}}e^{-xt}\,dt \tag{2.3}$$

for $x>0$ and $k\in\mathbb{N}=\{1,2,\cdots\}$, see Abramowitz and Stegun (1972, p. 260, 6.4.1), yields

$$\begin{aligned}
f_{a_i,b_i}(x)&=\theta(x)-a_i\left[(-1)^{i+1}\psi^{(i)}(x+b_i)\right]\\
&=\int_0^\infty\left[\frac{1}{t}\left(\frac{1}{e^t-1}-\frac{1}{t}+\frac{1}{2}\right)-\frac{a_it^ie^{-b_it}}{1-e^{-t}}\right]e^{-xt}\,dt\\
&=\int_0^\infty\left[\frac{1-e^{-t}}{t^{i+1}}\left(\frac{1}{e^t-1}-\frac{1}{t}+\frac{1}{2}\right)-a_ie^{-b_it}\right]\frac{t^ie^{-xt}}{1-e^{-t}}\,dt\\
&=\int_0^\infty\left[\frac{1}{t^{i+1}}\left(\frac{e^{-t}+1}{2}+\frac{e^{-t}-1}{t}\right)-a_ie^{-b_it}\right]\frac{t^ie^{-xt}}{1-e^{-t}}\,dt
\end{aligned}$$

For $t>0$ and $i\ge 2$, let

$$h_i(t)=\frac{1}{t^{i+1}}\left(\frac{e^{-t}+1}{2}+\frac{e^{-t}-1}{t}\right) \tag{2.4}$$

Completely monotonic functions were characterized by Widder (1946, p. 160, Theorem 12a) which reads that a necessary and sufficient condition that $f(x)$ should be completely monotonic in $0\le x<\infty$ is that

$$f(x)=\int_0^\infty e^{-xt}\,d\alpha(t) \tag{2.5}$$

where $\alpha(t)$ is bounded and non-decreasing and the integral converges for $0\le x<\infty$. Therefore, in order to prove complete monotonicity of $f_{a_i,b_i}(x)$ on $(0,\infty)$, it suffices to show the positivity or negativity of the function $h_i(t)-a_ie^{-b_it}$ on $(0,\infty)$, which is equivalent to

$$h_i(t) \gtrless a_i e^{-b_i t}, \quad \frac{1}{t}\ln\frac{h_i(t)}{a_i} \gtrless -b_i t, \quad \frac{1}{t}\ln\frac{h_i(t)}{a_i} \gtrless -b_i, b_i \gtrless -\frac{1}{t}\ln\frac{h_i(t)}{a_i}$$

Because

$$\frac{1}{t}\ln\frac{h_i(t)}{a_i} = \frac{1}{t}\ln\frac{h_1(t)}{a_i t^{i-1}} = \frac{\ln[12h_1(t)]}{t} - \frac{\ln\left(12a_i t^{i-1}\right)}{t}$$

$$\frac{\ln[12h_1(t)]}{t} \to \begin{cases} -\frac{1}{2}, & t \to 0^+, \\ 0, & t \to \infty, \end{cases} \quad \text{and} \quad \frac{\ln\left(12a_i t^{i-1}\right)}{t} \to \begin{cases} -\infty, & t \to 0^+ \\ 0, & t \to \infty \end{cases} \tag{2.6}$$

it follows that

$$\frac{1}{t}\ln\frac{h_i(t)}{a_i} \to \begin{cases} -\infty, & t \to 0^+ \\ 0, & t \to \infty \end{cases}$$

for $i \geq 2$. Thus, if $b_i > 0$ the function $h_i(t) - a_i e^{-b_i t}$ for $i \geq 2$ and any positive number a_i is neither constantly positive nor constantly negative on $(0, \infty)$, so, by Widder (1946, p. 160, Theorem 12a) mentioned above, neither the function $f_{a_i,b_i}(x)$ for $i \geq 2$, $a_i > 0$ and $b_i > 0$ nor its negative is completely monotonic on $(0, \infty)$.

It is easy to calculate that

$$\lim_{t \to 0^+} h_i(t) = \infty \quad \text{and} \quad \lim_{t \to \infty} h_i(t) = 0$$

for $i \geq 2$. Hence, if $b_i = 0$ and $a_i > 0$, the function $h_i(t) - a_i$ for $i \geq 2$ is still neither constantly positive nor constantly negative on $(0, \infty)$. Consequently, neither the function $f_{a_i,0}(x)$ for $i \geq 2$ and $a_i > 0$ nor its negative is completely monotonic on $(0, \infty)$. The proof of Theorem 1.1 is complete.

3. Remarks

In this section, we list several remarks about some functions related to the logarithm of the gamma function $\Gamma(x)$.

Remark 3.1 For $a_i \leq 0$ and $b_i \geq 0$, how about complete monotonicity of the function $f_{a_i,b_i}(x)$ on $(0, \infty)$?

Since the function

$$\frac{1}{e^t - 1} - \frac{1}{t} + \frac{1}{2} \tag{3.1}$$

is positive and increasing on $(0, \infty)$, see Guo and Qi (2009; 2010), Zhang, Guo, and Qi (2009) and closely related references therein, the remainder $\theta(x)$, defined by (2.2), of Binet's first formula for the logarithm of the gamma function $\Gamma(x)$ is clearly completely monotonic on $(0, \infty)$. The formula (2.3) reveals that the functions $(-1)^{i+1}\psi^{(i)}(x)$ are completely monotonic on $(0, \infty)$. As a result of the facts that the sum of finitely many completely monotonic functions and the product of any positive number of completely monotonic functions are both completely monotonic, we obtain that for all real numbers $a_i \leq 0$ and $b_i \geq 0$ the function $f_{a_i,b_i}(x)$ defined by (1.8) is trivially completely monotonic on $(0, \infty)$.

Remark 3.2 Is the constant $\frac{1}{12}$ in (1.7) the best possible? I believe it is.

Repeating the process in the proof of Theorem 1.1 for $i = 1$, we obtain that

(1) if $a_1 > \frac{1}{12}$,

$$\frac{1}{t}\ln\frac{h_1(t)}{a_1} = \frac{\ln[12h_1(t)]}{t} - \frac{\ln(12a_1)}{t} \to \begin{cases} -\infty, & t \to 0^+ \\ 0, & t \to \infty \end{cases} \tag{3.2}$$

(2) if $a_1 < \frac{1}{12}$,

$$\frac{1}{t}\ln\frac{h_1(t)}{a_1} = \frac{\ln[12h_1(t)]}{t} - \frac{\ln(12a_1)}{t} \to \begin{cases} \infty, & t\to 0^+ \\ 0, & t\to\infty \end{cases} \tag{3.3}$$

Combining these with (2.6), we see that only the function $f_{a_1,b_1}(x)$ for $a_1 \le \frac{1}{12}$ and $b_1 \ge 0$ is possible to be completely monotonic on $(0,\infty)$. It is apparent that

$$f_{a_1,b_1}(x) = f_{1/12,b_1}(x) + \left(\frac{1}{12} - a_1\right)\psi'(x+b_1)$$

This means that, for $a_1 \le \frac{1}{12}$ and $b_1 \ge \frac{1}{2}$, the function $f_{a_1,b_1}(x)$ is completely monotonic on $(0,\infty)$. Consequently, since $f_{1/12,b_1} = -f_{b_1}(x)$, the coefficient $\frac{1}{12}$ in (1.7) is the best possible in the sense that the scalar $\frac{1}{12}$ cannot be replaced by a bigger one. This sharpens the main result in Qi (2013; 2015), Şevli and Batir (2011).

Remark 3.3 Is the scalar $\frac{1}{2}$ in the last term of (1.5) the best possible? I believe it is.

For $a \ge 0$ and $b \ne 0$, let

$$G_{a,b}(x) = \theta(x) + b[\psi(x+a) - \ln x] \tag{3.4}$$

on $(0,\infty)$. It is clear that $G_{a,1/2}(x) = G_a(x)$. An easy calculation leads to

$$\begin{aligned} G'_{a,b}(x) &= \theta'(x) + b\left[\psi'(x+a) - \frac{1}{x}\right] \\ &= \int_0^\infty \left[b\left(\frac{te^{-at}}{1-e^{-t}} - 1\right) - \left(\frac{1}{e^t-1} - \frac{1}{t} + \frac{1}{2}\right)\right]e^{-xt}\,dt \\ &= \int_0^\infty \left\{2b\frac{t[te^{(1-a)t} - e^t + 1]}{(t-2)e^t + t + 2} - 1\right\}\left(\frac{1}{e^t-1} - \frac{1}{t} + \frac{1}{2}\right)e^{-xt}\,dt \end{aligned}$$

Let

$$q_a(t) = \frac{t[te^{(1-a)t} - e^t + 1]}{(t-2)e^t + t + 2}$$

for $a \ge 0$ and $t > 0$. By L'Hôspital rule, we have

$$\begin{aligned} \lim_{t\to\infty} q_a(t) &= -\lim_{t\to\infty}\frac{e^{-at}\{(t+1)e^{(1+a)t} - e^{at} + te^t[(a-1)t-2]\}}{(t-1)e^t+1} \\ &= -\lim_{t\to\infty}\frac{e^{-at}\left[(t+2)e^{at} - (a-1)^2t^2 + 4(a-1)t - 2\right]}{t} \\ &= -\lim_{t\to\infty}\left\{1 + \left[a^3t^2 - 2a^2t(t+3) + a\left(t^2+8t+6\right) - 2(t+2)\right]e^{-at}\right\} \\ &= \begin{cases} -1, & a>0 \\ \infty, & a=0 \end{cases} \end{aligned}$$

and $\lim_{t\to 0^+} q_a(t) = 3(1-2a)$. For the function $G_{a,b}(x)$ to be completely monotonic on $(0,\infty)$, it is necessary that $6b(1-2a) - 1 \le 0$ which implies that $a < \frac{1}{2}$ and $b \le \frac{1}{6(1-2a)}$ are both necessary. Since $\frac{1}{6(1-2a)}$ has a minimum $\frac{1}{2}$ for $\frac{1}{3} \le a < \frac{1}{2}$ and $G_{1/2,b}(x)$ is completely monotonic on $(0,\infty)$ if and only if $a \ge \frac{1}{3}$, the coefficient $\frac{1}{2}$ in front of the first bracket in (1.5) is the best possible in the sense that it cannot be substituted by a larger number.

Remark 3.4 Now we outline an alternative proof of the main result in Alzer and Batir (2007, p. 779).

When $b=\frac{1}{2}$, the function $G_{a,1/2}(x)=G_a(x)$. In order to prove that $G_{a,1/2}(x)$ or its negative is completely monotonic on $(0,\infty)$, it is necessary that the function $-G'_{a,1/2}(x)$ or its negative is completely monotonic on $(0,\infty)$. The latter is equivalent to the inequality $q_a(t)\lesseqgtr 1$ on $(0,\infty)$ and this inequality may be rewritten as

$$1-a\lesseqgtr\frac{1}{t}\ln\frac{2[(t-1)e^t+1]}{t^2}$$

The function

$$\frac{1}{t}\ln\frac{2[(t-1)e^t+1]}{t^2}$$

is strictly increasing on $(0,\infty)$, with limits

$$\lim_{t\to0^+}\frac{1}{t}\ln\frac{2[(t-1)e^t+1]}{t^2}=\frac{2}{3}\quad\text{and}\quad\lim_{t\to\infty}\frac{1}{t}\ln\frac{2[(t-1)e^t+1]}{t^2}=1$$

This implies that the function $G_{a,1/2}(x)$ is completely monotonic on $(0,\infty)$ if and only if $a\ge\frac{1}{3}$ and so is the function $-G_{a,1/2}(x)$ if and only if $a=0$.

Remark 3.5 By the way, some complete monotonicity properties of functions involving the gamma and polygamma functions and relating to differences between two remainders $\theta(x)$ of Binet's first formula for the logarithm of the gamma function $\Gamma(x)$ have been investigated in Chen, Qi, and Srivastava (2010, p. 162–163, Section 5), Qi and Guo (2010, p. 464, Section 1.4, Theorem 3) and Chen and Srivastava (2011), Chen and Srivastava, Li, and Manyama (2011), Guo, Qi, and Srivastava (2010), Guo and Qi (2009), Guo, Qi, and Srivastava (2012), Guo, Qi, and Srivastava (2007; 2008), Guo and Srivastava (2009; 2008), Qi, Niu, and Guo (2007), Srivastava, Guo, and Qi (2012). For more information, please refer to the survey articles Qi (2010), Qi and Luo (2012) and plenty of references therein.

Remark 3.6 Finally, we pose an open problem. Discuss complete monotonicity of the functions

$$\ln\Gamma(x)-\left(x-\frac{1}{2}\right)\ln x+x-\frac{1}{2}\ln(2\pi)-a_i\left[(-1)^i\psi^{(i)}(x+b_i)+\frac{(i-1)!}{x^i}\right] \tag{3.5}$$

and their negatives on $(0,\infty)$, where $a_i>0$ and $b_i\ge0$ are real numbers and $i\ge1$ are integers.

Funding
Feng Qi was partially supported by the Natural Science Foundation under grant number 2014JQ1006 of Shaanxi Province of China and by the National Natural Science Foundation under grant number 11361038 of China.

Author details
Bai-Ni Guo[1]
E-mail: bai.ni.guo@gmail.com; bai.ni.guo@hotmail.com
Feng Qi[2,3]
E-mail: qifeng618@gmail.com; qifeng618@hotmail.com; qifeng618@qq.com
[1] School of Mathematics and Informatics, Henan Polytechnic University, Jiaozuo City, Henan Province, 454010, China.
[2] College of Mathematics, Inner Mongolia University for Nationalities, Tongliao City, Inner Mongolia Autonomous Region, 028043, China.
[3] Department of Mathematics, College of Science, Tianjin Polytechnic University, Tianjin City, 300387, China.

References

Abramowitz, M., & Stegun, I. A. (Eds.). (1972). *Mathematical functions with formulas, graphs, and mathematical tables.* Vol. 55, applied mathematics series, Tenth printing, with corrections. Washington, DC: National Bureau of Standards.

Alzer, H., & Batir, N. (2007). Monotonicity properties of the gamma function. *Applied Mathematics Letters, 20,* 778–781. Retrieved from http://dx.doi.org/10.1016/j.aml.2006.08.026

Chen, C.-P., Qi, F., & Srivastava, H. M. (2010). Some properties of functions related to the gamma and psi functions. *Integral Transforms and Special Functions, 21,* 153–164. Retrieved from http://dx.doi.org/10.1080/10652460903064216

Chen, C.-P., & Srivastava, H. M. (2011). Some inequalities and monotonicity properties associated with the gamma and psi functions and the Barnes G-function. *Integral Transforms and Special Functions, 22,* 1–15. Retrieved from http://dx.doi.org/10.1080/10652469.2010.483899

Chen, C.-P., Srivastava, H. M., Li, L., & Manyama, S. (2011). Inequalities and monotonicity properties for the psi (or digamma) function and estimates for the Euler-Mascheroni constant. *Integral Transforms and Special Functions, 22*, 681–693. Retrieved from http://dx.doi.org/10.1080/10652469.2010.538525

Guo, B.-N., Qi, F., & Srivastava, H. M. (2010). Some uniqueness results for the non-trivially complete monotonicity of a class of functions involving the polygamma and related functions. *Integral Transforms and Special Functions, 21*, 849–858. Retrieved from http://dx.doi.org/10.1080/10652461003748112

Guo, S., & Qi, F. (2009). A class of completely monotonic functions related to the remainder of Binet's formula with applications. *Tamsui Oxford Journal of Mathematical Sciences, 25*, 9–14.

Guo, S., Qi, F., & Srivastava, H. M. (2012). A class of logarithmically completely monotonic functions related to the gamma function with applications. *Integral Transforms and Special Functions, 23*, 557–566. Retrieved from http://dx.doi.org/10.1080/10652469.2011.611331

Guo, S., Qi, F., & Srivastava, H. M. (2007). Necessary and sufficient conditions for two classes of functions to be logarithmically completely monotonic. *Integral Transforms and Special Functions, 18*, 819–826. Retrieved from http://dx.doi.org/10.1080/10652460701528933

Guo, S., Qi, F., & Srivastava, H. M. (2008). Supplements to a class of logarithmically completely monotonic functions associated with the gamma function. *Applied Mathematics and Computation, 197*, 768–774. Retrieved from http://dx.doi.org/10.1016/j.amc.2007.08.011

Guo, S., & Srivastava, H. M. (2009). A certain function class related to the class of logarithmically completely monotonic functions. *Mathematical and Computer Modelling, 49*, 2073–2079. Retrieved from http://dx.doi.org/10.1016/j.mcm.2009.01.002

Guo, S., & Srivastava, H.M. (2008). A class of logarithmically completely monotonic functions. *Applied Mathematics Letters, 21*, 1134–1141. Retrieved from http://dx.doi.org/10.1016/j.aml.2007.10.028

Magnus, W., Oberhettinger, F., & Soni, R. P. (1966). *Formulas and theorems for the special functions of mathematical physics*. Berlin: Springer. Retrieved from http://dx.doi.org/10.1216/rmjm/1181071755

Merkle, M. (1998). Convexity, Schur-convexity and bounds for the gamma function involving the digamma function. *Rocky Mountain Journal of Mathematics, 28*, 1053–1066.

Mitrinović, D. S., Pečarić, J. E., & Fink, A. M. (1993). *Classical and new inequalities in analysis*. Dordrecht: Kluwer Academic Publishers.

Qi, F. (2010). Bounds for the ratio of two gamma functions. *Journal of Inequalities and Applications*. Article ID 493058. 84 pp. Retrieved from http://dx.doi.org/10.1155/2010/493058

Qi, F. (2013). *A completely monotonic function involving the gamma and tri-gamma functions*. Retrieved from http://arxiv.org/abs/1307.5407

Qi, F. (2015, in press). A completely monotonic function involving the gamma and tri-gamma functions. *Applied Mathematics & Information Sciences, 8*.

Qi, F., & Guo, B.-N. (2010). Some properties of extended remainder of Binet's first formula for logarithm of gamma function. *Mathematica Slovaca, 60*, 461–470. Retrieved from http://dx.doi.org/10.2478/s12175-010-0025-7

Qi, F., & Luo, Q.-M. (2012). Bounds for the ratio of two gamma functions-From Wendel's and related inequalities to logarithmically completely monotonic functions. *Banach Journal of Mathematical Analysis, 6*, 132–158.

Qi, F., Niu, D.-W., & Guo, B.-N. (2007). Monotonic properties of differences for remainders of psi function. *International Journal of Pure and Applied Mathematical Sciences, 4*, 59–66.

Şevli, H., & Batir, N. (2011). Complete monotonicity results for some functions involving the gamma and polygamma functions. *Mathematical and Computer Modelling, 53*, 1771–1775. Retrieved from http://dx.doi.org/10.1016/j.mcm.2010.12.055

Srivastava, H. M., Guo, S., & Qi, F. (2012). Some properties of a class of functions related to completely monotonic functions. *Computers & Mathematics with Applications, 64*, 1649–1654. Retrieved from http://dx.doi.org/10.1016/j.camwa.2012.01.016

Widder, D. V. (1946). *The Laplace transform*. Princeton, NJ: Princeton University Press.

Zhang, S.-Q., Guo, B.-N., & Qi, F. (2009). A concise proof for properties of three functions involving the exponential function. *Applied Mathematics E-Notes, 9*, 177–183.

2

On some inequalities involving Turán-type inequalities

Piyush Kumar Bhandari[1*§] and S.K. Bissu[2§]
*Corresponding author: Piyush Kumar Bhandari, Department of Mathematics, Shrinathji Institute of Technology & Engineering, Nathdwara, Rajasthan 313301, India
E-mail: bhandari1piyush@gmail.com
Reviewing editor: Regina S. Burachik, University of South Australia, Australia

Abstract: Using a new form of the Cauchy–Bunyakovsky–Schwarz inequality, we prove inequalities involving Turán-type inequalities for some special functions.

Subjects: Applied mathematics; Mathematics & statistics; Science

Keywords: a form of Cauchy–Bunyakovsky–Schwarz inequality; Turán-type inequalities; polygamma functions; exponential integral function; Abramowitz's function
2010 Mathematics subject classifications: Primary 26D07; Secondary 33B15

1. Introduction

The integral representation of well-known Cauchy–Bunyakovsky–Schwarz inequality (see, for instance, Mitrinović, Pečarić, & Fink, 1993) in the space of continuous real-valued functions $C([a,b],\ \mathbb{R})$ is given by:

$$\left(\int_a^b u^{\frac{1}{2}}(t)v^{\frac{1}{2}}(t)\,dt\right)^2 \le \left(\int_a^b u(t)\,dt\right)\left(\int_a^b v(t)\,dt\right) \tag{1}$$

It is well known that the Cauchy–Bunyakovsky–Schwarz inequality plays an important role in different branches of modern mathematics such as Hilbert space theory, classical real and complex analysis, numerical analysis, qualitative theory of differential equations and probability and statistics. To date, a large number of generalisations and refinements of this inequality have been investigated in the literature, e.g. (Alzer, 1999; Callebaut, 1965; Masjed-Jamei, 2009; Masjed-Jamei, Dragomir, & Srivastava, 2009; Steiger, 1969; Zheng, 1998).

ABOUT THE AUTHORS

Piyush Kumar Bhandari is working as an assistant professor in the department of Mathematics, Shrinathji Institute of Technology & Engineering, Nathdwara, Rajasthan, India. He received his MSc degree in Mathematics from M.L. Sukhadia University, Udaipur in 2000, cleared CSIR-UGC NET in June 2001. He is pursuing his PhD in the field of "Inequalities and Special Function".

S.K. Bissu is working as an associate professor in department of Mathematics, Government College, Ajmer. He received his MSc degree in 1987 from University of Rajasthan, Jaipur and PhD degree in 1992 from M.L. Sukhadia University, Udaipur. He has published 18 research papers in national and international journals and has written several books of Board of Secondary Education, Rajasthan, for secondary and senior secondary levels. His area of interest is "Inequalities and Special Function, Fractional calculus".

PUBLIC INTEREST STATEMENT

In this paper, we prove inequalities involving Turán-type inequalities for some special functions using a new form of the Cauchy–Bunyakovsky–Schwarz inequality. These inequalities play an important role in different branches of modern mathematics such as Hilbert space theory, classical real and complex analysis, numerical analysis, probability and statistics. Also, Turán-type inequalities have important applications in complex analysis, number theory, theory of mean values or statistics and control theory.

Also, the importance, in many fields of mathematics, of the inequalities of the type:

$$f_n(x)f_{n+2}(x) - f_{n+1}^2(x) \le 0 \tag{2}$$

$n = 0, 1, 2, \ldots$ is well known. They are named, by Karlin and Szegö, Turán-type inequalities because the first of this type of inequalities was proved by Turán, 1950).

Laforgia and Natalini, 2006) used the following form of the Schwarz inequality (1):

$$\left(\int_a^b g(t)\, f^{\frac{m+n}{2}}(t)\, dt\right)^2 \le \left(\int_a^b g(t) f^m(t)\, dt\right)\left(\int_a^b g(t) f^n(t)\, dt\right) \tag{3}$$

to establish some new Turán-type inequalities involving the special functions as gamma, polygamma functions and Riemann's zeta function. Here, f and g are non-negative functions of a real variable and m and n belong to a set S of real numbers, such that the involved integrals in Equation (3) exist.

In this context, we have the idea to replace $u(t)$ and $v(t)$ in (1) by $g(t)h^{\alpha x}(t)f^{\nu}(t)$ and $g(t)h^{(2-\alpha)x}(t)f^{\mu}(t)$, respectively, to introduce the following new inequality:

$$\left(\int_a^b g(t)h^x(t)f^{\frac{\nu+\mu}{2}}(t)\, dt\right)^2 \le \left(\int_a^b g(t)h^{\alpha x}(t)f^{\nu}(t)\, dt\right)\left(\int_a^b g(t)h^{(2-\alpha)x}(t)f^{\mu}(t)\, dt\right) \tag{4}$$

in which $\alpha, \nu, \mu \in \mathbb{R}$ and g, h, f are real integrable functions, such that the involved integrals in Equation (4) exist.

For $h(t) = 1$, or $x = 0$, our new inequality Equation (4) reduces to the inequality Equation (3).

The aim of this paper is to apply the inequality (4) for some well-known special functions in order to get inequalities involving Turán-type inequalities.

2. The results

In this section, we apply the inequality Equation (4) to prove inequalities involving Turán-type inequalities for n-th derivative of gamma function and the Remainder of the Binet's first formula for $\ln \Gamma(x)$, polygamma functions, exponential integral function, Abramowitz's function and modified Bessel function of second kind.

2.1. An inequality for the n-th derivative of gamma function

THEOREM 2.1 *For every real number $x \in (0, \infty)$, $\alpha \in (0, 2)$ and for every integer $\nu, \mu \ge 1$, such that $\nu + \mu$ is even, it holds for the n-th derivative of gamma function:*

$$\left(\Gamma^{\left(\frac{\nu+\mu}{2}\right)}(x)\right)^2 \le \Gamma^{(\nu)}(\alpha x)\, \Gamma^{(\mu)}\left((2-\alpha)x\right)$$

Proof The classical Euler gamma function is defined for $x > 0$ as:

$$\Gamma(x) = \int_0^\infty t^{x-1} e^{-t}\, dt \tag{5}$$

By differentiating Equation (5), we obtain, for $n = 1, 2, 3, \ldots$

$$\Gamma^{(n)}(x) = \int_0^\infty e^{-t}\, t^{x-1} \log^n(t)\, dt \tag{6}$$

Hence, if we replace $g(t) = e^{-t}t^{-1}$, $h(t) = t$, $f(t) = \log t$ and $[a, b] = [0, \infty)$ in the inequality Equation (4), we get:

$$\left(\int_0^\infty t^{-1}e^{-t}t^{x} \log^{\frac{\nu+\mu}{2}}(t)\, dt\right)^2 \le \left(\int_0^\infty t^{-1}e^{-t}t^{\alpha x} \log^{\nu}(t)\, dt\right)\left(\int_0^\infty t^{-1}e^{-t}t^{(2-\alpha)x} \log^{\mu}(t)\, dt\right)$$

$$\Rightarrow \left(\int_0^\infty e^{-t}t^{x-1} \log^{\frac{\nu+\mu}{2}}(t)\, dt\right)^2 \le \left(\int_0^\infty e^{-t}t^{\alpha x-1} \log^{\nu}(t)\, dt\right)\left(\int_0^\infty e^{-t}t^{(2-\alpha)x-1} \log^{\mu}(t)\, dt\right)$$

By applying Equation (6) in the above inequality, the following result will eventually be obtained:

$$\left(\Gamma^{\left(\frac{\nu+\mu}{2}\right)}(x)\right)^2 \le \Gamma^{(\nu)}(\alpha x)\Gamma^{(\mu)}\left((2-\alpha)x\right) \tag{7}$$

$\forall \alpha \in (0, 2)$, $x > 0$ and for every integer $\nu, \mu \ge 1$, such that $\nu + \mu$ is even.

In particular, for $\alpha = 1$ and $\mu = \nu + 2$, it obtains the Turán-type inequality for $\nu \in \mathbb{N}$:

$$\left(\Gamma^{(\nu+1)}(x)\right)^2 \le \Gamma^{(\nu)}(x)\Gamma^{(\nu+2)}(x)$$

For instance, substituting $\alpha = \frac{1}{2}$, $\nu = 4$ and $\mu = 2$ in Equation (7), we get:

$$\left(\Gamma^{(3)}(x)\right)^2 \le \Gamma^{(4)}\left(\frac{1}{2}x\right)\Gamma^{(2)}\left(\frac{3}{2}x\right) \quad \forall\, x > 0 \tag{8}$$

2.2. An inequality for the polygamma function

THEOREM 2.2 *For every real number $x \in (0, \infty)$, $\alpha \in (0, 2)$ and for every integer $\nu, \mu \ge 1$, such that $\nu + \mu$ is even, it holds for the polygamma functions:*

$$\left(\psi^{\left(\frac{\nu+\mu}{2}\right)}(x)\right)^2 \le \psi^{(\nu)}(\alpha x)\psi^{(\mu)}\left((2-\alpha)x\right).$$

Proof As we know, the polygamma functions $\psi^{(n)}(x) = \frac{d^n\psi(x)}{dx^n}$, where $n = 1, 2, 3, \ldots$, are defined as the n-th derivative of the Psi function $\psi(x) = \frac{d}{dx} \ln \Gamma(x) = \frac{\Gamma'(x)}{\Gamma(x)}$, $(x > 0)$ with the usual notation for the gamma function and has an integral representation (Nikiforov & Uvarov, 1988) as:

$$\psi^{(n)}(x) = (-1)^{n+1} \int_0^\infty \frac{t^n}{1-e^{-t}} e^{-xt}\, dt \quad (n = 1, 2, \ldots; x > 0). \tag{9}$$

Now, if $g(t) = \frac{1}{1-e^{-t}}$, $h(t) = e^{-t}$ and $f(t) = t$ are substituted in inequality Equation (4) for $[a, b] = [0, \infty)$, the following inequality is derived:

$$\left(\int_0^\infty \frac{t^{\frac{\nu+\mu}{2}}}{1-e^{-t}} e^{-xt}\, dt\right)^2 \le \left(\int_0^\infty \frac{t^{\nu}}{1-e^{-t}} e^{-\alpha xt}\, dt\right)\left(\int_0^\infty \frac{t^{\mu}}{1-e^{-t}} e^{-(2-\alpha)xt}\, dt\right)$$

By the definition Equation (9), this is equivalent to:

$$\left(\psi^{\left(\frac{\nu+\mu}{2}\right)}(x)\right)^2 \le \psi^{(\nu)}(\alpha x)\,\psi^{(\mu)}\big((2-\alpha)x\big) \tag{10}$$

$\forall \alpha \in (0,\ 2)$, $x > 0$ and for every integer $\nu, \mu \ge 1$, such that $\nu + \mu$ is even.

In the particular case, for $\alpha = 1$ and $\mu = \nu + 2$, it obtains the Turán-type inequality for $\nu \in \mathbb{N}$:

$$\left(\psi^{(\nu+1)}(x)\right)^2 \le \psi^{(\nu)}(x)\ \psi^{(\nu+2)}(x).$$

2.3. An inequality for the n-th derivative of the remainder of the Binet's first formula for ln $\Gamma(x)$

THEOREM 2.3 *For every real number $x \in (0,\ \infty)$, $\alpha \in (0,\ 2)$ and for every integer $\nu, \mu \ge 1$, such that $\nu + \mu$ is even, it holds for the n-th derivative of the remainder of the Binet's first formula for the logarithm of the gamma function, i.e.* ln $\Gamma(x)$*:*

$$\theta^{\frac{\nu+\mu}{2}}(x) \le \theta^{\nu}(\alpha x)\,\theta^{\mu}\big((2-\alpha)x\big).$$

Proof Binet's first formula for ln $\Gamma(x)$ is given by:

$$\log \Gamma(x) = (x - 1/2)\ \log x - x + \log\ \sqrt{2\pi} + \theta(x)$$

For $x > 0$, where the function:

$$\theta(x) = \int_0^\infty \left(\frac{1}{e^t - 1} - \frac{1}{t} + \frac{1}{2}\right)\frac{e^{-xt}}{t}\,dt \tag{11}$$

is known as the remainder of the Binet's first formula for the logarithm of the gamma function; see (Abramowitz & Stegun, 1965).

By differentiating Equation (11), we obtain, for every positive integer $n \ge 1$.

$$\theta^{(n)}(x) = (-1)^n \int_0^\infty \left(\frac{1}{e^t - 1} - \frac{1}{t} + \frac{1}{2}\right)t^{n-1}e^{-xt}\,dt \tag{12}$$

Hence, if $g(t) = \frac{1}{t}\left(\frac{1}{e^t-1} - \frac{1}{t} + \frac{1}{2}\right)$, $h(t) = e^{-t}$, $f(t) = t$ and $[a,\ b] = [0,\ \infty)$, are considered in inequality Equation (4), then we get:

$$\left\{\int_0^\infty \frac{1}{t}\left(\frac{1}{e^t - 1} - \frac{1}{t} + \frac{1}{2}\right)t^{\frac{\nu+\mu}{2}}e^{-xt}\,dt\right\}^2$$

$$\le \left\{\int_0^\infty \frac{1}{t}\left(\frac{1}{e^t - 1} - \frac{1}{t} + \frac{1}{2}\right)t^{\nu}e^{-\alpha xt}\,dt\right\}\left\{\int_0^\infty \frac{1}{t}\left(\frac{1}{e^t - 1} - \frac{1}{t} + \frac{1}{2}\right)t^{\mu}e^{-(2-\alpha)xt}\,dt\right\}$$

$$\Rightarrow \left\{\int_0^\infty \left(\frac{1}{e^t - 1} - \frac{1}{t} + \frac{1}{2}\right)\ t^{\frac{\nu+\mu}{2}-1}e^{-xt}\,dt\right\}^2$$

$$\le \left\{\int_0^\infty \left(\frac{1}{e^t - 1} - \frac{1}{t} + \frac{1}{2}\right)t^{\nu-1}e^{-\alpha xt}\,dt\right\}\left\{\int_0^\infty \left(\frac{1}{e^t - 1} - \frac{1}{t} + \frac{1}{2}\right)t^{\mu-1}e^{-(2-\alpha)xt}\,dt\right\}$$

By Equation (12), this is transformed to:

$$\left(\theta^{\left(\frac{\nu+\mu}{2}\right)}(x)\right)^2 \le \theta^{(\nu)}(\alpha x)\theta^{(\mu)}\left((2-\alpha)x\right) \tag{13}$$

$\forall \alpha \in (0,\ 2), x > 0$ and for every integer $\nu, \mu \ge 1$, such that $\nu + \mu$ is even.

In particular, for $\alpha = 1$ and $\mu = \nu + 2$, it obtains the Turán-type inequality for $\nu \in \mathbb{N}$:

$$\left(\theta^{(\nu+1)}(x)\right)^2 \le \theta^{(\nu)}(x)\theta^{(\nu+2)}(x)$$

2.4. An inequality for the exponential integral function

THEOREM 2.4 *For every real number $x \in (0,\ \infty)$, $\alpha \in (0,\ 2)$ and for every integer $\nu, \mu \ge 0$, such that $\nu + \mu$ is even, it holds for the exponential integral function:*

$$\left(E_{\frac{\nu+\mu}{2}}(x)\right)^2 \le E_\nu(\alpha x)\, E_\mu\left((2-\alpha)x\right)$$

Proof If we consider the exponential integral function [11, p. 228, 5.1.4] with the following integral representation:

$$E_n(x) = \int_1^\infty e^{-xt} t^{-n}\, dt,\ (n = 0, 1, \ldots; x > 0) \tag{14}$$

and then replace $g(t) = 1$, $h(t) = e^{-t}$ and $f(t) = t^{-1}$ for $[a,\ b] = [1,\ \infty)$ in inequality Equation (4), we obtain:

$$\left(\int_1^\infty e^{-xt} t^{-\frac{\nu+\mu}{2}}\, dt\right)^2 \le \left(\int_1^\infty e^{-\alpha x t} t^{-\nu}\, dt\right)\left(\int_1^\infty e^{-(2-\alpha)xt} t^{-\mu}\, dt\right)$$

Using Equation (14), this is in fact equivalent to:

$$\left(E_{\frac{\nu+\mu}{2}}(x)\right)^2 \le E_\nu(\alpha x)\, E_\mu\left((2-\alpha)x\right) \tag{15}$$

$\forall \alpha \in (0,\ 2)$, $x > 0$ and for every integer $\nu, \mu \ge 0$, such that $\nu + \mu$ is even.

In particular, for $\alpha = 1$ and $\mu = \nu + 2$, it obtains the Turán-type inequality for $\nu \in \mathbb{N}$:

$$\left(E_{\nu+1}(x)\right)^2 \le E_\nu(x)\, E_{\nu+2}(x)$$

2.5. An inequality for the Abramowitz's function

THEOREM 2.5 *For every real number $x \ge 0$, $\alpha \in [0, 2]$ and for every non-negative integer ν and μ, such that $\nu + \mu$ is even, it holds for the Abramowitz function:*

$$\left(f_{\frac{\nu+\mu}{2}}(x)\right)^2 \le f_\nu(\alpha x)\, f_\mu\left((2-\alpha)x\right)$$

Proof The Abramowitz's function (Abramowitz & Stegun, 1965) which has been used in many fields of physics, as the theory of the field of particle and radiation transform, is defined as:

$$f_n(x) = \int_0^\infty t^n e^{-t^2 - xt^{-1}}\, dt \tag{16}$$

where n is a non-negative integer and $x \ge 0$.

Now, applying inequality Equation (4) for $g(t) = e^{-t^2}$, $h(t) = e^{-t^{-1}}$, $f(t) = t$ and $[a, b] = [0, \infty)$ results in:

$$\left(\int_0^\infty t^{\frac{\nu+\mu}{2}} e^{-t^2-xt^{-1}}\,dt\right)^2 \le \left(\int_0^\infty t^\nu e^{-t^2-\alpha xt^{-1}}\,dt\right)\left(\int_0^\infty t^\mu e^{-t^2-(2-\alpha)xt^{-1}}\,dt\right)$$

Therefore, according to Equation (16), one can finally arrive at:

$$\left(f_{\frac{\nu+\mu}{2}}(x)\right)^2 \le f_\nu(\alpha x) f_\mu((2-\alpha)x) \tag{17}$$

$\forall \alpha \in [0, 2]$, $x \ge 0$ and for every non-negative integer ν and μ, such that $\nu + \mu$ is even.

In particular, for $\alpha = 1$ and $\mu = \nu + 2$, it obtains the Turán-type inequality for $\nu \in \mathbb{N}$:

$$(f_{\nu+1}(x))^2 \le f_\nu(x)\, f_{\nu+2}(x)$$

2.6. An inequality for modified Bessel function of second kind

THEOREM 2.6 *For every real number $x \in (0, \infty)$, $\alpha \in (0, 2)$, $\nu > -1/2$ and $\mu > -1/2$, it holds for the modified Bessel function of second kind:*

$$K^2\left\{\frac{\nu+\mu}{2}; x\right\} \le \frac{\Gamma\left\{\nu+\frac{1}{2}\right\}\Gamma\left\{\mu+\frac{1}{2}\right\}}{(\alpha)^\nu(2-\alpha)^\mu\Gamma^2\left\{\frac{\nu+\mu}{2}+\frac{1}{2}\right\}} K\{\nu; \alpha x\}\, K\{\mu; (2-\alpha)x\}$$

Proof It is known that the modified Bessel function of second kind (Nikiforov & Uvarov, 1988) can be represented by the following relations for $x > 0$ and $\upsilon > -1/2$:

$$K_\upsilon(x) = K(\upsilon; x) = \frac{\sqrt{\pi}(x/2)^\upsilon}{\Gamma(\upsilon+1/2)} \int_1^\infty e^{-xt}\left(t^2-1\right)^{\upsilon-1/2} dt \tag{18}$$

By substituting $g(t) = \left(t^2-1\right)^{-1/2}$, $h(t) = e^{-t}$ and $f(t) = \left(t^2-1\right)$ in inequality Equation (4) for $[a, b] = [1, \infty)$, we obtain:

$$\left(\int_1^\infty e^{-xt}\left(t^2-1\right)^{\frac{\nu+\mu}{2}-\frac{1}{2}} dt\right)^2 \le \left(\int_1^\infty e^{-\alpha xt}\left(t^2-1\right)^{(\nu-1/2)} dt\right)\left(\int_1^\infty e^{-(2-\alpha)xt}\left(t^2-1\right)^{(\mu-1/2)} dt\right)$$

Corresponding to definition Equation (18), the following result after simplification eventually yields:

$$K^2\left\{\frac{\nu+\mu}{2}; x\right\} \le \frac{\Gamma\left\{\nu+\frac{1}{2}\right\}\Gamma\left\{\mu+\frac{1}{2}\right\}}{(\alpha)^\nu(2-\alpha)^\mu\Gamma^2\left\{\frac{\nu+\mu}{2}+\frac{1}{2}\right\}} K\{\nu; \alpha x\}\, K\{\mu; (2-\alpha)x\} \tag{19}$$

provided that $x > 0$, $\alpha \in (0, 2)$, $\nu > -1/2$ and $\mu > -1/2$.

In the particular case for $\alpha = 1$ and $\mu = \nu + 2$, it obtains the Turán-type inequality:

$$K^2\{\nu+1; x\} \le \frac{\left(\nu+\frac{3}{2}\right)}{\left(\nu+\frac{1}{2}\right)} K\{\nu; x\}\, K\{\nu+2; x\}, \quad \forall\, \nu > -1/2 \tag{20}$$

Acknowledgement
The authors appreciate the anonymous referees for their careful corrections to and valuable comments on the original version of this paper.

Funding
The authors received no direct funding for this research.

Author details
Piyush Kumar Bhandari[1]
E-mail: bhandari1piyush@gmail.com
ORCID ID: http://orcid.org/0000-0001-9656-5441
S.K. Bissu[2]
E-mail: susilkbissu@gmail.com
ORCID ID: http://orcid.org/0000-0001-8856-9872
[1] Department of Mathematics, Shrinathji Institute of Technology & Engineering, Nathdwara, Rajasthan 313301, India.
[2] Department of Mathematics, Government College of Ajmer, Ajmer, Rajasthan 305001, India.
§ All authors contributed equally and significantly in writing this article. All authors read and approved the final manuscript.

References

Abramowitz, M., & Stegun, I. A. (Eds.). (1965). *Handbook of mathematical functions with formulas, graphs and mathematical tables*. New York, NY: Dover Publications.

Alzer, H. (1999). On the Cauchy-Schwarz inequality. *Journal of Mathematical Analysis and Applications, 234*, 6–14. http://dx.doi.org/10.1006/jmaa.1998.6252

Callebaut, D. K. (1965). Generalization of the Cauchy-Schwarz inequality. *Journal of Mathematical Analysis and Applications, 12*, 491–494. http://dx.doi.org/10.1016/0022-247X(65)90016-8

Laforgia, A., & Natalini, P. (2006). Turán-type inequalities for some special functions. *Journal of Inequalities in Pure and Applied Mathematics, 27*, Article no: 32.

Masjed-Jamei, M. (2009). A functional generalization of the Cauchy–Schwarz inequality and some subclasses. *Applied Mathematics Letters, 22*, 1335–1339. http://dx.doi.org/10.1016/j.aml.2009.03.001

Masjed-Jamei, M., Dragomir, S. S., & Srivastava, H. M. (2009). Some generalizations of the Cauchy–Schwarz and the Cauchy–Bunyakovsky inequalities involving four free parameters and their applications. *Mathematical and Computer Modelling, 49*, 1960–1968. http://dx.doi.org/10.1016/j.mcm.2008.09.014

Mitrinović, D. S., Pečarić, J. E., & Fink, A. M. (1993). *Classical and new inequalities in analysis*. Dordrecht: Kluwer Academic. http://dx.doi.org/10.1007/978-94-017-1043-5

Nikiforov, A. F., & Uvarov, V. B. (1988). *Special functions of mathematical physics*. Basel: Birkhäuser. http://dx.doi.org/10.1007/978-1-4757-1595-8

Steiger, W. L. (1969). On a generalization of the Cauchy-Schwarz inequality. *The American Mathematical Monthly, 76*, 815–816. http://dx.doi.org/10.2307/2317882

Turán, P. (1950). On the zeros of the polynomials of Legendre. *Casopis Pro Pestovani Matematiky, 75*, 113–122.

Zheng, L. (1998). Remark on a refinement of the Cauchy–Schwarz inequality. *Journal of Mathematical Analysis and Applications, 218*, 13–21. http://dx.doi.org/10.1006/jmaa.1997.5720

3

Effect of viscous dissipation on hydromagnetic fluid flow and heat transfer of nanofluid over an exponentially stretching sheet with fluid-particle suspension

M.R. Krishnamurthy[1], B.C. Prasannakumara[2], B.J. Gireesha[1,3]* and Rama S.R. Gorla[3]

*Corresponding author: B.J. Gireesha, Department of Studies and Research in Mathematics, Kuvempu University, Shankaraghatta, Shimoga 577 451, Karnataka, India; Department of Mechanical Engineering, Cleveland State University, Cleveland, OH 44115, USA

E-mail: g.bijjanaljayanna@csuohio.edu

Reviewing editor: Antoinette Tordesillas, University of Melbourne, Australia

Abstract: This paper considers the problem of steady, boundary layer flow and heat transfer of a nanofluid with fluid-particle suspension over an exponentially stretching surface in the presence of transverse magnetic field and viscous dissipation. The stretching velocity and wall temperature are assumed to vary according to specific exponential form. The governing equations in partial forms are reduced to a system of coupled non-linear ordinary differential equations using suitable similarity transformations. An effective Runge–Kutta–Fehlberg (RKF-45) is used to solve the obtained differential equations with the help of a symbolic software MAPLE. The effects of flow parameters—such as nanofluid interaction parameter, magnetic parameter, solid volume fraction of nanoparticle parameter, Prandtl number and Eckert number—on the flow field and heat-transfer characteristics were obtained and are tabulated. Useful discussions were carried out with the help of plotted graphs and tables. Under the limiting cases, comparison with the existing results was made and found to be in good

ABOUT THE AUTHORS

B.J. Gireesha is an assistant professor in Department of Mathematics, Kuvempu University, Karnataka, India. He has received MPhil (1999) and PhD (2002) in Fluid Mechanics from Kuvempu University, Shimoga, India. Currently, he is working as a visiting research scholar in the Department of Mechanical Engineering, Cleveland State University, Cleveland, USA. His research interests include the areas of Fluid Mechanics particularly, boundary layer flows, Newtonian/non-Newtonian fluids, heat and mass transfer, Nanofluid and dusty fluid flow problems.

B.C. Prasannakumara is faculty member in Department of Mathematics, Government First Grade College, Koppa, Karnataka and M.R.Krishnamurthy is working for his doctoral degree under the guidance of Dr. B.J.Gireesha.

Rama S.R. Gorla is Fenn Distinguished Research Professor in the Department of Mechanical Engineering at Cleveland State University. He received the PhD degree in Mechanical Engineering from the University of Toledo in 1972. His primary research areas are combustion, heat transfer and fluid dynamics.

PUBLIC INTEREST STATEMENT

This paper considers the problem of steady, boundary layer flow and heat transfer of a nanofluid with fluid-particle suspension over an exponentially stretching surface in the presence of transverse magnetic field and viscous dissipation. The governing equations in partial forms are reduced to a system of coupled non-linear ordinary differential equations using suitable similarity transformations. The heat-transfer analysis is carried for two heating process, namely (1) prescribed exponential-order surface temperature and (2) prescribed exponential-order heat flux. Useful discussions were carried out with the help of plotted graphs. Under the limiting cases, comparison with the existing results was found to be in good agreement.

agreement. The results demonstrate that the skin friction coefficient increases for both magnetic and solid volume fraction nanoparticle parameters. However, dusty fluid with copper (Cu) nanoparticles has the appreciable cooling performance than other fluids.

Subjects: Applied Mathematics; Mathematical Modeling; Mathematics & Statistics; Non-Linear Systems; Science; Thermodynamics

Keywords: boundary layer flow; dusty fluid; nanofluid; viscous dissipation; heat transfer; exponentially stretching surface

1. Introduction

During the past few decades, the study of boundary layer flow and heat transfer over a stretching surface has achieved a lot of success because of its large number of applications in industry and technology. Few of these applications are materials manufactured by polymer extrusion, drawing of copper wires, continuous stretching of plastic films, artificial fibres, hot rolling, wire drawing, glass fibre, metal extrusion and metal spinning etc. After the pioneering work by Sakiadis (1961), a large amount of literature is available on boundary layer flow of Newtonian and non-Newtonian fluids over linear and non-linear stretching surfaces. Crane (1970) investigated the flow caused by the stretching of a sheet. Many researchers such as Gupta and Gupta (1977), Dutta, Roy, and Gupta (1985), Chen and Char (1988), Andersson (2002) extended the work of Crane (1970) to study the effect of heat- and mass-transfer analysis under different physical situations. On the other hand, Gupta and Gupta (1977) stressed that realistically, stretching surface is not necessarily continuous. Most of the available literature deals with the study of boundary layer flow over a stretching surface where the velocity of the stretching surface is assumed to be linearly proportional to the distance from the fixed origin. However, it is often argued that (Gupta & Gupta, 1977) stretching of plastic sheet realistically may not necessarily be linear. This situation was effectively dealt by Kumaran and Ramanaiah (1996) in their work on boundary layer flow of fluid where a general quadratic stretching sheet has been assumed for the first time.

In determining the particle accumulation and impingement on the surface, the study on boundary layer flow of fluid-particle suspension flow finds its importance. In view of this, Saffman (1962) formulated governing equations for the flow of dusty fluid and has discussed the stability of the laminar flow of a dusty gas in which dust particles are uniformly distributed. Based on this model, Chakrabarti (1974) analysed the boundary layer flow for a dusty gas. Datta and Mishra (1982) investigated the boundary layer flow of a dusty fluid over a semi-infinite flat plate. Numerical investigations carried out by Vajravelu and Nayfeh (1992) to analyse the hydromagnetic flow of a dusty fluid over a stretching sheet. Recently, Gireesha, Ramesh, and Bagewadi (2012) have critically analysed the magnetohydrodynamics (MHDs) flow and heat transfer of a dusty fluid over a stretching sheet using numerical technique.

Ali (1995) has investigated the thermal boundary layer flow by considering the non-linear stretching surface. Later, Magyari and Keller (1999) focused on heat- and mass-transfer analysis on boundary layer flow due to an exponentially continuous stretching sheet. Partha, Murthy, and Rajasekhar (2005) investigated the effect of viscous dissipation on the mixed convection heat transfer from an exponentially stretching surface. Sajid and Hayat (2008) studied the influence of thermal radiation on the boundary layer flow due to an exponentially stretching sheet using homotopy analysis method. The study of MHD has important applications, and may be used to deal with problems such as cooling of nuclear reactors by liquid sodium and induction flow metre, which depends on the potential difference in the fluid in the direction perpendicular to the motion and to the magnetic field. Elbashbeshy (2001) added new dimension to the study on exponentially continuous stretching surface. Khan and Pop (2010) and Sanjayan and Khan (2006) have studied the viscoelastic boundary layer flow and heat transfer due to an exponentially stretching sheet. Recently, Bidin and Nazar (2009) numerically analysed the effect of thermal radiation on the steady laminar two-dimensional boundary layer flow and heat transfer over an exponentially stretching sheet, which was earlier solved analytically by Sajid and Hayat (2008). Pal (2010) reported the mixed convection flow past an exponentially stretching surface in the presence of a magnetic field. Nadeem, Zaheer, and Fang (2011) addressed the flow of Jeffrey fluid and heat transfer past an exponentially stretching sheet. Ishak (2011) studied the effect of radiation on MHD boundary layer flow of a viscous fluid over an exponentially stretching sheet. It was found that the local heat-transfer rate at the surface decreases with increasing values of the magnetic and radiation parameters. Sahoo and Poncet (2011) addressed the flow of third-grade fluid past an exponentially stretching sheet with slip condition.

Nanofluids, a new class of nano-engineered liquid solutions of colloidal particles with a diameter of 1–100 nm, have been shown great energy saving potentials and attractive properties for applications such as energy, bio and pharmaceutical industry, and chemical, electronic, environmental, material, medical and thermal engineering, among others. The concept of nanofluid was first introduced by Choi (1995) in the article enhancing thermal conductivity of fluids with nanoparticles. Based on this pioneer work, Mabood, Khan, and Ismail (2015a, 2015b) focused on the study of combined heat and mass transfer of electrically conducting nanofluid over a non-linear stretching surface in the presence of a first-order chemical reaction and viscous dissipation. Further, they (Mabood & Khan, 2014) have obtained the series solution for MHD stagnation point flow in porous medium for different values of Prandtl number and suction/injection parameter. Abbasi, Shehzad, Hayat, Alsaedi, and Obid (2015) presented an analysis to address the MHD two-dimensional boundary layer flow of Jeffrey nanofluid over a stretching sheet with thermal radiation. Hussain, Hayat, Shehzad, Alsaedi, and Chen (2015) examined the flow problem resulting from the stretching of a surface with convective conditions in a MHD third-grade nanofluid in the presence of thermal radiation. Hayat, Hussain, Shehzad, and Alsaedi (2015) investigated the Brownian motion and thermophoresis effects on two-dimensional boundary layer flow of an Oldroyd-B nanofluid in the presence of thermal radiation and heat generation.

Recently, Pavithra and Gireesha (2013) discussed the effect of internal heat generation/ absorption on dusty fluid flow over an exponentially stretching sheet with viscous dissipation. Mukhopadhyay and Gorla (2012) analysed the effects of partial slip on flow past an exponentially stretching sheet. Rudraswamy and Gireesha (2014) investigated the influence of chemical reaction and thermal radiation on MHD boundary layer flow and heat transfer of a nanofluid over an exponentially stretching sheet. To the best of author's knowledge, studies on hydromagnetic fluid flow and heat transfer of nanofluid over an exponentially stretching sheet with fluid particle suspension in the presence of viscous dissipation have not been considered so far. The governing boundary layer equations of the flow problem are transformed into ordinary differential equations, using similarity transformations. The resulting ordinary differential equations are then solved numerically, using Runge–Kutta–Fehlberg (RKF-45) method. The effects of flow pertinent parameters on the flow- and heat-transfer characteristics are discussed and are presented in detail. Obtained results are compared with the known results available in the literature and the comparison shows good agreement. It is believed that the results of this study can be used in the design of an effective cooling system for electronic machinery to help ensure effective and safe operational conditions, which includes nuclear power plants, gas turbines, aircraft, missiles, satellites, space vehicles, etc.

2. Mathematical formulation and solution of the problem

Consider a steady, two-dimensional, boundary layer flow and heat transfer of an incompressible nanofluid with fluid-particle suspension over an exponentially stretching surface. Cartesian coordinate system is considered in such a way that x-axis is taken along the stretching surface in the direction of motion and y-axis is normal to it. The plate is stretched in the x-direction with a velocity $U_w(x) = U_0 e^{\frac{x}{L}}$, defined at $y = 0$. The flow is generated as a consequence of exponential stretching of the sheet caused by simultaneous application of equal and opposite forces along x-axis keeping the origin fixed. A uniform magnetic field B_0 is assumed to be applied in the y-direction. The geometry of the flow configuration is as shown in Figure 1.

The flow- and heat-transfer characteristics under the boundary layer approximations are governed by the following equations,

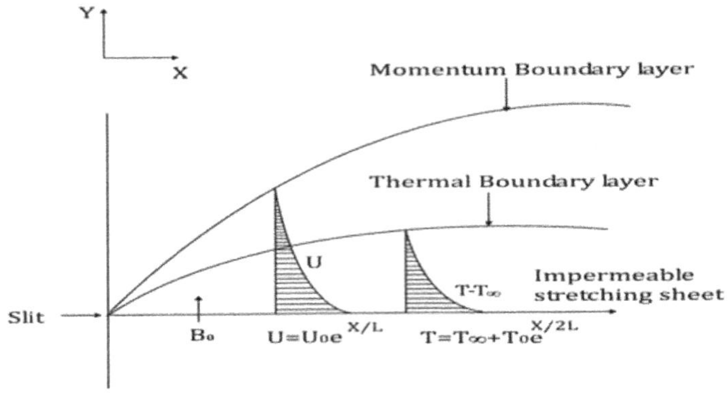

Figure 1. Schematic diagram of the flow geometry.

$$\frac{\partial u}{\partial x}+\frac{\partial v}{\partial y}=0 \tag{2.1}$$

$$u\frac{\partial u}{\partial x}+v\frac{\partial u}{\partial y}=\frac{\mu_{\text{nf}}}{\rho_{\text{nf}}}\frac{\partial^2 u}{\partial y^2}+\frac{KN}{\rho_{\text{nf}}}\left(u_p-u\right)-\frac{\sigma B_0^2}{\rho_{\text{nf}}}u \tag{2.2}$$

$$\frac{\partial u_p}{\partial x}+\frac{\partial v_p}{\partial y}=0 \tag{2.3}$$

$$u_p\frac{\partial u_p}{\partial x}+v_p\frac{\partial u_p}{\partial y}=\frac{K}{m}\left(u-u_p\right) \tag{2.4}$$

where x and y, respectively, represents coordinate axes along the continuous surface in the direction of motion and perpendicular to it. (u, v) and (u_p, v_p) denotes the velocity components of the nanofluid and dust phases along x and y directions respectively, N denotes number density of the dust particles, m is the mass concentration of dust particles and σ symbolises electrical conductivity.

The associated boundary conditions are,

$$\begin{aligned} &u=U_w(x), \quad v=0, \quad T=T_w \ \text{at} \ y=0\\ &u=0, \quad u_p\to 0, \quad v_p\to v, \ \text{as}\ y\to\infty \end{aligned} \tag{2.5}$$

where $U_w(x)=U_0e^{\frac{x}{L}}$ is the sheet velocity, U_0 is reference velocity and L is the reference length. The effective density of the nanofluid is given by,

$$\rho_{\text{nf}}=(1-\phi)\rho_f+\phi\rho_s \tag{2.6}$$

where ϕ is the solid volume fraction of nanoparticles and the heat capacitance of the nanofluid is obtained as,

$$(\rho C_p)_{\text{nf}}=(1-\phi)(\rho C_p)_f+\phi(\rho C_p)_s \tag{2.7}$$

and the thermal conductivity of the nanofluid k_{nf} for spherical nanoparticles can be written as Maxwell (1904),

$$\frac{k_{\text{nf}}}{k_f}=\frac{k_s+2k_f-2\phi(k_f-k_s)}{k_s+2k_f-2\phi(k_f-k_s)} \tag{2.8}$$

Table 1. Thermo-physical properties of water and nanoparticles

	ρ (kg/m^3)	C_p (J/kg K)	k (W/mk)
Pure water (H_2O)	997.1	4,179	0.613
Copper (Cu)	8,933	385	401
Copper oxide (CuO)	6,320	531.8	76.500
Aluminium oxide (Al_2O_3)	3,970	765.0	40.000
Silver (Ag)	10,500	235.0	429.00

Also the effective dynamic viscosity of the nanofluid given by Brinkman (1952) as,

$$\mu_{nf} = \frac{\mu_f}{(1-\phi)^{2.5}} \tag{2.9}$$

Here the subscripts nf, f and s represent the thermophysical properties of the nanofluids, base fluid and the nano-solid particles, respectively.

We are interested in similarity solution of the above boundary value problem; therefore, introduce the following similarity transformations;

$$u = U_0 e^{\frac{x}{L}} f'(\eta), \quad v = -\sqrt{\frac{U_0 v_f}{2L}} e^{\frac{x}{2L}} \left[f(\eta) + \eta f'(\eta)\right]$$

$$u_p = U_0 e^{\frac{x}{L}} F'(\eta), \quad v_p = -\sqrt{\frac{U_0 v_f}{2L}} e^{\frac{x}{2L}} \left[F(\eta) + \eta F'(\eta)\right]$$

$$\eta = \sqrt{\frac{U_0}{2v_f L}} e^{\frac{x}{2L}} y \tag{2.10}$$

Making use of the transformations (2.10), Equation 2.1 and 2.3 are identically satisfied and Equations 2.2 and 2.4 will take the following form,

$$f''' + (1-\phi)^{2.5}\left[(1-\phi) + \phi\frac{\rho_s}{\rho_f}\right](ff'' - 2f'^2) + (1-\phi)^{2.5}\left[2l\beta(F' - f') - Mf'\right] = 0 \tag{2.11}$$

$$FF'' - 2[F'^2] + 2\beta[f' - F'] = 0 \tag{2.12}$$

where a prime denotes differentiation with respect to η, $l = \frac{mN}{\rho_f}$ is the mass concentration, v_f is the kinematic viscosity of the fluid, $\tau_v = m/K$ is the relaxation time of the particle phase, $\beta = \frac{L}{\tau_v U_0}$ is the fluid particle interaction parameter for velocity and $M = \frac{2\sigma B_0^2 L}{\rho_f U_0}$ is the magnetic parameter.

The boundary conditions defined as in Equation 2.5 will become;

$$f'(\eta) = 1, \quad f(\eta) = 0 \ \text{at} \ = 0$$

$$f'(\eta) = 0, \quad F'(\eta) = 0 \quad F(\eta) = f(\eta) \ \text{as} \ \eta \to \infty \tag{2.13}$$

3. Heat-transfer analysis

The governing steady, boundary layer, heat transport equations for both fluid and dust phases with viscous dissipation are given by,

$$(\rho C_p)_{nf}\left[u\frac{\partial T}{\partial x}+v\frac{\partial T}{\partial y}\right]=k_{nf}\frac{\partial^2 T}{\partial y^2}+\frac{NC_{pf}}{\tau_T}\left(T_p-T\right)+\frac{N}{\tau_v}\left(u_p-u\right)^2+\mu_f\left(\frac{\partial u}{\partial y}\right)^2 \tag{3.1}$$

$$u_p\frac{\partial T_p}{\partial x}+v_p\frac{\partial T_p}{\partial y}=\frac{C_{pf}}{C_{mf}\tau_T}\left(T-T_p\right) \tag{3.2}$$

where T and T_p are the temperatures of the fluid and dust particle inside the boundary layer, c_{pf} and c_{mf} are the specific heat of fluid and dust particles, τ_T is the thermal equilibrium time i.e. the time required by a dust cloud to adjust its temperature to the fluid, k_{nf} is the thermal conductivity, τ_v is the relaxation time of the dust particle, that is, the time required by a dust particle to adjust its velocity relative to the fluid.

Heat-transfer phenomenon are solved for two types of heating process, namely,

- Prescribed exponential-order surface temperature (PEST) and
- Prescribed exponential-order heat flux (PEHF).

3.1. Case-1: Prescribed exponential-order surface temperature (PEST)

For this heating process, the following boundary conditions are employed;

$$\begin{aligned}&T=T_w(x)\ \text{ at }\ y=0\\&T\to T_\infty,\quad T_p\to T_\infty\ \text{ as }\ y\to\infty\end{aligned} \tag{3.3}$$

where $T_w=T_\infty+T_0e^{\frac{c_1x}{2L}}$ is the temperature distribution in the stretching surface, T_0 is the reference temperature and c_1 is a constant.

Defining the non-dimensional variables for fluid-phase temperature $\theta(\eta)$ and dust-phase temperature $\theta_p(\eta)$ as,

$$\theta(\eta)=\frac{T-T_\infty}{T_w-T_\infty},\quad \theta_p(\eta)=\frac{T_p-T_\infty}{T_w-T_\infty} \tag{3.4}$$

where $T=T_\infty=T_0e^{\frac{c_1x}{2L}}\theta(\eta)$. Using the similarity variable η and Equation 3.4 into Equations 3.1–3.2 and on equating the coefficient of $\left(\frac{x}{L}\right)^0$ on both sides, one can arrive at the following system of equations:

$$\begin{aligned}&\frac{k_{nf}}{k_f}\theta''+Pr\left[(1-\phi)+\phi\frac{(\rho C_p)_s}{(\rho C_p)_f}\right](f\theta'-c_1f'\theta)\\&\qquad+\frac{2NPr\beta_T}{\rho_f}\left[\theta_p-\theta\right]+\frac{2NPrEc\beta}{\rho_f}[F'-f']^2+EcPr[f'']^2=0\end{aligned} \tag{3.5}$$

$$F\theta_p'-c_1F'\theta_p+2\gamma\beta_T\left[\theta-\theta_p\right]=0 \tag{3.6}$$

where $Pr=\frac{(\mu C_p)_f}{\rho_f}$ is the Prandtl number, $Ec=\frac{U_0^2}{T_0C_{pf}}$ is the Eckert number, $\beta_T=\frac{L}{U_0\tau_T}$ is the fluid particle interaction parameter for temperature and $\gamma=\frac{C_{pf}}{C_{mf}}$ is the ratio of specific heat.

The corresponding thermal boundary conditions becomes,

$$\begin{aligned}&\theta(\eta)=1\ \text{ at }\ \eta=0\\&\theta(\eta)\to0,\quad \theta_p(\eta)\to0\ \text{ as }\ \to\infty\end{aligned} \tag{3.7}$$

3.2. Case-2: Prescribed exponential-order heat flux (PEHF)

For this heating process, consider the boundary conditions as follows,

$$\frac{\partial T}{\partial y} = -\frac{q_w}{k_{nf}} \text{ at } y = 0$$

$$T \to T_\infty, \quad T_p \to T_\infty \text{ as } y \to \infty \tag{3.8}$$

where $q_w(x) = T_1 e^{\frac{(c_2+1)x}{2L}}$, $T_w = T_\infty + \frac{T_1}{k_{nf}} e^{\frac{c_2 x}{2L}} \sqrt{\frac{2v_f L}{U_0}}$, T_1 is reference temperature, c_2 is constant and $T - T_\infty = \frac{T_1}{k_{nf}} e^{\frac{c_2 x}{2L}} \sqrt{\frac{2v_f L}{U_0}} \theta(\eta)$. Using the similarity variable η and Equation 3.4 into Equation 3.1–3.2 and on equating the coefficient of $\left(\frac{x}{L}\right)^0$ on both sides, one can arrive the following system of equations

$$\frac{k_{nf}}{k_f}\theta'' + Pr\left[(1-\phi) + \phi\frac{(\rho C_p)_s}{(\rho C_p)_f}\right](f\theta' - c_1 f'\theta) + \frac{2NPr\beta_T}{\rho_f}\left[\theta_p - \theta\right]$$
$$+ \frac{2NPrEc\beta}{\rho_f}[F' - f']^2 + EcPr[f'']^2 = 0 \tag{3.9}$$

$$F\theta_p' - c_1 F'\theta_p + 2\gamma\beta_T[\theta - \theta_p] = 0 \tag{3.10}$$

where $Ec = \frac{k_{nf} U_0^2}{T_1 C_{pf}} \sqrt{\frac{U_0}{2v_f L}}$ is the Eckert number and the corresponding thermal boundary conditions are written as

$$\theta'(\eta) = -1 \text{ at } \eta = 0$$

$$\theta(\eta) \to 0, \quad \theta_p(\eta) \to 0 \text{ as } \eta \to \infty \tag{3.11}$$

The important physical parameters for the boundary layer flow are the skin friction coefficient and heat-transfer coefficient which are, respectively, defined as,

$$C_f = \frac{\tau_w}{\rho_f U_w^2}, \quad Nu_x = \frac{x q_w}{k_{nf}(T_w - T_\infty)} \tag{3.12}$$

where the skin friction τ_w and the heat transfer from the sheet q_w are given by,

$$\tau_w = \mu_f\left(\frac{\partial u}{\partial y}\right)_{y=0}, \quad q_w = -k_{nf}\left(\frac{\partial T}{\partial y}\right)_{y=0} \tag{3.13}$$

Using the non-dimensional variables, one obtains,

$$\sqrt{2Re}\, C_f = f''(0)$$

$$\frac{Nu_x}{\sqrt{2Re}} = -\frac{x}{2L}\theta'(0) \text{ (PEST Case) and } \frac{Nu_x}{\sqrt{2Re}} = \frac{x}{2L}\frac{1}{\theta(0)} \text{(PEHF Case)} \tag{3.14}$$

where $Re = \frac{U_0 L}{v_f}$ is the Reynolds number.

4. Numerical solution

A two-dimensional hydromagnetic fluid flow and heat transfer of nanofluid at an exponentially stretching sheet with fluid-particle suspension in presence of viscous dissipation is considered. The non-linear differential Equation 2.11 and 2.12, 3.5 and 3.6 or 3.9 and 3.10 together with the boundary conditions 2.13, 3.7 and 3.11 for both PEST and PEHF cases are solved numerically using Runge–Kutta–Fehlberg (RKF-45) scheme with the help of Maple software. We have chosen suitable finite values of $\eta \to \infty$ say $\eta = 5$.

The results for $f''(0)$ obtained in the present work and those by Vajravelu and Nayfeh (1992) and Das (2012) are recorded in Table 2, for different values of ϕ. Further, Table 3 shows the comparative values of $\theta'(0)$ with Magyari and Keller (1999), Bidin and Nazar (2009), Ishak (2011) and Abd El-Aziz (2009)

Table 2. Comparison of the results of $f''(0)$ for Cu-water for various values of solid volume fraction of nanoparticles ϕ when $\beta = M = 0$

ϕ	$f''(0)$		
	Vajravelu and Nayfeh (1992)	**Das (2012)**	**Present result**
0.0	−1.001411	−1.001411	−1.001396
0.1	−1.175209	−1.175251	−1.175394
0.2	−1.218301	−1.218315	−1.218580

Table 3. Values of $\theta'(0)$ for several values of Prandtl number and in the absence of β, β, Ec, M and ϕ

Pr	$\theta'(0)$				
	Magyari and Keller (1999)	**Bidin and Nazar (2009)**	**Ishak (2011)**	**El-Aziz (2009)**	**Present result**
1	−0.9548	−0.9547	−0.9548	−0.9548	−0.95764
2	-	−1.4714	−1.4715	-	−1.47084
3	−1.8691	−1.8691	−1.8691	−1.8691	−1.86854
4	−2.5001	-	−2.5001	−2.5001	−2.49972
5	−3.6604	-	−3.6604	−3.6604	−3.66005

Table 4. Values of wall temperature gradient $[-\theta'(0)]$ in PEST case and wall temperature $\theta(0)$ in PEHF case for different values of the parameters Pr, Ec, M, N, ϕ and β

β	M	Pr	Ec	N	ϕ	$-\theta'(0)$ (PEST)	$\theta(0)$ (PEHF)
0.2	0.5	6.2	0.5	0.5	0.2	1.15302	0.92894
0.6						1.07472	0.95149 ($\beta = 0.4$)
1						1.04198	0.96496 ($\beta = 0.6$)
0.6	0	6.2	0.5	0.5	0.2	1.16853	0.92179
	0.1					1.06747	0.96496
	0.15					0.98521	1.00699
0.6	0.5	2.2	0.5	0.5	0.2	0.68437	1.28081
		4.2				0.92198	1.04621
		6.2				1.07472	0.96496
0.6	0.5	6.2	0	0.5	0.2	2.13308	0.46880
			0.5			1.07472	0.96496
			1			0.16366	1.46113
0.6	0.5	6.2	0.5	0.5	0.2	1.07472	0.96496
				1		1.42719	0.82803
				2		2.00800	0.67278
0.6	0.5	6.2	0.5	0.5	0.05	1.45603	0.84217
					0.1	1.28518	0.89053
					0.2	1.07472	0.96496

for various values of *Pr* in the absence of magnetic field. From these tables, one can notice that there is a close agreement between these approaches and thus verifies the accuracy of the method used. The results of thermal characteristics at the wall are examined for the values of the temperature gradient function $\theta'(0)$ in PEST case and $\theta(0)$ in PEHF case, and are tabulated in Table 4. The effects of various physical parameters such as nanofluid interaction parameter (β), magnetic parameter (M), Prandtl number (Pr) Eckert number (Ec) number density parameter (N) and solid volume fraction parameter (ϕ) are examined and discussed in detail.

5. Results and discussion

The hydromagnetic boundary layer flow and heat transfer of a dusty nanofluid over an exponentially stretching sheet are investigated in the presence of viscous dissipation. Similarity transformations are used to convert the governing time-independent non-linear boundary layer equations into a system of non-linear ordinary differential equations. The obtained highly non-linear ordinary differential equations are then solved numerically by means of most efficient numerical technique, fourth–fifth-order Runge–Kutta–Fehlberg scheme with the help of Maple software. The temperature profile $\theta(\eta)$ and $\theta_p(\eta)$ for both PEST and PEHF cases are depicted graphically. A parameter of interest for the present study is the nanofluid interaction parameter (β), magnetic parameter (M) Prandtl number (Pr) Eckert number (Ec) number density parameter (N) and solid volume fraction of nanoparticle parameter (ϕ).

The variation of velocity profiles with η for different values of the nanofluid interaction parameter for both nanofluid and dust phases are illustrated in Figure 2. It is noticed from this figure that the velocity profiles decrease with increasing values of β for nanofluid phase and increase for dust phase in the boundary layer. The increasing values of β reduce the velocity $f'(\eta)$ and thereby increase the boundary layer thickness.

The effects of magnetic parameter on velocity profiles for both nanofluid and dust phases are illustrated graphically through Figure 3. It is interesting to note that as strength of the magnetic field increases, velocity profiles for both the phases decrease. This is due to the fact that, the introduction of transverse magnetic field has a tendency to create a drag, known as the Lorentz force which tends to resist the flow. This behaviour is even true in the case of increasing values of nanofluid interaction parameter for fluid phase.

Figure 4 depicts the temperature profiles $\theta(\eta)$ and $\theta_p(\eta)$ vs. η for different values of nanofluid interaction parameter β. We infer from these figures that the temperature increases with increase in

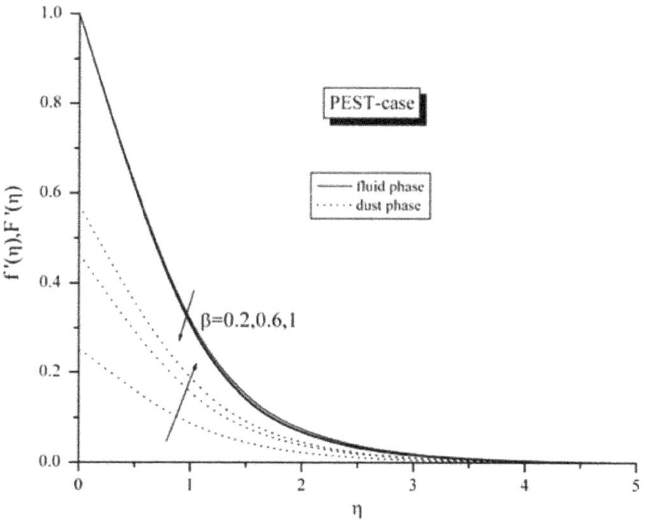

Figure 2. Effect of β on velocity profiles.

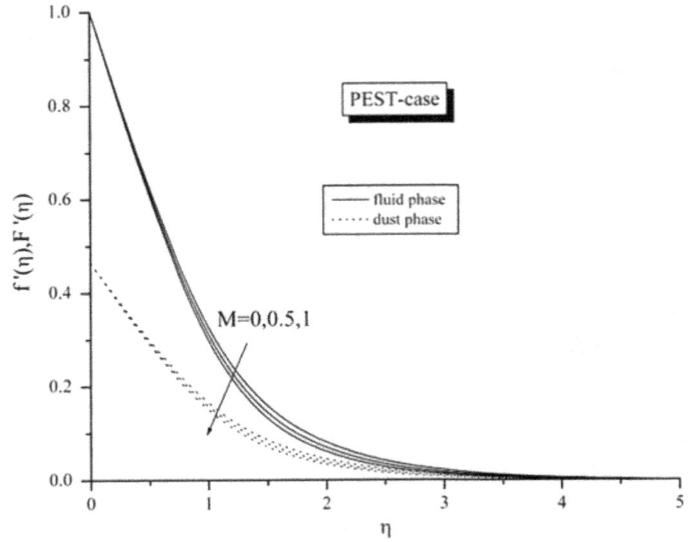

Figure 3. Effect of *M* on velocity profiles.

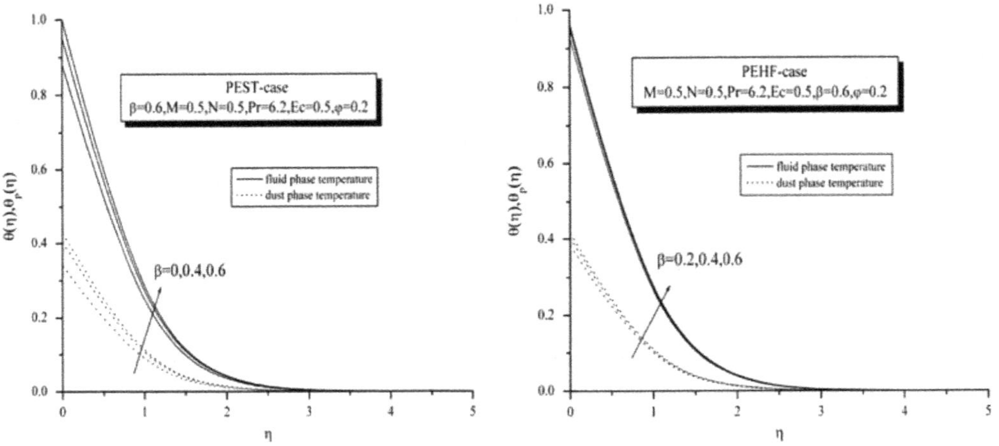

Figure 4. Effect of β on temperature profile for both PEST and PEHF cases.

nanofluid interaction parameter for both PEST and PEHF cases. Also one can observed that nanofluid phase temperature is higher than that of dust phase.

The effects of magnetic field parameter (*M*) on temperature profiles $\theta(\eta)$ and $\theta_p(\eta)$ for both PEST and PEHF cases are depicted in Figure 5. We infer from this figure that the temperature profiles increases with increase in magnetic parameter and it also indicates that both the nanofluid and the dust particle temperature are parallel to each other. This is true for both PEST and PEHF cases.

The temperature profile for various values of Prandtl number (*Pr*) is represented for both PEST and PEHF cases in Figure 6. From this figure, it reveals that the temperature decreases with increase in the value of *Pr*. Hence Prandtl number can be used to increase the rate of cooling.

Figure 7 explains the effect of viscous dissipation in terms of Eckert number on temperature profiles. Viscous dissipation can changes the temperature distribution by playing a role like an energy source, which leads to affect heat transfer rates. Here the temperature of both nanofluid and dust phase increase with increase in the value of *Ec*. It is because heat energy is stored in the liquid due to frictional heating and this is true in both the cases.

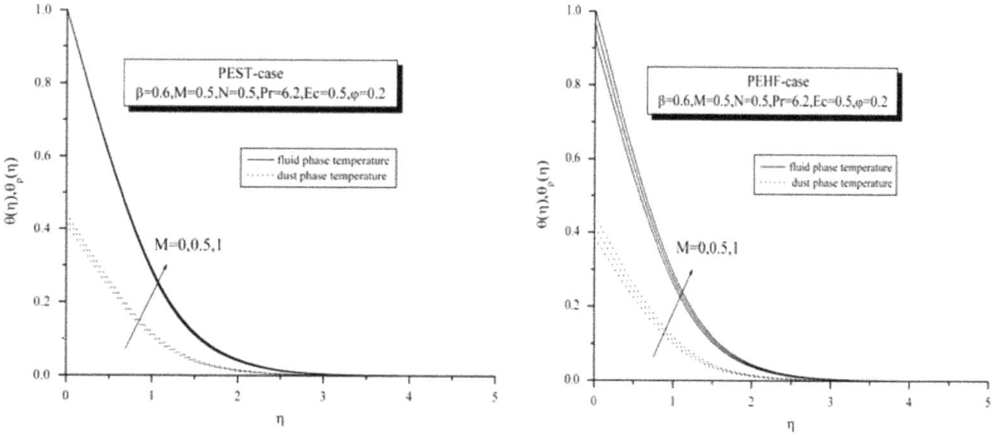

Figure 5. Effect of *M* on temperature profiles for both PEST and PEHF cases.

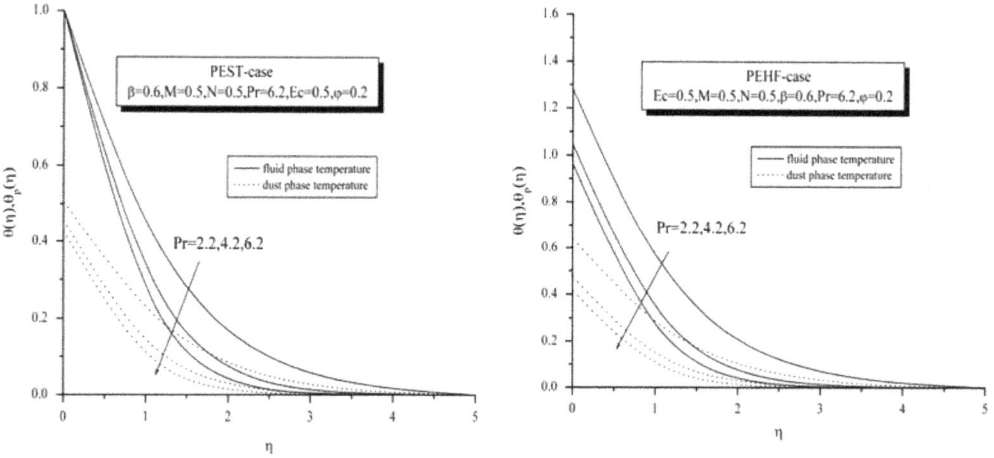

Figure 6. Effect of *Pr* on temperature profiles for both PEST and PEHF cases.

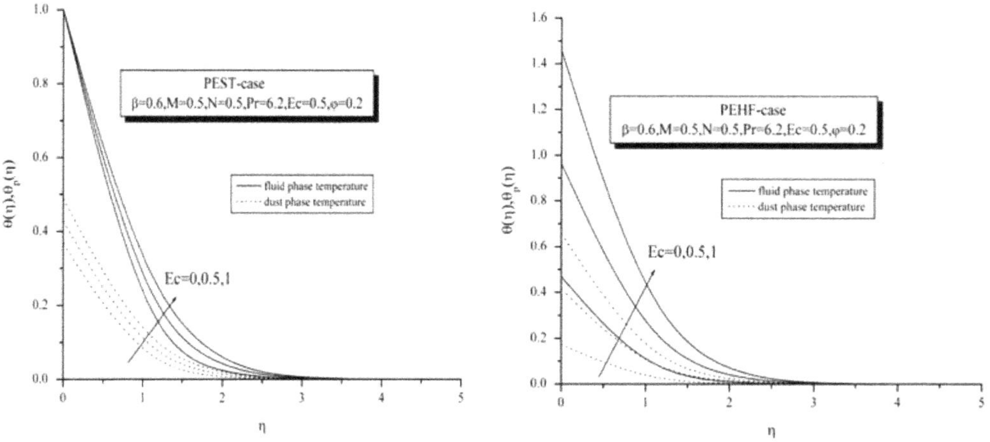

Figure 7. Effect of *Ec* on temperature profiles for both PEST and PEHF cases.

Figure 8 shows the temperature distribution $\theta(\eta)$ and $\theta_p(\eta)$ v. η for different values of number density (N). We infer from this figure that the temperature decreases with increase in N for both PEST and PEHF cases.

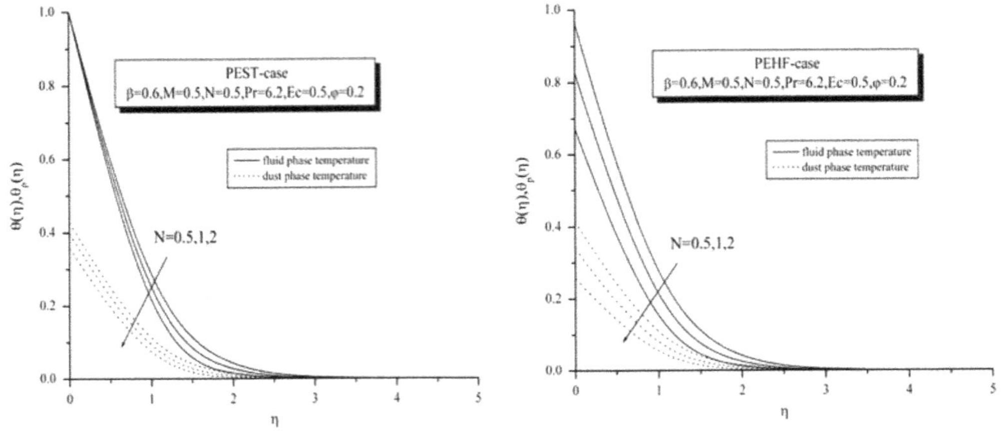

Figure 8. Effect of *N* on temperature profiles for both PEST and PEHF cases.

The effect of solid volume fraction of nanoparticle parameter ϕ on velocity profiles for both fluid and dust phases are illustrated graphically through Figure 9. From this figure, it is observed that as the volume fraction of nanoparticles increases from 0 to 0.2, velocity profile for both fluid and dust phases decrease inside the boundary layer, while they increase outside.

Figure 10, is plotted over temperature profiles of different values of nanoparticles volume fraction ϕ for both PEST and PEHF cases. It is observed that, by increasing nanoparticles volume fraction, the thermal boundary layer increases for both fluid and dust phases.

Figure 11 shows the velocity profile for different type of fluids in the region of uniform magnetic field. It is observed that the velocity profile for regular fluid is much higher than that of nanofluid, dusty fluid and dusty nanofluid, respectively, in their order. This result is well evident to say that dusty nanofluid has high thermal conductivity than dusty fluid, nanofluid and regular fluid.

The effect of nanoparticles (Ag, Cu, CuO and Al_2O_3) on velocity profiles for both fluid and dust phases are depicted graphically in Figure 12. It is interesting to note that, velocity profile increases in the order of silver (Ag), copper (Cu), copper oxide (CuO) and aluminium oxide (Al_2O_3). From this, it is concluded that velocity profile for aluminium oxide (Al_2O_3) is much higher than that of others (see Table 1).

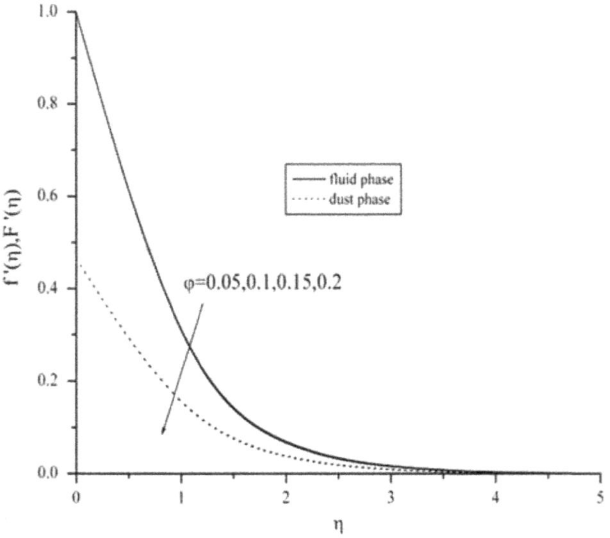

Figure 9. Effect of ϕ on velocity profile.

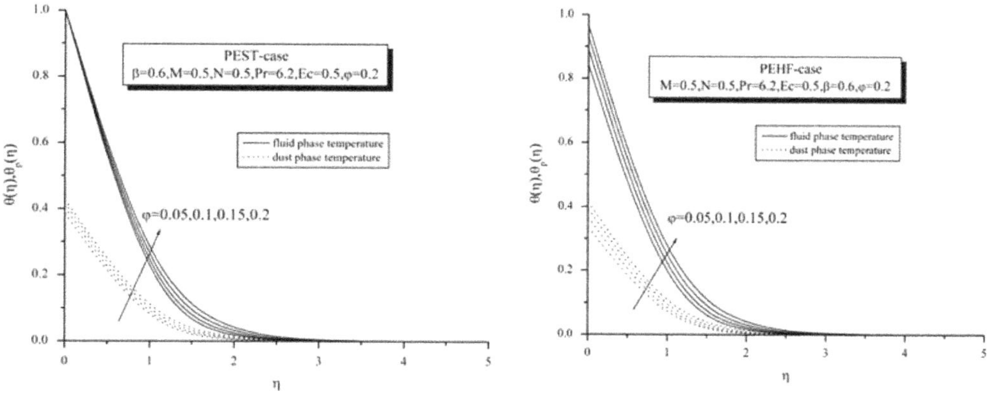

Figure 10. Effect of ϕ on temperature profiles for both PEST and PEHF cases.

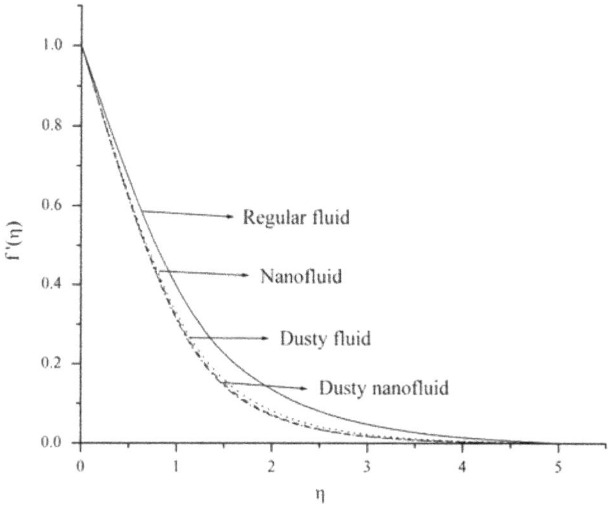

Figure 11. Effect of all fluids on velocity profiles.

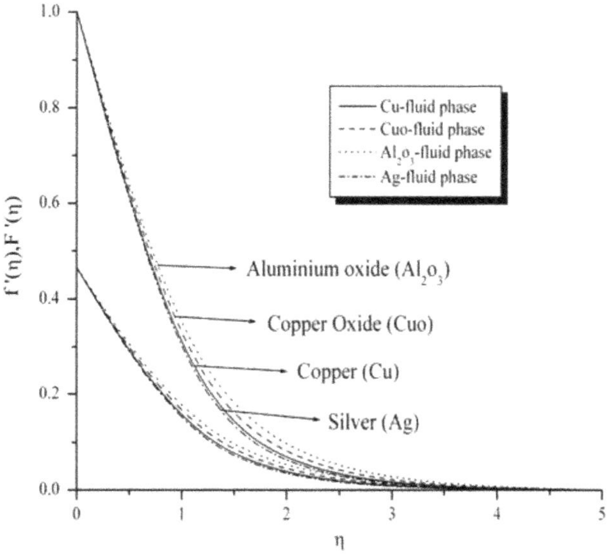

Figure 12. Effect of nanoparticles on velocity profiles.

Figure 13 exhibit the temperature profile for few nanoparticles in both PEST and PEHF cases. It is observed that temperature profile for silver (Ag) is much higher than that of copper (Cu), copper oxide (CuO) and Aluminium oxide (Al_2O_3).

Figure 14 displays the variation of skin friction ($-f''(0)$) for increasing values of nanofluid particle interaction parameter (β), magnetic parameter (M) as well as solid volume fraction nanoparticle parameter (ϕ). It can be noticed that the nanofluid particle interaction parameter increases the skin friction. The same effect can also be found with the magnetic parameter and solid volume fraction nanoparticle parameter.

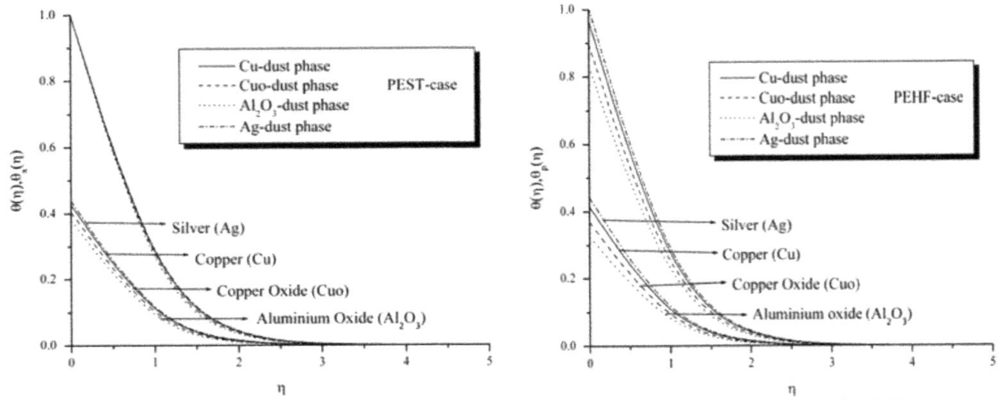

Figure 13. Effect of different nanoparticles on temperature profiles for both PEST and PEHF cases.

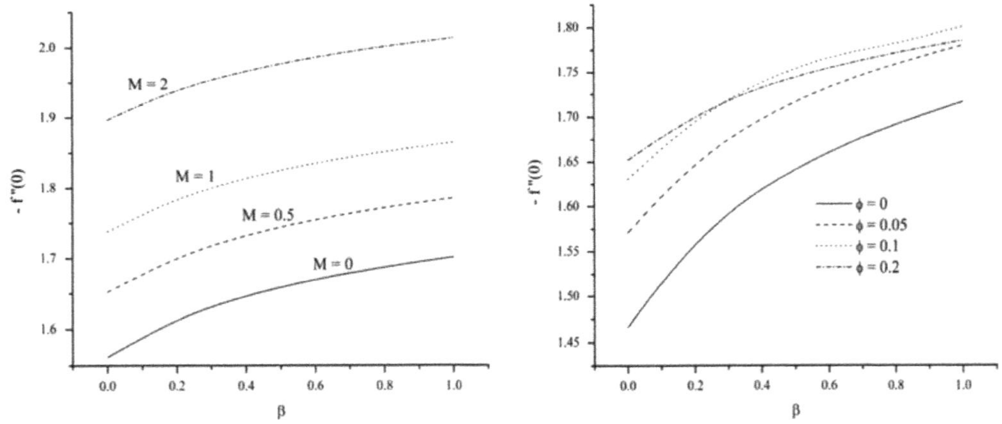

Figure 14. Effect of M, ϕ and β on skin friction coefficient.

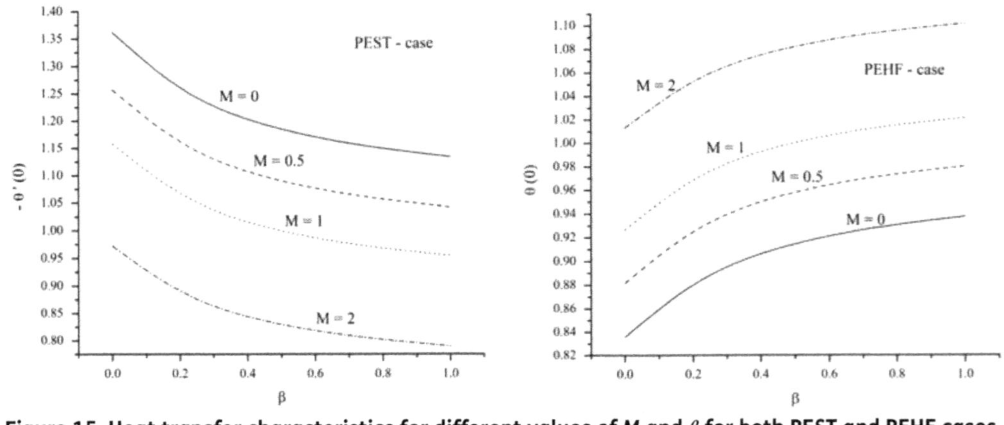

Figure 15. Heat transfer characteristics for different values of M and β for both PEST and PEHF cases.

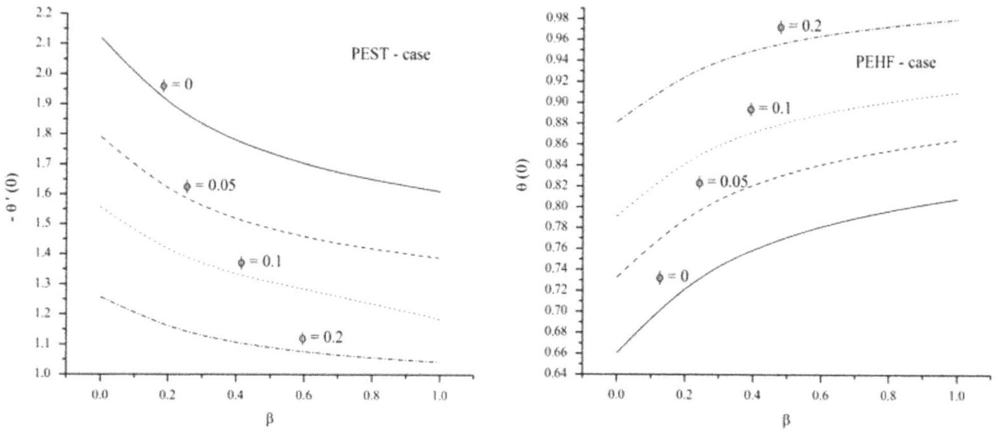

Figure 16. Heat transfer characteristics for different values of ϕ and β for both PEST and PEHF cases.

Figures 15 and 16 depict the nature of heat-transfer coefficient ($-\theta'(0)$) in PEST case and $\theta(0)$ in PEHF case with nanofluid particle interaction parameter (β) for different values of magnetic parameter (M) and solid volume fraction nanoparticle parameter (ϕ), respectively. It is very clear that the heat transfer increases with nanofluid particle interaction parameter (β) where as it decreases with magnetic parameter (M) and solid volume fraction nanoparticle parameter (ϕ) in PEST case but it increases with both (β) and (ϕ) in PEHF case. The temperature gradient function $\theta'(0)$ in PEST case is negative, by this we mean that the heat is transferred from fluid to the stretching surface. Further, the temperature function $\theta(0)$ in PEHF case at the sheet is positive means, heat is absorbing from the fluid.

6. Conclusions

A mathematical analysis has been carried out to study the effect of viscous dissipation on hydromagnetic flow and heat transfer of nanofluid over an exponentially stretching sheet with fluid particle suspension. Some of the important observations of our analysis obtained by the graphical representation are reported as follows.

The effect of magnetic parameter is to increase the temperature distribution in the flow region for both the cases of PEST and PEHF for both nanofluid and dust particle phases. The nanofluid phase temperature is higher than that of the dust phase. The temperature of nanofluid and dust phase increases with increasing values of Ec in both PEST and PEHF cases. The velocity of nanofluid and dust phases decreases while the temperature of nanofluid and dust phases increases as solid volume fraction of nanoparticles (ϕ) increases. For the aluminium oxide nanoparticles, the velocity is higher than that of other nanoparticles, and for silver (Ag) nanoparticles, the thermal conductivity is higher than other particles such as copper (Cu), copper oxide (CuO) and aluminium oxide (Al_2O_3). It is found that the dusty fluid with copper (Cu) nanoparticles have the appreciable cooling performance. The dusty nanofluids are found to have higher thermal conductivity, when compared with dusty fluid and nanofluid.

Funding
The author B.J. Gireesha gratefully acknowledges the financial support of University Grants Commission, India under Raman Fellowship for Post Doctoral Research for Indian Scholars in USA [grant number 5-1/2013(IC)].

Author details
M.R. Krishnamurthy[1]
E-mail: kittysa.mr@gmail.com
http://orcid.org/0000-0002-9752-2790
B.C. Prasannakumara[2]
E-mail: dr.bcprasanna@gmail.com
http://orcid.org/0000-0003-1950-4666
B.J. Gireesha[1,3]
E-mail: g.bijjanaljayanna@csuohio.edu
http://orcid.org/0000-0003-2995-6981
Rama S.R. Gorla[3]
E-mail: r.gorla@csuohio.edu
[1] Department of Studies and Research in Mathematics, Kuvempu University, Shankaraghatta, Shimoga 577 451, Karnataka, India.
[2] Government First Grade College, Koppa, Chikkamagaluru 577126, Karnataka, India.
[3] Department of Mechanical Engineering, Cleveland State University, Cleveland, OH 44115, USA.

References

Abbasi, F. M., Shehzad, S. A., Hayat, T., Alsaedi, A., & Obid, M. A. (2015). Influence of heat and mass flux conditions in hydromagnetic flow of Jeffrey nanofluid. *AIP Advances, 5*, 037111. doi:10.1063/1.4914549.

Abd El-Aziz, M. (2009). Viscous dissipation effect on mixed convection flow of a micropolar fluid over an exponentially stretching sheet. *Canadian Journal of Physics, 87*, 359–368. http://dx.doi.org/10.1139/P09-047

Ali, M. E. (1995). On thermal boundary layer on a power law stretched surface with suction or injection. *International Journal of Heat and Fluid Flow, 16*, 280–290. http://dx.doi.org/10.1016/0142-727X(95)00001-7

Andersson, H. I. (2002). Slip flow past a stretching surface. *Acta Mechanica, 158*, 121–125. http://dx.doi.org/10.1007/BF01463174

Bidin, B., & Nazar, B. (2009). Numerical solution of the boundary layer flow over an exponentially stretching sheet with thermal radiation. *European Journal of Scientific Research, 33*, 710–717.

Brinkman, H. C. (1952). The viscosity of concentrated suspensions and solutions. *The Journal of Chemical Physics, 20*, 571–581. http://dx.doi.org/10.1063/1.1700493

Chakrabarti, K. M. (1974). Note on boundary layer in a dusty gas. *AIAA Journal, 12*, 1136–1137. http://dx.doi.org/10.2514/3.49427

Chen, C. K., & Char, M. I. (1988). Heat transfer of a continuous stretching surface with suction or blowing. *Journal of Mathematical Analysis and Applications, 135*, 568–580. http://dx.doi.org/10.1016/0022-247X(88)90172-2

Choi, S. U. S. (1995). Enhancing thermal conductivity of fluids with nanoparticles. *Proceedings of the ASME International Mechanical Engineering Congress and Exposition, 66*, 99–105.

Crane, L. J. (1970). Flow past a stretching plate. *Zeitschrift für angewandte Mathematik und Physik ZAMP, 21*, 645–647. http://dx.doi.org/10.1007/BF01587695

Das, K. (2012). Slip flow and convective heat transfer of nanofluids over a permeable stretching surface. *Computers & Fluids, 64*, 34–42. http://dx.doi.org/10.1016/j.compfluid.2012.04.026

Datta, N., & Mishra, S. K. (1982). Boundary layer flow of a dusty fluid over a semi infinite flat plate. *Acta Mechanica, 42*, 71–83. http://dx.doi.org/10.1007/BF01176514

Dutta, B. K., Roy, P., & Gupta, A. S. (1985). Temperature field in flow over a stretching sheet with uniform heat flux. *International Communications in Heat and Mass Transfer, 12*, 89–94. http://dx.doi.org/10.1016/0735-1933(85)90010-7

Elbashbeshy, E. M. A. (2001). Heat transfer over an exponentially stretching continuous surface with suction. *Archives of Mechanics, 53*, 643–651.

Gireesha, B. J., Ramesh, G. K., & Bagewadi, C. S. (2012). Heat transfer in MHD flow of a dusty fluid over a stretching sheet with viscous dissipation. *Journal of Applied Sciences Research, 3*, 2392–2401.

Gupta, P. S., & Gupta, A. S. (1977). Heat and mass transfer on a stretching sheet with suction or blowing. *The Canadian Journal of Chemical Engineering, 55*, 744–746. http://dx.doi.org/10.1002/cjce.v55:6

Hayat, T., Hussain, T., Shehzad, S. A., & Alsaedi, A. (2015). Flow of oldroyd-B fluid with nanoparticles and thermal radiation. *Applied Mathematics and Mechanics, 36*, 69–80. http://dx.doi.org/10.1007/s10483-015-1896-9

Hussain, T., Hayat, T., Shehzad, S. A., Alsaedi, A., & Chen, B. (2015). A model of solar radiation and joule heating in flow of third grade nanofluid. *Zeitschrift für Naturforschung A, 70*, 177–184.

Ishak, A. (2011). MHD boundary layer flow due to an exponentially stretching sheet with radiation effect. *Sains Malaysiana, 40*, 391–395.

Khan, W. A., & Pop, I. (2010). Boundary-layer flow of a nanofluid past a stretching sheet. *International Journal of Heat and Mass Transfer, 53*, 2477–2483. http://dx.doi.org/10.1016/j.ijheatmasstransfer.2010.01.032

Kumaran, V., & Ramanaiah, G. (1996). A note on the flow over a stretching sheet. *Acta Mechanica, 116*, 229–233. http://dx.doi.org/10.1007/BF01171433

Mabood, F., & Khan, W. A. (2014). Approximate analytic solutions for influence of heat transfer on MHD stagnation point flow in porous medium. *Computer & Fluids, 100*, 72–78.

Mabood, F., Khan, W. A., & Ismail, A. I. M. (2015a). MHD stagnation point flow and heat transfer impinging on stretching sheet with chemical reaction and transpiration. *Chemical Engineering Journal, 273*, 430–437. http://dx.doi.org/10.1016/j.cej.2015.03.037

Mabood, F., Khan, W. A., & Ismail, A. I. M. (2015b). MHD boundary layer flow and heat transfer of nanofluids over a nonlinear stretching sheet: A numerical study. *Journal of Magnetism and Magnetic Materials, 374*, 569–576. http://dx.doi.org/10.1016/j.jmmm.2014.09.013

Magyari, E., & Keller, B. (1999). Heat and mass transfer in the boundary layers on an exponentially stretching continuous surface. *Journal of Physics D: Applied Physics, 32*, 577–585. http://dx.doi.org/10.1088/0022-3727/32/5/012

Maxwell, J. (1904). *A treatise on electricity and magnetism* (2nd ed.). Cambridge: Oxford University Press.

Mukhopadhyay, S., & Gorla, R. S. R. (2012). Effects of partial slip on boundary layer flow past a permeable exponential stretching sheet in presence of thermal radiation. *Heat Mass Transfer, 48*, 1773–1781. http://dx.doi.org/10.1007/s00231-012-1024-8

Nadeem, S., Zaheer, S., & Fang, T. (2011). Effects of thermal radiation on the boundary layer flow of a Jeffrey fluid over an exponentially stretching surface. *Numerical Algorithms, 57*, 187–205. http://dx.doi.org/10.1007/s11075-010-9423-8

Pal, D. (2010). Mixed convection heat transfer in the boundary layers on an exponentially stretching surface with magnetic field. *Applied Mathematics and Computation, 217*, 2356–2369. http://dx.doi.org/10.1016/j.amc.2010.07.035

Partha, M. K., Murthy, P. V. S. N., & Rajasekhar, G. P. (2005). Effect of viscous dissipation on the mixed convection heat transfer from an exponentially stretching surface. *Heat and Mass Transfer, 41*, 360–366. http://dx.doi.org/10.1007/s00231-004-0552-2

Pavithra, G. M., & Gireesha, B. J. (2013). Effect of internal heat generation/absorption on dusty fluid flow over an exponentially stretching sheet with viscous dissipation. *Journal of Mathematics, 2013*, 1–10.

Rudraswamy, N. G., & Gireesha, B. J. (2014). Influence of chemical reaction and thermal radiation on MHD boundary layer flow and heat transfer of a nanofluid over an exponentially stretching sheet. *Journal of Applied Mathematics and Physics, 02*, 24–32. http://dx.doi.org/10.4236/jamp.2014.22004

Saffman, P. G. (1962). On the stability of laminar flow of a dusty gas. *Journal of Fluid Mechanics, 13*, 120–128. http://dx.doi.org/10.1017/S0022112062000555

Sahoo, B., & Poncet, S. (2011). Flow and heat transfer of a third grade fluid past an exponentially stretching sheet with partial slip boundary condition. *International Journal of Heat and Mass Transfer, 54*, 5010–5019. http://dx.doi.org/10.1016/j.ijheatmasstransfer.2011.07.015

Sajid, M., & Hayat, T. (2008). Influence of thermal radiation on the boundary layer flow due to an exponentially stretching sheet. *International Communications in Heat and Mass Transfer, 35*(3), 1–19.

Sakiadis, B. C. (1961). Boundary-layer behavior on continuous solid surfaces: II. The boundary layer on a continuous flat surface. *AIChE Journal, 7*, 221–225. http://dx.doi.org/10.1002/(ISSN)1547-5905

Sanjayanand, E., & Khan, S. K. (2006). On heat and mass transfer in a viscoelastic boundary layer flow over an exponentially stretching sheet. *International Journal of Thermal Sciences, 45*, 819–828. http://dx.doi.org/10.1016/j.ijthermalsci.2005.11.002

Vajravelu, K., & Nayfeh, J. (1992). Hydromagnetic flow of a dusty fluid over a stretching sheet. *International Journal of Non-Linear Mechanics, 27*, 937–945. http://dx.doi.org/10.1016/0020-7462(92)90046-A

Initial value problem of fractional order

A. Guezane-Lakoud[1]*

*Corresponding author: A. Guezane-Lakoud, Faculty of Sciences, Department of Mathematics, Laboratory of Advanced Materials,Badji Mokhtar-Annaba University, P.O. Box 12, 23000 Annaba, Algeria
E-mail: a_guezane@yahoo.fr

Reviewing editor: Yong Hong Wu, Curtin University of Technology, Australia

Abstract: In this work, we discuss the existence of positive solutions for a class of fractional initial value problems. For this, we rewrite the posed problem as a Volterra integral equation, then, using Guo–Krasnoselskii theorem, positivity of solutions is established under some conditions. An example is given to illustrate the obtained results.

Subjects: Science; Mathematics & Statistics; Applied Mathematics

Keywords: initial value problem; fixed point theorem; positivity of solution

Mathematics Subject classifications: 05C38; 15A15; 15A18; 34B10; 26A33; 34B15

1. Introduction

This work is devoted to the study of positive solutions for the following fractional differential equation with initial conditions

$$(P):\begin{cases} D^q_{0^+}u(t) = f\left(t, u(t), u'(t)\right), & 0 < t \le 1, \\ u\left(0\right) = u'(0) = u''\left(0\right) = 0. \end{cases}$$

Where $f\colon [0,1] \times \mathbb{R} \times \mathbb{R} \to \mathbb{R}$ is a given function, $2 < q < 3$, $D^q_{0^+}$ denotes the Riemann's fractional derivative. We note that few papers dealing with fractional differential equations, considered the nonlinearity f in (P) depending on the derivative of u, due to this fact we need more assumptions on f and the problem becomes more complicated.

Fractional initial value problems have been studied recently by many authors. In the paper of Yoruk, Gnana Bhaskar, and Agarwal (2013), Krasnoselskii-Krein, Nagumo's type uniqueness result and successive approximations have been extended to differential equations of fractional order $0 < q < 1$. Some results in literature are given for boundary value problems for ordinary differential equation, by Webb (2009) and Graef, Kong, and Wang (2008) in the case where the Green function associated to the posed problem is vanishing on a set of zero measure. By means of Guo–Krasnosel'skii fixed point theorem the existence of nontrivial positive solution is proved.

ABOUT THE AUTHOR

A. Guezane-Lakoud is a professor in Mathematics at Badji Mokhtar Annaba University, Algeria. She received her PhD degree of Science in Mathematics from this University. Her research interests are on partial differential equations, ordinary and fractional differential equations and inequalities. For more information, please see http://fbedergi.sdu.edu.tr/docs/GuezaneLakoud.pdf

PUBLIC INTEREST STATEMENT

Under suitable conditions on the nonlinearity term, we prove the existence of positive solutions for an initial fractional value problem. The proofs are based on a fixed point theorem.

Existence and positivity of solutions for boundary value problems have been studied by using different methods, such as fixed point theory, topological degree methods, upper and lower solutions... (see Agarwal, O'Regan, & Stanek, 2010; Ahmad & Nieto, 2009; Cabada & Infante, 2013; Graef et al., 2008; Guezane-Lakoud & Khaldi, 2012a; 2012b; 2012c; Guo & Lakshmikantham, 1988; Henderson & Thompson, 2000; Infante & Webb, 2002; Lakshmikantham & Vatsala, 2008; Ntouyas, Wang, & Zhang, 2011; Webb, 2009; 2001; Webb & Infante, 2008).

In this work, we discuss the existence of positive solutions for the problem (P). To prove our results, we assume some conditions on the nonlinear term f, then we use a cone fixed point theorem due to Guo–Krasnoselskii.

2. Preliminaries

We present some definitions from fractional calculus theory which will be needed later (see Kilbas, Srivastava, & Trujillo, 2006; Podlubny, 1999).

Definition 2.1 The Riemann–Liouville fractional integral of order $\alpha > 0$ of a function g is defined by

$$I_{a^+}^{\alpha} g(t) = \frac{1}{\Gamma(\alpha)} \int_a^t \frac{g(s)}{(t-s)^{1-\alpha}} ds.$$

Definition 2.2 The Riemann fractional derivative of order q of g is defined by

$$D_{a^+}^{q} g(t) = \frac{1}{\Gamma(n-q)} \left(\frac{d}{dt}\right)^n \int_a^t \frac{g(s)}{(t-s)^{q-n+1}} ds,$$

where $n = [q] + 1$. ($[q]$ is the integer part of q).

LEMMA 2.3 *The homogenous fractional differential equation* $D_{a^+}^{q} g(t) = 0$ *has a solution*

$$g(t) = c_1 t^{q-1} + c_2 t^{q-2} + \cdots + c_n t^{q-n}$$

where $c_i \in \mathbb{R}, i = 1, \ldots, n$ *and* $n = [q] + 1$.

LEMMA 2.4 *Let* $p, q \geq 0$, $f \in L_1[a, b]$. *Then* $I_{0^+}^{p} I_{0^+}^{q} f(t) = I_{0^+}^{p+q} f(t) = I_{0^+}^{q} I_{0^+}^{p} f(t)$ *(properties of semi-groups) and* $D_{a^+}^{q} I_{0^+}^{q} f(t) = f(t)$, *for all* $t \in [a, b]$.

We start by solving an auxiliary problem which allows us to get the expression of the solution, let us consider the following linear problem (P_0):

$$\begin{aligned} &D_{0^+}^{q} u(t) = y(t), \quad 0 < t \leq 1, \\ &u\left(0\right) = u'(0) = u''\left(0\right) = 0. \end{aligned} \tag{2.1}$$

LEMMA 2.5 *Assume that* $y \in C([0, 1], \mathbb{R})$, *then the problem* (P_0) *has a unique solution given by:*

$$u(t) = \frac{1}{\Gamma(q)} \int_0^t (t-s)^{q-1} y(s) ds. \tag{2.2}$$

Proof Using Lemmas 2.3 and 2.4, we get :

$$u(t) = I_{0^+}^{q} y(t) + at^{q-1} + bt^{q-2} + ct^{q-3}. \tag{2.3}$$

The condition $u(0) = 0$ implies that $c = 0$. Differentiating both sides of (2.5) and using the initial condition $u'(0) = 0$, it yields $b = 0$. The condition $u''(0) = 0$ implies $a = 0$. Substituting a, b and c by their values in (2.5), we obtain

$$u(t) = I_{0^+}^{q} y(t) = \frac{1}{\Gamma(q)} \int_0^t (t-s)^{q-1} y(s) ds. \tag{2.4}$$

Let E be the Banach space of all function $u \in C^1([0,1])$ into $\mathbb{R}$ with the norm $||u|| = ||u||_\infty + ||u'||_\infty$ where $||u||_\infty = \max_{t\in[0,1]} |u(t)|$. Define the operator $T: E \to E$ as follows:

$$Tu(t) = \frac{1}{\Gamma(q)} \int_0^t (t-s)^{q-1} f(s, u(s), u'(s))\, ds. \tag{2.5}$$

LEMMA 2.6 *The function $u \in E$ is solution of the initial value problem (P) if and only if $Tu(t) = u(t)$, for all $t \in [0,1]$.*

From here we see that to solve the FIVP (P) it remains to prove that the map T has a fixed point in E.

3. Main results

First, we state the assumptions that will be used to prove the existence of positive solutions:

($\mathbf{H_1}$) $f(t,u,v) = g(t) f_1(u,v)$ where $g \in L^1([0,1], \mathbb{R}_+)$, $f_1 \in C(\mathbb{R}_+ \times \mathbb{R}_+, \mathbb{R}_+)$, $f_1(0,0) \neq 0$.

($\mathbf{H_2}$) There exists two positive constants g_1 and g_2 such that $0 < g_1 \le g(t) \le g_2$ for all $t \in [0,1]$.

The operator $T: E \to E$ becomes

$$Tu(t) = \frac{1}{\Gamma(q)} \int_0^t (t-s)^{q-1} g(s) f_1(u(s), u'(s)) ds.$$

Let us introduce the following notations

$$A_\delta = \lim_{(|u|+|v|)\to\delta} \frac{f_1(u,v)}{u+v}, \ (\delta = 0^+ \text{or} + \infty).$$

Let K be the classical cone

$$K = \{u \in E, u(t) \ge 0, u'(t) \ge 0, \text{ for all } t \in [0,1]\}.$$

Recall the definition of a positive solution:

Definition 3.1 A function u is called positive solution of problem (P) if $u(t) \ge 0, \forall t \in [0,1]$ and it satisfies the differential equation and the initial conditions in (P).

Now, we give the main result of this paper

THEOREM 3.2 *Under the assumptions (H_1) and (H_2) and if f_1 is convex and decreasing to each variables (i.e. for u fix, $f_1(u,.)$ is decreasing according to the second variable and for v fix the function $f_1(.,v)$ is decreasing according to the first variable), then the problem (P) has at least one nontrivial positive solution in the cone K, in the case $A_0 = +\infty$ and $A_\infty = 0$.*

Recall that a function $F: \Delta = [a,b] \times [c,d] \to \mathbb{R}$ is convex on Δ if

$$F(\lambda x + (1-\lambda)z, \lambda y + (1-\lambda)w) \le \lambda F(x,y) + (1-\lambda)F(z,w)$$

holds for all $(x, y),(z, w) \in \Delta$ and $\lambda \in [0, 1]$.

Jensen's inequality for a convex function is given by:

THEOREM 3.3 (Zabandan & Kılıçman, 2012) *Let p be a non-negative continuous function on* $[a, b]$ *such that* $\int_a^b p(x)dx > 0$. *If g and h are real-valued continuous functions on* $[a, b]$ *and* $m_1 \le g(x) \le M_1$, $m_2 \le h(x) \le M_2$ *for all* $x \in [a, b]$, *and F is convex on* $\Delta = [m_1, M_1] \times [m_2, M_2]$, *then*

$$F\left(\frac{\int_a^b g(t)p(t)dt}{\int_a^b p(t)dt}, \frac{\int_a^b h(t)p(t)dt}{\int_a^b p(t)dt}\right) \le \frac{\int_a^b F\left(g(t), h(t)\right) p(t)dt}{\int_a^b p(t)dt}.$$

The inequalities hold in reversed order if f is concave on Δ.

For the proof of Theorem 3.2, we need the following results:

LEMMA 3.4 (Wang, 2003) *If* f_1 *is continuous then* $A_0^* = A_0$ *and* $A_\infty^* = A_\infty$, *where* $A^*: \mathbb{R}_+ \to \mathbb{R}_+$, $A^*(r) = \max\left\{f_1(u, v), 0 \le u + v \le r\right\}$ *and*

$$A_\delta^* = \lim_{r \to \delta} \frac{A^*(r)}{r} \quad , \left(\delta = 0^+ or + \infty\right).$$

For the proof of Theorem 3.2, we use the following version of Guo–Krasnoselskii fixed point theorem Guo and Lakshmikantham (1988):

THEOREM 3.5 *Let E be a Banach space, and let* $K \subset E$ *be a cone. Assume* Ω_1 *and* Ω_2 *are open-bounded subsets of E with* $0 \in \Omega_1$, $\overline{\Omega_1} \subset \Omega_2$ *and let*

$$\mathcal{A}: K \cap \left(\overline{\Omega_2} \backslash \Omega_1\right) \to K$$

be a completely continuous operator such that

(i) $||\mathcal{A}u|| \le ||u||, u \in K \cap \partial\Omega_1$, *and* $||\mathcal{A}u|| \ge ||u||, u \in K \cap \partial\Omega_2$; *or*

(ii) $||\mathcal{A}u|| \ge ||u||, u \in K \cap \partial\Omega_1$, *and* $||\mathcal{A}u|| \le ||u||, u \in K \cap \partial\Omega_2$.

Then $\mathcal{A}$ *has a fixed point in* $K \cap \left(\overline{\Omega_2} \backslash \Omega_1\right)$.

Proof of Theorem 3.2. Using Ascoli Arzela Theorem, we prove that T is a completely continuous operator. From $A_0 = +\infty$, we deduce that for $M \ge \frac{\Gamma(q+2)}{g_1}$, there exists $r_1 > 0$, such that if $0 < u + v \le r_1$ then $f_1(u, v) \ge M(u + v)$. Let $\Omega_1 = \left\{u \in E, ||u|| < r_1\right\}$, we should prove the first statement of Theorem 3.5. Assume that $u_1 \in K \cap \partial\Omega_1$, then the mean value theorem implies

$$\begin{aligned} ||Tu_1|| \ge ||Tu_1||_\infty &\ge \int_0^1 Tu_1(t)dt \\ &= \frac{1}{\Gamma(q)} \int_0^1 \left(\int_0^t (t-s)^{q-1}dt\right) g(s)f_1(u_1(s), u_1'(s))ds \\ &= \frac{1}{\Gamma(q+1)} \int_0^1 (1-s)^q g(s)f_1(u_1(s), u_1'(s))ds. \end{aligned}$$

Now from the convexity of f_1, then Jensen's inequality and the assumption (H_2), it yields

$$\begin{aligned}||Tu_1|| &\geq \frac{1}{\Gamma(q+1)}\int_0^1 (1-s)^q g(s) f_1(u_1(s), u_1'(s))ds \\ &\geq \frac{1}{\Gamma(q+1)}\left(\int_0^1 (1-s)^q g(s)ds\right) \\ &\quad\times f_1\left(\frac{\int_0^1 (1-s)^q g(s)u_1(s)ds}{\int_0^1 (1-s)^q g(s)ds}, \frac{\int_0^1 (1-s)^q g(s)u_1'(s)ds}{\int_0^1 (1-s)^q g(s)ds}\right) \\ &\geq \frac{g_1}{\Gamma(q+2)} f_1\left(\frac{\int_0^1 (1-s)^q g(s)u_1(s)ds}{\int_0^1 (1-s)^q g(s)ds}, \frac{\int_0^1 (1-s)^q g(s)u_1'(s)ds}{\int_0^1 (1-s)^q g(s)ds}\right),\end{aligned}$$

consequently

$$||Tu_1|| \geq \frac{g_1}{\Gamma(q+2)} f_1\left(\frac{\int_0^1 (1-s)^q g(s)u_1(s)ds}{\int_0^1 (1-s)^q g(s)ds}, \frac{\int_0^1 (1-s)^q g(s)u_1'(s)ds}{\int_0^1 (1-s)^q g(s)ds}\right). \tag{3.1}$$

Since $\frac{\int_0^1 (1-s)^q g(s)u_1(s)ds}{\left(\int_0^1 (1-s)^q g(s)ds\right)} \leq ||u_1||_\infty, \frac{\int_0^1 (1-s)^q g(s)u_1'(s)ds}{\left(\int_0^1 (1-s)^q g(s)ds\right)} \leq ||u_1'||_\infty$ and f_1 is decreasing in each variables, then (3.1) becomes

$$\begin{aligned}||Tu_1|| &\geq \frac{g_1}{\Gamma(q+2)} f_1\left(\frac{\int_0^1 (1-s)^q g(s)u_1(s)ds}{\int_0^1 (1-s)^q g(s)ds}, \frac{\int_0^1 (1-s)^q g(s)u_1'(s)ds}{\int_0^1 (1-s)^q g(s)ds}\right) \\ &\geq \frac{g_1}{\Gamma(q+2)} f_1\left(||u_1||_\infty, ||u_1'||_\infty\right) \geq \frac{g_1}{\Gamma(q+2)} M\,||u_1|| \geq ||u_1||.\end{aligned}$$

Secondly, taking into account Lemma 3.4 and the fact that $A_\infty = 0$, it results that $A_\infty^* = 0$, so for $0 < \varepsilon \leq \frac{\Gamma(q+1)}{(q+1)g_2}$, there exists $R > 0$, such that if $r \geq R$ then $A^*(r) \leq \varepsilon r$. Let $r_2 < \max\left(r_1, R\right)$ and set $\Omega_2 = \left\{u \in E, ||u|| < r_2\right\}$, it is easy to see that $\overline{\Omega_1} \subset \Omega_2$. Assume that $u_2 \in K \cap \partial\Omega_2$, then

$$\begin{aligned}||Tu_2|| &= \frac{1}{\Gamma(q)} \max_{t\in[0,1]} \int_0^t \left((t-s)^{q-1} + (q-1)(t-s)^{q-2}\right) g(s) f_1(u_2(s), u_2'(s))ds \\ &\leq \frac{g_2 A^*(r_2)}{\Gamma(q)} \max_{t\in[0,1]} \int_0^t \left((t-s)^{q-1} + (q-1)(t-s)^{q-2}\right) ds \\ &\leq \frac{g_2(q+1)\varepsilon r_2}{\Gamma(q+1)} \leq ||u||.\end{aligned}$$

then from the second statement of Theorem 3.5, T has a fixed point in $K \cap \left(\overline{\Omega_2}\backslash\Omega_1\right)$.

Example 3.6 Let us consider the problem (P) with $q = \frac{5}{2}, f_1(u,v) = \frac{1}{1+u+v}, f_1(0,0) \neq 0, g(t) = 1 + t^2$, $g_1 = 1, g_2 = 2$. We check easily that $A_0 = +\infty$ and $A_\infty = 0$ and that the assumptions $\left(H_1\right) - \left(H_2\right)$ are satisfied. Theorem 3.2 implies that there exists at least one nontrivial positive solution in the cone K.

Acknowledgements
The author would like to thank the anonymous referees for their valuable remarks.

Funding
This work is done under the research project CNEPRU Code B01120120002.

Author details
A. Guezane-Lakoud[1]
E-mail: a_guezane@yahoo.fr
[1] Faculty of Sciences, Department of Mathematics, Laboratory of Advanced Materials, Badji Mokhtar-Annaba University, P.O. Box 12, 23000 Annaba, Algeria.

References

Agarwal, R. P., O'Regan, D., & Stanek, S. (2010). Positive solutions for Dirichlet problems of singular nonlinear fractional differential equations. *Journal of Mathematical Analysis and Applications, 371*, 57–68.

Ahmad, B., & Nieto, J. J. (2009). Existence results for a coupled system of nonlinear fractional differential equations with three-point boundary conditions. *Computers & Mathematics with Applications, 58*, 1838–1843.

Cabada, A., & Infante, G. (2013, July 29). Positive solutions of a nonlocal Caputo fractional BVP. arXiv 1307.7651 v1 [math.CA].

Graef, J. R., Kong, L., & Wang, H. (2008). A periodic boundary value problem with vanishing Green's function. *Applied Mathematics Letters, 21*, 176–180.

Guezane-Lakoud, A., & Khaldi, R. (2012a). Solvability of a two-point fractional boundary value problem. *The Journal of Nonlinear Science and Applications, 5*, 64–73.

Guezane-Lakoud, A., & Khaldi, R. (2012b). Positive solution to a higher order fractional boundary value problem with fractional integral condition. *Romanian Journal of Mathematics and Computer Sciences, 2*, 28–40.

Guezane-Lakoud, A., & Khaldi, R. (2012c). Solvability of a three-point fractional nonlinear boundary value problem. *Differerential Equations and Dynamical Systems, 20*, 395–403.

Guo, D. J., & Lakshmikantham, V. (1988). Nonlinear problems in abstract cones. In *Notes and reports in mathematics in science and engineering* (Vol. 5). Boston, MA: Academic Press.

Henderson, J., & Thompson, H. B. (2000). Multiple symmetric positive solutions for a second order boundary value problem. *Proceedings of the American Mathematical Society, 128*, 2373–2379.

Infante, G., & Webb, J. R. L. (2002). Nonzero solutions of Hammerstein integral equations with discontinuous kernels. *Journal of Mathematical Analysis and Applications, 272*, 30–42.

Kilbas, A. A., Srivastava, H. M., & Trujillo, J. J. (2006). *Theory and applications of fractional differential equations*. Amsterdam: B. V. Elsevier.

Lakshmikantham, V., & Vatsala, A. S. (2008). Basic theory of fractional differential equations. *Nonlinear Analysis, 69*, 2677–2682.

Ntouyas, K., Wang, G., & Zhang, L. (2011). Positive solutions of arbitrary order nonlinear fractional differential equations with advanced arguments. *Opuscula Mathematica, 31*, 433–442.

Podlubny, I. (1999). *Fractional differential equations mathematics in sciences and engineering*. New York, NY: Academic Press.

Wang, H. (2003). On the numbers of positive solutions of a nonlinear systems. *Journal of Mathematical Analysis and Applications, 281*, 287–306.

Webb, J. R. L. (2001). Positive solutions of some three point boundary value problems via fixed point index theory. *Nonlinear Analysis, 47*, 4319–4332.

Webb, J. R. L. (2009). Boundary value problems with vanishing Green's function. *Communications in Applied Analysis, 13*, 587–596.

Webb, J. R. L., & Infante, G. (2008). Positive solutions of nonlocal boundary value problems involving integral conditions. *Nonlinear Differential Equations and Applications NODEA, 15*, 45–67.

Yoruk, F., Gnana Bhaskar, T., & Agarwal, R. P. (2013). New uniqueness results for fractional differential equations. *Applicable Analysis, 92*, 259–269.

Zabandan, G., & Kılıçman, A. (2012). A new version of Jensen's inequality and related results. *Journal of Inequalities and Applications, 2012*, p. 238.

5

Kernel techniques for generalized audio crossfades

William A. Sethares[1]* and James A. Bucklew[1]

*Corresponding author: William A. Sethares, Department of Electrical and Computer Engineering, University of Wisconsin, Madison, WI, USA

E-mail: sethares@wisc.edu

Reviewing editor: Kok Lay Teo, Curtin University, Australia

Abstract: This paper explores a variety of density and kernel-based techniques that can smoothly connect (crossfade or "morph" between) two functions. When the functions represent audio spectra, this provides a concrete way of adjusting the partials of a sound while smoothly interpolating between existing sounds. The approach can be applied to both interpolation crossfades (where the crossfade connects two different sounds over a specified duration) and to repetitive crossfades (where a series of sounds are generated, each containing progressively more features of one sound and fewer of the other). The interpolation surface can be thought of as the two dimensions (time and frequency) of a spectrogram, and the kernels can be chosen so as to constrain the surface in a number of desirable ways. When successful, the timbre of the sounds is changed dynamically in a plausible way. A series of sound examples demonstrate the strengths and weaknesses of the approach.

Subjects: Acoustical Engineering; Arts; Arts Humanities; Audio; Audio Engineering; Engineering Technology; Music; Music Technology; Technology

Keywords: audio crossfade; morphing; kernel techniques; Poisson process; interpolation

1. Introduction

Crossfading between two sounds can be simple: one sound decreases in volume as the second sound increases in volume. More interesting crossfades may attempt to maintain common aspects of the sounds while smoothly changing dissimilar aspects. For example, it may be desirable to gradually transform one sound into another while requiring that nearby partials sweep between nearby partials or it may be advantageous to require that the sound retains its harmonic integrity over the duration of the crossfade. Sometimes called audio morphing, such generalized crossfades are an area of investigation in the computer music field (Hatch, 2004; Slaney, Covell, & Lassiter, 1996;

ABOUT THE AUTHORS

William A Sethares and James A Bucklew are both with the Department of Electrical and Computer Engineering at the University of Wisconsin-Madison. Their research interests include signal processing as applied to audio, images, and telecommunications.

PUBLIC INTEREST STATEMENT

A common cinematic effect is the morphing of one image to another: a person transforms smoothly into a werewolf or the features of one person change fluidly into those of another. The analogous effect in audition is sometimes called a *crossfade*, and this paper examines two kinds of generalized crossfades that allow one sound to smoothly transform into another. Using ideas from differential equations and probability theory, the "kernel" of the crossfade is defined, and its structure helps to determine the behavior of the resulting sound in terms of audible ridges. A number of sound examples present the uses and limitations of the method.

Tellman, Haken, & Holloway, 1995) and the techniques may also find use in speech synthesis, where smoothly connecting speech sounds is not a trivial operation (Farnetani & Recasens, 2010).

Two kinds of crossfades may be distinguished based on the information used and the desired time over which the fade is to be conducted. In *interpolation crossfades*, two sounds A and B are separated in time by some interval t. The goal of the fade is to smoothly and continuously change from A (the source) to B (the destination) over the time t. The fade "fills in" the time between a single (starting) frame in A and a single (ending) frame in B. Figure 1(a) shows this schematically. In a *repetitive crossfade*, the goal is to create a series of intermediate sounds $M_i, i = 1, 2, \dots, n$ each of which exhibits progressively more aspects of B and fewer aspects of A, as shown in Figure 1(b). Observe that repetitive crossfading is formally analogous to image morphing since it creates a series of intermediaries between the specified start and end points. Interpolation crossfades, by filling in a silence between two sounds, can be thought of as a time-stretching procedure where the start and end sounds may be chosen arbitrarily. In both cases, kernel-based techniques can be used to place constraints on and guide the crossfade.

Perhaps the most common strategy for creating audio morphings is to:

(1) derive sets of features f_A and f_B,
(2) create a correspondence where features in sound A are assigned to features in sound B,
(3) interpolate between the corresponding features over the specified time of the morph, and
(4) synthesize the morphed sound from the interpolated features.

Most current approaches to morphing follow the general plan (1–4). For example, Boccardi and Drioli (2001) model the sound as a Gaussian Mixture which is trained on notes from the same instrument played with different intensities or on notes from different instruments. Other approaches exploit the sinusoidal plus noise decomposition of Serra (1994) or use the bandwidth-enhanced sinusoidal approach (Fitz, Haken, Lefvert, & O'Donnell, 2002) to allow for the more faithful reproduction of nonsinusoidal elements in the sound. A variety of spectral manipulations including audio morphings are suggested by Erbe (1994) and Polansky and McKinney (1991). Our previous work (Sethares, Milne, Tiedje, Prechtl, & Plamondon, 2009) separated the noise part of the sound from the tonal part using a median filter, then morphed the two parts independently. Most such methods incorporate peak-finding routines [as may be familiar from McAulay and Quatieri's tracking method (1986)] in the choice of features and use some kind of ad hoc assignment method for creating the correspondences. Tellman et al. (1995) describes some of the issues that arise when carrying out complex assignments.

This paper suggests an alternative procedure for the construction of smooth audio connections that generalizes to any sensible kernel function. An advantage of this method is that two of the common problems in the general scheme (1–4) are avoided. First, no choice of specific features is made and there is no need to locate significant partials or features in the sound. Hence, there can be no mistakes made in identifying such features. Second, since the crossfade is defined by a PDE or, in a probabilistic sense, as a density or kernel function, no correspondence of features is required, and hence there is no possibility of error in the assignment of such correspondences.

Section 2 presents the conceptual and analytical foundations of the method, which reside in the specification of a pair of density-like functions $f_{z|L}$ and $f_{z|R}$ that describe how the left and right spectra of the sound are propagated and a pair of mixing functions G_L and G_R that describe how the spectra are combined. Section 3 presents a number of crossfades between sinusoids that are simple enough to approach analytically, and the idea of a *ridge* able to connect nearby partials is introduced and analyzed. Section 4 then presents several sound examples that demonstrate the basic functioning of the generalized crossfading process and a selection of examples are conducted between both instrumental and environmental sounds, including a set of fades between clarinet multiphonics.

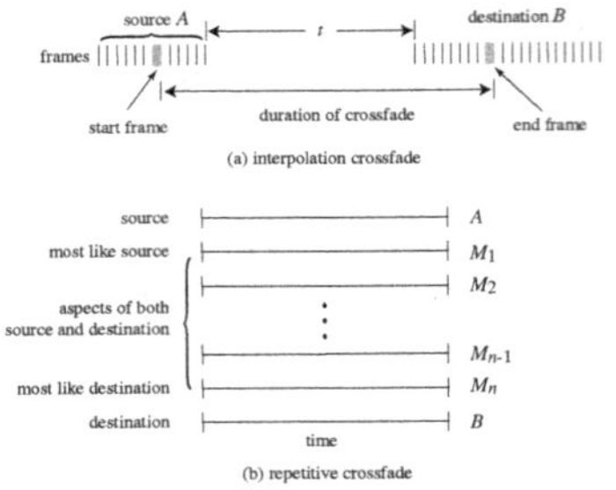

Figure 1. Audio crossfades generate sounds that change smoothly between a source and a

Notes: In interpolation crossfades (a), the sound begins as *A* and over time smoothly becomes like *B*. The total duration of the output sound is independent of the duration of *A* and *B* and the cross only depends on the sound in the starting and ending frames. The overall effect is one of stretching time under the constraint that the sound must emerge continuously from *A* and merge continuously into *B*. In repetitive crossfades (b), a series of intermediate sounds M_i merge aspects of *A* and *B*, analogous to the intermediary photographs of an image morph that merges various aspects of the starting and ending photographs. The duration of each output sound M_i is equal to the common duration of *A* and *B*. Thus, interpolation crosses begin as one sound and end as another, while in a repetitive cross, each M_i contains features of both of the original sounds. F or instance, an interpolation crossfade might start with the attack portion of a cymbal and end with the final moments of a lion's roar. The interpolation crossfade is the transition that occurs over a user specified time. In contrast, each intermediate sound in a repetitivecrossfade merges aspects of both the complete lion sound (from start to end) with those of the complete cymbal (from attack to decay).

Section 4.2 then provides details on the repetitive crossfades along with corresponding sound demonstrations.

2. Crossfading, potentials, and probability theory

Given two functions of a real variable, $S_0(y)$ and $S_d(y)$, the solution to the mathematical crossfade problem may be defined to be a real-valued function of two real variables $S(x,y)$ with domain $D = \{(x,y) \in \Re^2{:}0 \le x \le d, y \in (-\infty, \infty)\}$ and such that $S(0,y) = S_0(y)$ and $S(d,y) = S_d(y)$. The domain D is an infinite strip of width d in the $\Re^2$ plane, with the strip extending from $x = 0$ to $x = d$ and extending infinitely in the positive and negative y directions. The two functions S_0 and S_d act as boundary conditions on the left and right margins (respectively) of the infinite strip. A solution to the crossfade problem is then any real-valued function over the strip that when restricted to the left (right) margin is S_0 (S_d). We often impose additional conditions in order to avoid useless and/or trivial answers. For example, in this paper, we always require that $S(x,y)$ have some sort of smoothness or differentiability on the interior of D to ensure that the surface $S(x,y)$ is smooth.

This is analogous in many ways to the Dirichlet problem which consists of finding a solution to Laplace's equation on some domain D where the solution on the boundary of D is equal to a given function. Perhaps the simplest field equation is Laplace's equation, which is the linear, second-order, steady-state elliptic PDE

$$\nabla^2 u = 0 \tag{1}$$

where ∇^2 is the Laplacian operator. For 2-D rectangular coordinates,

$$\nabla^2 \equiv \frac{\partial^2}{\partial x^2} + \frac{\partial^2}{\partial y^2}. \tag{2}$$

Problems of great physical diversity can be studied using this equation. For example, in the thermal case, the field potential function $u(x,y)$ represents the temperature; in gravitational problems, it is the gravitational potential; in hydrodynamics, it is the velocity potential; and in electrostatics, it is the voltage.

Laplace's equation is the condition required from a variational analysis for minimizing the field energy of a surface "stretched across" the boundaries (Gustafson, 1980). Imagine a rectangular wire frame where the contour of the left-hand side is specified by the spectrum of the sound *A* (given by the function $S_0(y)$), the contour of the right-hand side is given by the spectrum of the sound *B* (given by $S_d(y)$), and where the top and bottom are set to zero as depicted in Figure 2. This is tantamount to an assumption

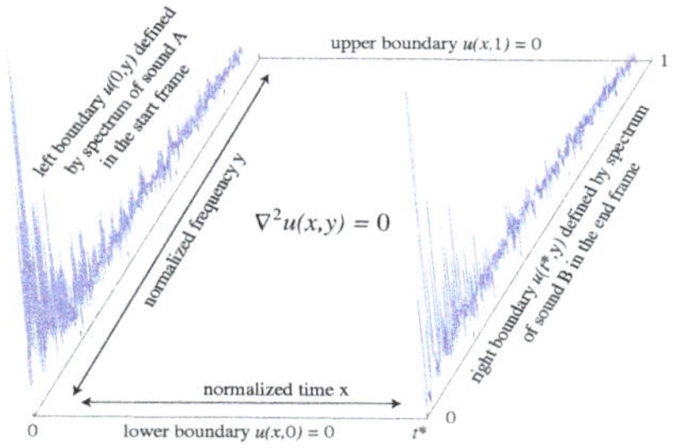

Figure 2. A crossfade surface can be defined by Laplace's equation $\nabla^2 u(x, y) = 0$ with boundary conditions given by the spectra of two sounds *A* and *B*.

Notes: The *x*-axis (representing time) proceeds from time 0 to time t^*, while the *y*-axis (representing frequency) covers the range from DC (at 0) to the Nyquist rate (at 1). The surface is formally analogous to a spectrogram and can be inverted back into the time domain using a variety of standard techniques.

that there is no sound energy at DC and none at high frequencies, for instance, those outside the normal range of hearing. If this wire frame is dipped into a pool of soapy water and carefully retracted, a smooth sheet forms that is characterized as the surface that minimizes the surface energy where the height of the sheet at each point is $u(x, y)$. Mathematically, this can be stated as the PDE (1) with the specified boundary conditions. Reinterpreting the contour of the soap film (i.e. the field values) as sound provides the audio output, which can be heard to smoothly interpolate from the left-hand spectrum to the right-hand spectrum. This views the crossfade function as the solution to a boundary value problem over a two-dimensional domain defined by the spectrum of the sound in the *y* dimension and the duration of the crossfade in the *x* direction. The soapy film is, in essence, reinterpreted as a spectrogram.

Close connections exist between potential theory and the theory of Markov processes (Doob, 1984); most famously, the solution to the Dirichlet problem can be expressed as a functional of the mean hitting time of a standard Brownian motion. Suppose that B_z is a standard two-dimensional Brownian motion whose value at time zero is $z = (x_z, y_z) \in D$. Let $E_z[\cdot]$ denote the expectation operator with respect to this Brownian motion and let $\tau_{\partial D}$ denote the time that the Brownian motion first hits the boundary of the strip $\partial D = \{x = 0\} \cup \{x = d\}$. The value of B_z at this time is $B_z(\tau_{\partial D})$. Defining the "initial condition" function over ∂D as $S_{\partial D}(x, y) = 1_{\{x=0\}} S_0(y) + 1_{\{x=d\}} S_d(y)$, the solution to the Dirichlet problem can be rewritten

$$S(z) = E_z[S_{\partial D}(B_z(\tau_{\partial D}))].$$

A Brownian motion that begins at the point *z* in the interior of D wanders about in *D* until (with probability one) it hits either the left $\{x = 0\}$ or the right $\{x = d\}$ boundary. It is true (and intuitive) that areas on the boundary closer to *z* have a greater chance of being hit than areas further away, and the probability distribution of the points hit on the boundary (the so-called hitting distribution) is

$$f_z(x, y) = \frac{1}{2d}\left(K\left(\frac{x_z \pi}{d}, \frac{(y - y_z)\pi}{d}\right)1_{\{x=0\}} + K\left(\pi - \frac{x_z \pi}{d}, \frac{(y - y_z)\pi}{d}\right)1_{\{x=d\}}\right), \tag{3}$$

where $K(x, y)$ is called the *kernel*. Making the specific choice of the Poisson kernel

$$K(x, y) = P(x, y) = \frac{\sin(x)}{\cosh(y) - \cos(x)} \tag{4}$$

allows a complete analogy with the heat equation. The indicator functions keep track of the hitting distributions on the left and right boundaries. $1_A = 1$ if *A* is true and is zero if *A* is false. Since

$$\int_{-\infty}^{\infty} P(x, y)dy = 2(\pi - x),$$

it can be shown that starting from the point $z = (x_z, y_z)$, the Brownian motion will hit the left boundary with a probability $1 - x_z/d$ and the right boundary with a probability x_z/d. Thus, the hitting distribution conditioned on the event that the left boundary is hit first is

$$f_{z|L}(y) = \frac{1}{2(d - x_z)} K\left(\frac{x_z \pi}{d}, \frac{(y - y_z)\pi}{d}\right) \tag{5}$$

and the hitting distribution conditioned on the event that the right boundary is hit first is

$$f_{z|R}(y) = \frac{1}{2x_z} K\left(\pi - \frac{x_z \pi}{d}, \frac{(y - y_z)\pi}{d}\right). \tag{6}$$

Observe that $f_{z|L}(y)$ converges to the Dirac delta function $\delta(y - y_z)$ as x_z approaches zero and $f_{z|R}(y)$ converges to $\delta(y - y_z)$ as x_z approaches d. Thus, the Dirichlet problem can be restated as

$$S(z) = G_L(x_z) \int_{-\infty}^{\infty} f_{z|L}(y) S_0(y) dy + G_R(x_z) \int_{-\infty}^{\infty} f_{z|R}(y) S_d(y) dy. \tag{7}$$

We call the functions G_L and G_R *crossover functions* because they control the relative hitting probabilities in the x direction. Making the specific choice that $G_L(x_z) = 1 - G(x_z)$ and $G_R(x_z) = G(x_z)$ where

$$G(x_z) = \frac{x_z}{d} \tag{8}$$

allows a complete analogy with the heat equation, that is, $S(z) = S(x_z, y_z)$ is equal to the field potential function $u(x_z, y_z)$ given by the heat Equation (2).

In the crossfade setting, there is no compelling reason that the kernel function $K(x, y)$ must have exactly the form of the Poisson kernel (4) or that the crossover function $G(z)$ must have the form (8). The essence of the audio morphing design problem is encapsulated in (7), which exists independently of the heat equation or reference to Brownian motions. The crucial factors (for audio morphing) are that $f_{z|L}(y)$ converges to $\delta(y - y_z)$, $G_L(x_z) \to 1$, and $G_R(x_z) \to 0$ as x_z converges to zero, and similarly that $f_{z|R}(y)$ converges to $\delta(y - y_z)$, $G_R(x_z) \to 1$, and $G_L(x_z) \to 0$ as x_z converges to d. Together, these allow $S(z)$ to converge to the desired boundary conditions (as z approaches the boundaries), and to smoothly connect the left-hand initial condition with the right-hand initial condition. Thus, the role of the hitting distributions may be played by any kernel that smoothly connects the boundaries, which, in the audio morphing application, are given by the spectra of the starting and ending sounds. By choosing kernel and crossover functions judiciously, fades with a variety of different (audible) properties may be selected.

Example 1 (Simple Linear Crossfade) Let $G(x) = x/d$, $f_{z|L}(y) = \delta(y - y_z)$, and $f_{z|R}(y) = \delta(y - y_z)$. Then, $S(z) = (1 - x_z/d)S_0(y_z) + (x_z/d)S_d(y_z)$.

This crossfade is the standard audio crossfade in which the volume of the first sound is lowered proportionally as the volume of the second is raised. Fortunately, there are more interesting forms of crossfades.

Example 2 (Heat Equation) With $f_{z|L}(y)$ and $f_{z|R}(y)$ chosen as in (5) and (6) with kernel (4) and with $G(x) = x/d$, this is the standard heat equation corresponding to the solution given by (Equation 2) (and the intuition of Figure 2).

The heat equation formulation is used in several of the sound examples as it gives a smooth fade that connects nearby partials at the two endpoints. For instance, a frequency f at the left boundary sweeps smoothly upwards to meet another frequency g at the right boundary. By its nature, the heat equation diffuses energy as it moves away from the boundaries, and this can sometimes be heard as a lowering of the volume of the sound toward the middle of the crossfade surface.

Example 3 *(Harmonic Integrity)* Since the human auditory apparatus perceives pitches (roughly) on a log scale, it makes sense to allow the hitting distribution to scale so that it is wider at higher frequencies. Let $f(z)$ be an arbitrary probability density function and choose a reference frequency y_0. For a point $z = (x_z, y_z)$, define the left hitting density

$$f_{z|L}(y) = \frac{1}{x_z}\frac{y_z}{y_0} f\left((y - y_z)\frac{y_z}{x_z y_0}\right)$$

and the right hitting density

$$f_{z|R}(y) = \frac{1}{d - x_z}\frac{y_z}{y_0} f\left((y - y_z)\frac{y_z}{(d - x_z)y_0}\right).$$

Note that as x_z approaches either zero or d, these hitting distributions collapse to the required boundary Dirac delta functions. This strategy tends to maintain the perceptual integrity of a harmonic collection. A number of other choices for the functional forms of $G_L(x)$, $f_{z|L}(y)$, $G_R(x)$, and $f_{z|R}(y)$ are investigated in the following sections.

3. Crossfades between sinusoids

The simplest setting is where the starting and ending sounds both consist of a small number of sinusoids. In the first example, a pair of sinusoids with normalized frequencies $\omega_{L_1} = 5$ and $\omega_{L_2} = 12$ at the left boundary are crossed with a pair of sinusoids with normalized frequencies $\omega_{R_1} = 6$ and $\omega_{R_2} = 11$ at the right boundary. Accordingly, the left boundary function is the (one-sided) Fourier transform $S_0(y) = \delta(y - \omega_{L_1}) + \delta(y - \omega_{L_2})$ and the right boundary function is $S_\pi(y) = \delta(y - \omega_{R_1}) + \delta(y - \omega_{R_2})$. For simplicity, the duration of the crossfade is scaled to be $d = \pi$ and the two boundary functions only consider positive frequencies (the negative frequencies proceed analogously). Because the boundary functions have a simple form (as a sum of $\delta()$ functions), the crossfade surface (7) can be integrated exactly as

$$S(x, y) = \frac{1}{2\pi}\left(\frac{x\sin(x)}{\cos(x) + \cosh(\omega_{R_2} - y)} + \frac{x\sin(x)}{\cos(x) + \cosh(\omega_{R_1} - y)} \cdots \right.$$
$$\left. + \frac{(\pi - x)\sin(x)}{\cosh(\omega_{L_2} - y) - \cos(x)} + \frac{(\pi - x)\sin(x)}{\cosh(\omega_{L_1} - y) - \cos(x)}\right),$$

when the kernels are chosen to mimic the heat equation as in Example 2 (with $K(x, y)$ given by (4) and $G(x)$ by (8)).

This is plotted in Figure 3(a). The boundaries at the left and right show the two sinusoids (as delta functions at their respective frequencies) while the surface gradually descends to the middle where they meet. Observe that there are two shapes that connect the nearby frequencies, ω_{L_1} to ω_{R_1} and ω_{L_2} to ω_{R_2}. These are local maxima (in the y direction) which form a connected set as x varies over its range; call these *ridges*. Observe that there is a significant loss of height in the ridges of Figure 3(a). Since the magnitude of the surface corresponds to the amplitude of the spectral components, this may be perceptible as a drop in the volume toward the middle of the crossfade region.

Figure 3(b) also uses the Poisson kernel (4) but chooses crossover functions $G_L(x) = (\pi - x)\sin(x)$ and $G_R(x) = x\sin(x)$. This tends to increase the total mass in the middle of the crossfade, and the ridge sags less than in (a). Figure 3(c) shows the results when using crossover functions

$$G_L(x) = G_R(x) = \sin(x). \tag{9}$$

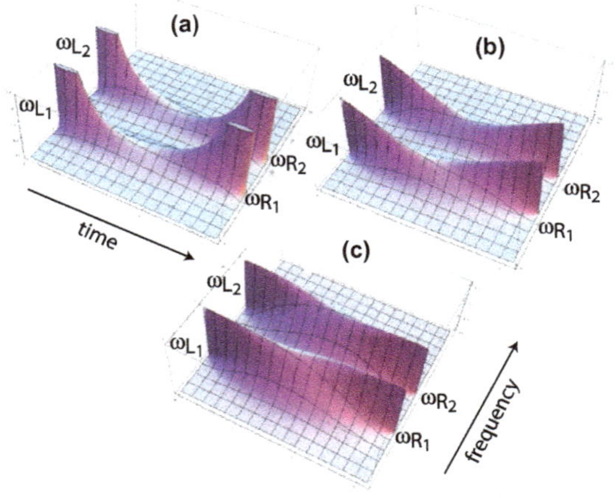

Figure 3. Sinusoids of frequencies $\omega_{L_1} = 5$ and $\omega_{L_2} = 12$ are crossed with frequencies $\omega_{R_1} = 6$ and $\omega_{R_2} = 11$ using the Poisson kernel and three different crossover functions (see text for details).

Notes: Though the ridges connecting the nearby frequencies appear in all three figures, the drop in (a) is likely to be heard as a drop in volume over the course of the first half of the crossfade.

This boosts the ridge to a (near) constant height as it spans the duration to connect the sinusoidal pairs on the two boundaries. Observe that in all three cases, the sinusoids sweep smoothly from their starting to ending frequencies. In contrast, a linear combination of the two sounds (as in the cross-fade of Example 1) has no ridges: the amplitudes of the two starting frequencies die away to zero over the duration of the fade while the amplitudes of the two ending frequencies slowly increase.

The kernels used in Figure 3 have the same width at all frequencies y, which may not be desirable when attempting to cross more complex sounds. Consider a source sound with partials at (relative) frequencies 8, 16, 32, and 64 and a destination sound with partials at 9, 18, 36, and 72. If these sounds are to be spectrally crossed, it is desirable to have $8 \rightarrow 9, 16 \rightarrow 18, 32 \rightarrow 36$, and $64 \rightarrow 72$. With an equal width between all pairs, this is impossible since the distance between 9 and 16 (two partials which should not be connected by a ridge) is less than 8, while the distance between 64 and 72 (two partials which should be connected by a ridge) is 8. This is shown in the left side of Figure 4. While the lower ridges appear as expected, the upper two pairs are not joined together by a ridge. Once again, the freedom to modify the kernels allows a solution. The right-hand side of Figure 4 shows a kernel, as suggested by Example 3, that is narrow at lower frequencies and wider at higher frequencies, allowing ridges to form for all the pairs. The specific kernel used is

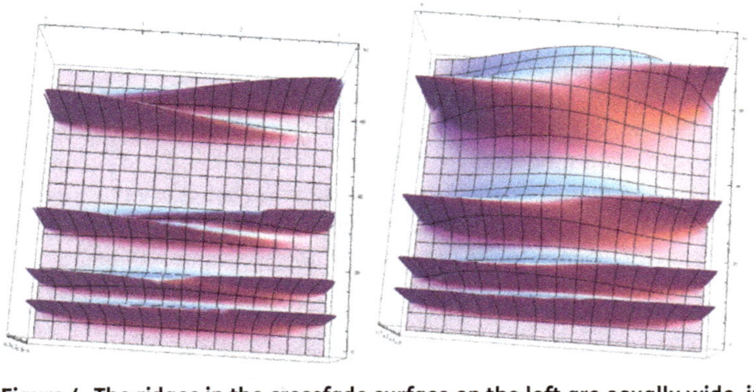

Figure 4. The ridges in the crossfade surface on the left are equally wide, irrespective of the absolute frequency. In some situations, it may be advantageous to allow the width of the ridges to become wider at higher frequencies, as shown on the right. This can be accomplished by defining the kernels as in (10).

$$K(x,y) = \frac{\sin(x)}{\cosh\left(\frac{y-y_0}{cy_0}\right) - \cos(x)}. \quad (10)$$

With $c < 1$, the $K(x,y)$ values stretch more for larger y, mimicking the sensitivity of the auditory system. In subsequent examples, $c = 0.12$ is chosen as a compromise. If c is chosen much larger, the desired ridges fail to exist; if chosen much smaller, the ridges in the low frequencies tend to merge together.

The above discussion emphasizes the importance of the ridges, and it is crucial to be able to make good choices of kernels that lead to desirable ridges. While it is difficult to prove in general when ridges will occur and how wide they are, in the simple case where the kernel is a rectangle function, the existence and behavior of ridges can be described analytically. Viewing the smooth kernels as having a support that can be approximated by an appropriate set of rectangle functions suggests that insights gained from studying the rectangle kernels may be useful in more general situations.

The rectangle function $\mathrm{rect}(x)$ is defined as one for $x \in (-1/2, 1/2)$ and zero otherwise. For $a > 0$, let $f(x) = a\,\mathrm{rect}(ax)$ and define the kernel as in Example 3. The support of the left boundary hitting density is $\left[y - \frac{xy}{ay_0}, y + \frac{xy}{ay_0}\right]$ and the support of the right boundary hitting density is $\left[y - \frac{(d-x)y}{ay_0}, y + \frac{(d-x)xy}{ay_0}\right]$. The support of the left density varies linearly in x (if y is held constant) from zero at $x = 0$ to a maximum of $2dy/y_0a$ at $x = d$ (and similarly for the support of the right density). Consider the crossfade between a pure frequency ω_L on the left boundary to a pure frequency ω_R on the right boundary. Thus, $S_0(y) = \delta(y - \omega_L)$ and $S_d(y) = \delta(y - \omega_R)$. The crossfade surface is

$$\begin{aligned} S(z) &= (1 - G(x))f_{z|L}(\omega_L) + G(x)f_{z|R}(\omega_R) \\ &= (1 - G(x))\frac{a}{x}\frac{y_0}{y}\,\mathrm{rect}\left(a(\omega_L - y)\frac{y_0}{xy}\right) \\ &\quad + G(x)\frac{a}{d-x}\frac{y_0}{y}\,\mathrm{rect}\left(a(\omega_R - y)\frac{y_0}{(d-x)y}\right) \end{aligned}$$

A ridge is said to exist whenever there is a trajectory $T = \{(x, y(x)) : \forall x \in [0,1]$ such that both terms in the above expression are nonzero$\}$.

THEOREM 1 *(The Ridge Theorem) Suppose that $\omega_R > \omega_L$ and that $d < 2ay_0$. A ridge exists if and only if*

$$\frac{\omega_R}{\omega_L} < 1 + \frac{d}{2ay_0}. \quad (11)$$

A proof is given in Appendix A.1.

4. Audio crossfades

This section presents a series of experiments that carry out generalized crossfades between a variety of sounds, including sinusoids, instrumental, and environmental sounds. The experiments demonstrate the ridge theorem concretely by showing the interaction between the width of the kernel and the frequencies joined by the ridges. To be practical, it is desirable to have ridges that connect partials of the starting and ending sounds when the frequencies are close and to *not* have ridges when the frequencies of the partials are distant.

In order to implement the crossfade procedure, it is necessary to discretize the two dimensions, to choose the size n of the FFTs that will be used to specify the boundary spectra, and to select a window that will extract the n samples from the sound waveforms. These choices are familiar from short-time Fourier transform (STFT) modeling (Oppenheim & Schafer, 2009), and the same trade-offs apply. In addition, n must be equal to the number of points in the vertical y direction. We have found $n = 2^{10}, 2^{11}$, and 2^{12} to be convenient and have used a standard Hann window. In the horizontal x direction, we have typically used between $m = 200$ and $m = 500$ points.

The inversion of the two-dimensional surface $S(x, y)$ of (7) into a sound waveform can be accomplished using any of the techniques that would invert an STFT image into sound. The sound examples of this section implement a “phase vocoder” strategy that is well known in applications such as time scaling and pitch transposition (Dolson, 1986; Laroche & Dolson, 1999; Sethares, 2007). This method synthesizes phase values for a given set of magnitude values, effectively choosing phase values that guarantee continuity across successive frames. To be explicit, suppose that the frequency f_i is to be mapped to some value g. Let k be the closest frequency bin in the FFT vector, i.e. the integer k that minimizes $\left|k\frac{sr}{n} - g\right|$ where sr is the sampling rate. Then, the kth bin of the output spectrum at time index $j+1$ has magnitude equal to the magnitude of the ith bin of the input spectrum with corresponding phase

$$\theta_k^{j+1} = \theta_k^j + 2\pi\, dt\, g \tag{12}$$

where dt is the time separation between consecutive frames. The phase values in (12) guarantee that the resynthesized partials are continuous across frame boundaries, reducing the likelihood of discontinuities and clicks. An advantage of this approach is that it allows the duration of the fade to be freely chosen after the solution to the crossfade surface has been obtained. Thus, the relationship between t^* in Figure 2 and real time can be freely adjusted even after the calculation of the surface $S(x, y)$.

A series of generalized crossfades demonstrate that the ridges of Figures 3 and 4 are perceived as pitch glides. Sound examples 220to230.wav, 220to240.wav, through 220to270.wav are available at the website (http://sethares.engr.wisc.edu/papers/audioMorph.html, date last viewed 27 September 2015, as are all other sound files discussed throughout the paper). All examples use the kernel $K(x, y)$ in (10) and the crossover function $G(x)$ of (9). In each case, the crossfade starts at the pitch corresponding to the first frequency and rises smoothly to the pitch corresponding to the second frequency, as shown graphically in Figure 5(a). The frequency values are calculated from the output of the phase vocoder using an analysis that interpolates three frequency bins in each FFT frame. In these graphs, the method is accurate to about 2 Hz (far better than the $\frac{44100}{2048} \approx 22$ Hz resolution of the FFT bins).

When the frequencies of the sinusoids at the start and end are far apart, there is less interaction. The sound example in 220to300.wav begins as a sine wave at 220 Hz and ends as a sine wave at 300 Hz. What happens is that the starting sinusoid decreases in amplitude and the ending sinusoid increases in amplitude throughout the process. Essentially, the kernel is no longer wide enough to form ridges and the connecting sound has become a simple crossfade. The instantaneous frequencies of the two sines are shown in Figure 5(b), which show that both sines are individually identifiable throughout the process. The pitches are not completely fixed at 220 and 300, but bend slightly toward each other. The final sinusoidal example shows how superposition applies to the crossfade

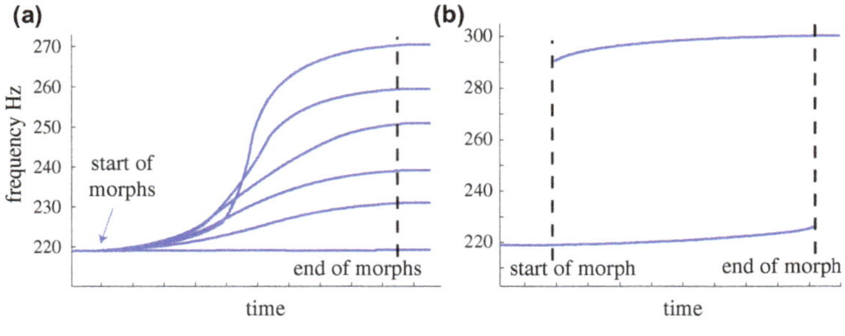

Figure 5. (a) Six different crossfades begin at 220 Hz and proceed to 220, 230, 240, 250, 260, and 270 Hz. Each sounds like a single sine wave that slowly increases in pitch up to the specified frequency. (b) A sinusoid at 220 Hz is crossfaded with a sinusoid at 300 Hz. Because the pitches only bend slightly, the process is almost indistinguishable from a simple amplitude crossfade.

process when the sine waves are far apart in frequency. In the example 220to260+440to400.wav, a sine at 220 glides smoothly to 260, while a sine at 440 glides smoothly to 400. The two are effectively independent. Indeed, the output to the two crossfaded pairs is (almost exactly) the sum of outputs to the two pairs crossfaded separately. Observe that the goal of this method is not to connect every frequency with every other but to gain control over the spreading of the kernel. Indeed, this is exactly what the ridge theorem quantifies: how to trade off the spreading of the energy to nearby frequencies vs. the continuity over the course of the crossfade.

4.1. Instrumental and environmental crossfades

The crossfades in this section are conducted as interpolation crossfades, which stretch time proportional to the x-width of the surface $S(x, y)$. Again, the kernel used is $K(x, y)$ of (10) and the crossover function $G(x)$ is (9). The first two examples cross between single-tone instrumental sounds. In morph-PianoClarinet.wav, an $A2$ attack on the piano changes slowly into a sustained $A2$ on the clarinet. Similarly, in morph-ViolinTrumpet.wav, both instruments play a $C4$ as the attack of the violin crossfades into the sustained portion of the trumpet. Two spectrally rich sounds, a chinese gong and a low C on a minimoog synth, are crossed in morph-GongMinimoog.wav. Several nonobvious effects can be heard including the rising and falling pitch contours, and the slow swelling of the low C toward the end. Then, in morph-GongLion.wav, the same gong recording is crossed with the roar of a lion. Spectrally rich sounds seem to crossfade particularly well.

Multiphonics occur in wind instruments when the coupling between the driver (the reed or lips) and the resonant tube evokes more than a single fundamental pitch. The sounds tend to be inharmonic and spectrally rich; the timbres range from soft and mellow to noisy and harsh. We recorded Paris-based instrumentalist Carol Robinson playing a large number (about 80) of multiphonics. These ranged in duration from brief (a few hundred milliseconds) to fully sustained (several seconds). The timbres ranged from soft and mellow to noisy and harsh. For the present application, a number of these were selected, and sustained crossfades were calculated between a variety of starting and ending multiphonics. These are

morph-MultiXMultiY.wav

where (X,Y) take on values (13, 23), (29, 66), (32, 14), (39, 28), (48, 64), and (74, 53). All of these can be heard (along with the original recordings of the multiphonics) on the website for the paper (http://sethares.engr.wisc.edu/papers/audioMorph.html, date last viewed 27 September 2015). Despite the variety of starting and ending timbres, the crossfades connect smoothly. There are partials that move in frequency (as suggested by the experiments of Section 3) and the basic level of noisiness in some of the samples also changes smoothly throughout the process.

4.2. Repetitive crossfades

Interpolation crossfades tend to change the timbre of the sounds in proportion to the amount time is stretched. Repetitive crossfades more closely parallel visual morphing since the output is a collection of sounds that are each the same duration as the sounds A and B. In this case, the sounds are not partitioned into frames and the boundaries of the crossfade surface are the complete spectra of the sounds. Each column of the solution $S(x, y)$ represents the spectrum of a different intermediate sound.

This distinction has several implications. First, the sounds cannot be too long since they must be analyzed (and inverted) all at once; at the normal CD sampling rate, this limits the duration to a few seconds. Second, the horizontal axis needs only as many points as the desired number of output (intermediate) sounds (recall that for the interpolation crossfades, there needs to be as many mesh points as there are frames in the duration t). Thus, while the frequency y dimension is significantly larger, the time dimension x is significantly smaller. It is possible to be clever. Appendix A.2 shows how, when using the Poisson kernel (4), it is possible to calculate the crossed signal at the midpoint $d/2$ without calculating the complete surface, that is, to calculate $S(d/2, y)$ in isolation. This can reduce the numerical complexity significantly. The method of the Appendix can also be iterated to yield the solutions for $S(d/4, y)$, $S(3d/4, y)$, etc.

Perhaps the greatest difference is in the reinterpretation of the $S(x, y)$ into sound. In the interpolation crossfade, it is necessary to reconstruct the phases of the spectra in some way (for instance, using the phase vocoder strategy as in (12)). In the repetitive crossfade, it is possible to use the complete complex-valued spectra as the boundary conditions; the surface $S(x, y)$ becomes complex valued and each column represents the complete spectrum of the sound.

The first two examples of the repetitive crossfade are between single-tone instrumental sounds. In repmorph-PianoClarinet.wav, an $A2$ attack on the piano is crossed with an $A2$ on the clarinet. Each of the sounds was truncated to about 2.5 seconds, and nine different intermediate sounds were generated. In the soundfile, each of the nine sounds is separated by about 0.25 seconds of silence. The first sound is the trumpet (sound *A*), the last is the clarinet (sound *B*), and the others are the intermediaries. Similarly, in repmorph-TrumpetViolin.wav, both instruments play a $C4$ as the attack of the trumpet is crossed into the violin.

Two spectrally rich sounds, a chinese gong and a low *C* on a minimoog synth, are crossed in repmorph-MinimoogGong.wav. The first 2.5-second sound is the minimoog note, and the next several slowly incorporate increasing amount of gong noise. The final segment is the pure gong sound. Observe that this is quite a different set of effects from the interpolation crossfades of the same sounds. In repmorph-Gong1Gong2.wav, two different gong sounds are faded together, creating a variety of "new" intermediate gong-like sounds. Finally, in repmorph-LionGong.wav, the same gong recording is crossed with the roar of a lion. Spectrally rich sounds cross easily, and the middle sounds are plausible hybrids.

5. Conclusion

By formalizing the idea of a crossfade function as one which smoothly connects two signals, this paper provides a basis for studying processes that underlie sound transitions. The use of a variety of kernels is key, as this specification connects a family of uninteresting transitions (such as simple crossfades) with more interesting transitions (such as spectral crossfades). The ridge theorem delineates in a simple setting when spectral peaks in one signal connect to those in another. The methodology (of regarding the spectrogram as a surface defined by hitting points of a stochastic process) provides some hope that similar questions can also be handled analytically. The mathematics is applied concretely to the problems of interpolation and repetitive crossfades, and each is demonstrated in a handful of sound examples where the strengths and weaknesses of the approach become apparent. In many of the examples, it is possible to clearly hear the ridges, indicating that these plausibly correspond (in an audio sense) to the smooth ridges that appear in Figures 3 and 4. Using this setup, interpolation crossfades appear to be more convincing (as spectrally rich audio morphs) than the repetitive crossfades, which tend to sound more like simple amplitude crossfades. There are two competing factors in the choice of the kernels and the crossover functions: the desire to have a level ridge and the desire to *not* have the energy spread out in the middle. We suspect it may be possible to find better functions using some kind of optimization procedure that trades off these two features, but we have been unable to formulate and solve this in a concrete way.

Acknowledgements
The authors would like to thank Howard Sharpe for extensive discussions during the early phases of this project.

Funding
The authors received no direct funding for this research.

Author details
William A. Sethares[1]
E-mail: sethares@wisc.edu
James A. Bucklew[1]
E-mail: bucklew@engr.wisc.edu
[1] Department of Electrical and Computer Engineering, University of Wisconsin, Madison, WI, USA.

References
Boccardi, F., &Drioli, C. (2001). Sound morphing with Gaussian mixture models. In *Proceedings on 4th COST-G6 Conference on Digital Audio Effects*. Limerick.

Dolson, M. (1986). The phase vocoder: A tutorial. *Computer Music Journal, 10*, 14–27.
Doob, J. L. (1984). *Classical potential theory and its probabilistic counterpart*. Berlin: Springer-Verlag.
Erbe, T. (1994). *Soundhack manual* (pp. 7–40). Lebanon, NH: Frog Peak Music.
Farnetani, E., & Recasens, D. (2010). Coarticulation and connected speech processes. In W. J. Hardcastle, J. Laver, & F. E. Gibbon (Eds.), *Handbook of phonetic sciences* (2nd ed., pp. 316–352). Chichester: Blackwell.
Fitz, K., Haken, L., Lefvert, S., & O'Donnell, M. (2002). Sound morphing using Loris and the reassigned bandwidth-enhanced additive sound model: Practice and applications. In *International Computer Music Conference*. Gotenborg.
Gustafson, K. E. (1980). *Introduction to partial differential equations and Hilbert space methods* (pp. 1–35). Hoboken, NJ: Wiley.
Hatch, W. (2004). *High-level audio morphing strategies* (MS thesis). McGill University, Montreal.
Laroche, J., & Dolson, M. (1999). Improved phase vocoder time-scale modification of audio. *IEEE Transactions on Audio, Speech and Language Processing, 7*, 323–332.
McAulay, R. J., & Quatieri, T. F. (1986). Speech analysis/synthesis based on a sinusoidal representation. *IEEE Transactions on Acoustics, Speech, and Signal Processing, ASSP, 34*, 744–754.
Oberhettinger, F. (1973). *Fourier transforms of distributions and their inverses* (pp. 15–17). New York, NY: Academic Press.
Oppenheim, A. V., & Schafer, R. W. (2009). *Discrete-time signal processing* (3rd ed., pp. 730–742).Upper Saddle River, NJ: Prentice-Hall.
Polansky, L., & McKinney, M. (1991). Morphological mutation functions: Applications to motivic transformations and to a new class of cross-synthesis techniques. *Proceedings of the International Computer Music Conference*. Montreal.
Serra, X. (1994). Sound hybridization based on a deterministic plus stochastic decomposition model. In *Proceedings of the International Computer Music Conference* (pp. 348-351). Aarhus.
Sethares, W. A. (2007). *Rhythm and transforms* (pp. 111–145). London: Springer-Verlag.
Sethares, W. A., Milne, A., Tiedje, S., Prechtl, A., & Plamondon, J. (2009). Spectral tools for dynamic tonality and audio morphing. *Computer Music Journal, 33*, 71–84.
Slaney, M., Covell, M., & Lassiter, B. (1996). Automatic audio morphing. In *Procceedings of the 1996 International Conference on Acoustics, Speech, and Signal Processing*. Atlanta, GA.
Tellman, E., Haken, L., & Holloway, B. (1995). Timbre morphing of sounds with unequal numbers of features. *Journal of the Audio Engineering Society, 43*, 678–689.

Appendix 1

Proof of the Ridge Theorem

Fix a value of x in the interval $[0, d]$. There is a nonzero contribution from both terms as long as the upper part of the rectangle for the first term extends further than the lower part of the rectangle for the second term. The y value for where the upper part of the rectangle for the first term terminates satisfies

$$(y - \omega_L)\frac{ay_0}{xy} = \frac{1}{2}$$
$$y = \frac{\omega_L}{1 - \frac{x}{2ay_0}}.$$

Similarly, the y value for where the lower part of the rectangle for the second term terminates satisfies

$$(y - \omega_R)\frac{ay_0}{((d - x)y} = -\frac{1}{2}$$
$$y = \frac{\omega_R}{1 - \frac{d-x}{2ay_0}}.$$

Thus, the condition for overlap is

$$\frac{\omega_L}{1 - \frac{x}{2ay_0}} > \frac{\omega_R}{1 - \frac{d-x}{2ay_0}}$$
$$\frac{\omega_R}{\omega_L} < \frac{2ay_0 + (d - x)}{2ay_0 - x}$$

It is easy to verify that the right-hand side of the above inequality is increasing in x and thus takes on its minimum value at $x = 0$. This gives the theorem statement. □

A computational simplification

Let $P(x, y)$ be the Poisson kernel (4). The line where $x = d/2 = \pi/2$ represents the center strip of the crossfade surface. A Brownian motion started on this center strip has the hitting distribution

$$
\begin{aligned}
f_{\pi/2}(y) &= \frac{1}{2\pi}(P(\pi/2,y)1_L + P(\pi-\pi/2,y)1_R) \\
&= \frac{1}{2\pi}(\frac{1}{\cosh(y)}1_L + \frac{1}{\cosh(y)}1_R) \\
&= \frac{1}{2\pi}(\frac{2\exp(|y|)}{\exp(2|y|)+1})(1_L + 1_R).
\end{aligned}
$$

To find the characteristic function or Fourier transform of this probability density,

$$
\begin{aligned}
z_x(y) &= \frac{P(x,y)}{2(\pi-x)} \\
&= \frac{1}{2(\pi-x)}\frac{\sin(x)}{\cosh(y)-\cos(x)}.
\end{aligned}
$$

The following transform pair can be found in Oberhettinger (1973), Table 1A, Even Functions, # 201:

$$f(x) \Longleftrightarrow g(y)$$

$$\frac{1}{2N}\frac{1}{\cosh(ax)+\cos(b)} \Longleftrightarrow \frac{1}{N}\frac{1}{a}\pi\csc(b)\frac{\sinh(\frac{by}{a})}{\sinh(\frac{\pi y}{a})}$$

where $N = \frac{b}{a}\csc(b) b < \pi$. Letting $a = 1$, $b = \pi - t$, and $N = (\pi - t)\csc(\pi - t)$ gives the transform relation

$$\frac{1}{\cosh(x)-\cos(t)} \Longleftrightarrow \frac{2\pi}{\sin(\pi-t)}\frac{\sinh[(\pi-t)y]}{\sinh[\pi y]}$$

Hence,

$$
\begin{aligned}
z_x(y) &= \frac{1}{2(\pi-x)}\frac{\sin(x)}{\cosh(y)-\cos(x)} \\
&\Longleftrightarrow \frac{\pi}{(\pi-x)}\frac{\sin(x)}{\sin(\pi-x)}\frac{\sinh[(\pi-x)\omega]}{\sinh[\pi\omega]} \\
&= \frac{\pi}{(\pi-x)}\frac{\sinh[(\pi-x)\omega]}{\sinh[\pi\omega]} \\
&= Z_x(\omega)
\end{aligned}
$$

where $Z_x(\omega)$ is the characteristic function (and Fourier transform since we are dealing with even functions) of $z_x(y)$.

6

Competitive facility location problem with attractiveness adjustment of the follower on the closed supply chain

Arsalan Rahmani[1]*

*Corresponding author: Arsalan Rahmani, Department of Mathematics, University of Kurdistan, Pasdaran Boulevard, P. O. Box 416, Sanandaj, Iran

E-mail: arsalan.rah@gmail.com

Reviewing editor: Akiko Yoshise, University of Tsukuba, Japan

Abstract: In this paper, the problem examined concerns a firm entering a new facility in the market where facilities belonging to the firm and another competitor firm exist. The market's reverse logistics network that has attracted growing attention with stringent pressures from environmental and social requirements is considered. The firm wants to maximize its profit by finding the best location and the most attractive facility to open. The other firm (the competitor) can counteract this situation and attempt to maximize its own profit by adjusting the attractiveness of its existing facilities. The demand is assumed to be aggregated at certain points in the plane and the facilities of the firm can be located at pre-determined candidate sites. To do this, Huff's gravity-based rule in modeling the behavior of the customers is employed where the fraction of customers at a demand point that visit a certain facility is proportional to the facility attractiveness and inversely proportional to the distance between the facility site and demand point. A mathematical bi-level mixed-integer nonlinear programming model where the firm entering the market is the leader and the competitor is the follower is delivered. Furthermore, for finding the optimal solution of this model, it is converted into a one-level mixed-integer nonlinear program so that it can be solved by global optimization methods.

ABOUT THE AUTHOR

He is a faculty member of the Mathematics Department at the University of Kurdistan, Sanandaj, Iran.

His field of specialty includes combinatorial optimization, mathematical modeling, and stochastic programming.

In 2015, he received his PhD in Applied Mathematics (Operations Research) in the Department of Mathematics and Computer Science, Amirkabir University of Technology, Tehran, Iran. In 2011, he completed his MSc degree in Applied Mathematics (Operations Research) in the Department of Mathematics and Computer Science, Amirkabir University of Technology, Tehran, Iran. In 2008, he completed his BSc degree in Pure Mathematics from the University of Kurdistan, Sanandaj, Iran. He is also the member of Operations Research Society, Iran.

PUBLIC INTEREST STATEMENT

Competitive facility location (CFL) problems differ from the classical facility location problems considered in the operations research area because they incorporate the competition among the facilities belonging to different firms. The new facility or facilities to be located by a firm have to compete with the facilities of the other firm(s) that are already (or will be) present in the market in order to capture market share. The nature of the competition is the primary factor determining the class of the CFL problem.

In this paper, the proposed model is applicable for sales and marketing systems which encompass at least two firms as competitors. The solution methodology and the proposed model can be used separately as a basic way for using in the same work in mathematics, engineering science, or even management branches.

The computational results on some examples obtained from random generated problems show that the method is able to solve the Bi-Level Programming Problem (BLPP) efficiently in a reasonable time.

Subjects: Applied Mathematics; Engineering & Technology; Industrial Engineering & Manufacturing; Mathematics & Statistics; Mathematics & Statistics for Engineers; Science; Technology

Keywords: competitive location problems; attractiveness adjustment; closed supply chain; leader–follower problem; convexify method; B&B method

1. Introduction

In many competitive location problems, a firm wants to open a set of new facilities to provide goods to customers of a given geographical area where other competing firms offering the same goods are already present. The new facility or facilities to be located by a firm have to compete with the facilities of other firm(s) that are already (or will be) present in the market in order to capture market share. Many factors influence such decisions, including the available budget and expansion strategy of the firm, the location of the firm's existing facilities and their performance, location of competitors' facilities, the market size, demographics (e.g. population, age, gender, purchasing power) of the demand-generating population, target customer segments, availability and logistical suitability of candidate locations, and existence of other nearby attractions for generating traffic.

Huff (1964, 1966) proposed one basic formulation for estimating the market share among the competing facilities. He assumes that the probability that a customer patronizes a certain facility is proportional to its floor area and inversely proportional to a certain power of the distance from it. Huff's model has been improved in (Cooper & Nakanishi, 1974) by replacing the floor area by a product of attraction factors. One application of this model has been found in (Jane & Mahajan, 1979) and then (Hadgson, 1981) which suggested to replace the power of the distance by an exponential function of the distance which accelerates the distance decay. Literature reviews encompass "the probability of consumers patronizing a facility" called attractiveness. It means that a customer patronizing certain facility is proportional to its attraction toward the facility. This paradigm is expressed as a parameter called the facility quality (depends on the characteristics of the facility), divided by a non-negative nondecreasing function of the distance from it. (For more details see (Drezner, 1994), (Fernández, Pelegrín, Plastria, & Tóth, 2007)) In their models, the market share captured by a facility depends on its quality and its location.

In this paper, it is supposed that the firm as leader wants to locate a single new facility in a given area of the plane. In this chain, already $m + s$ facilities exist which offer the same good or product. It is further supposed that the first m facilities belong to the leader and the next s facilities belong to the competitor firm as follower. This problem includes competitive facility location CFL problem. It was assumed that the demand is inelastic and concentrated at n demand points; these demand points include both demand points for new products (delivered from facilities to customers) and used products (delivered from customers to facilities).

The primary works comprising of the CFL problem are (Hakimi, 1983) and (Revelle, 1986). In these problems, the new facilities must be located on a network space to compete with a number of existing (competitor) facilities. The authors use the assumption that a customer always visits the nearest open facility. According to Huff, customers' patronization of different facilities depends on the attractiveness of each facility and is also inversely proportional to the distance to the facility. This means that customers may not always choose the nearest facility; they may sometimes visit a distant facility because it is more attractive. This idea has been extended a step further by introducing additional attributes for GIS-Based by (Nakanishi & Cooper, 1974) (a geographic information system or geographical information system (GIS) is a system designed to capture, store, manipulate, analyze, manage, and present all types of spatial or geographical data).The optimization attitude for

Competitive Multi-Facility Location-Routing Problem is calculating facility attractiveness in the form of a multiplicative interaction model. In this work, this rule is applied. Moreover, our considered chain includes forward and reverses logistics operations via three kinds of facilities; forward processing facility, backward processing facility, and hybrid processing facility. New products are delivered to a number of geographically dispersed customers from plants via forward processing facilities. Used products are taken back from the customers and shipped to the plants via backward processing facilities for the purpose of recovery or safe disposal. In this paper, instead of only handling separate forward processing and collection facilities, a type of intermediate depot, namely hybrid processing facility, is also taken into account. Both new products and used products can be transferred via hybrid processing facilities. Advantages of building such hybrid processing facilities might include cost savings and pollution reduction as a result of sharing material handling equipment and infrastructure.

We found how to manipulate the B&B method by reviewing existing B&B methods in (Dempe & Gadhi, 2007) for solving Bi-level programming problem (BLPP). However, B&B methods require linear or convex quadratic nonlinear functions in the lower level problem of the bi-level programming (BP) models. In our bi-level model, the objective function of the follower's problem is nonlinear and non-quadratic. Thus, for applying a solution method, the mixed-integer nonlinear BP formulation is firstly converted into a one-level mixed-integer nonlinear programming (MINLP) problem by using the fact that the follower's problem is a continuous concave maximization problem. Therefore, the Karush–Kuhn–Tucker optimality conditions make it necessary and efficient. Although the NLP relaxation of resulting equivalent one-level MINLP problem is not a concave programming problem, we transformed it by introducing new variables so that it can be solved for global optimality by applying the General Structure Mixed-Integer Nonlinear a BB (GMIN-aBB) algorithm which is based on parameters and branch and bound method.

The rest of this paper is organized as follows: Section 2 describes the model formulation. In Section 3, an insight into the solution method is provided. In Section 4, concluding remarks and related examples are summarized and finally in Section 5 some conclusions are drawn.

2. Model description

In this section, the problem faced by the firm during the course of entering a single facility into the market is at first described. Then, BP formulation is delivered in which the market entrant firm that attempts to enter (locate) a facility is the leader and the other competing firms (referred to as the competitor) existing in the market are the followers. The firm wishes to determine the optimal location and attractiveness of the new kind of facility to be opened so as to maximize its profit. In addition, there are m existing facilities belonging to the leader and $s$existing facility belongs to the competitor. It is assumed that the follower reacts to the market entry of the new facility (by leader) via adjusting the attractiveness of some or all of its existing facilities for maximizing its profit. Attractiveness level of a facility is determined by a function of its attributes related to the facility. For example, when the considered facility is a shopping mall, these factors include the variety of stores, availability of food, court/restaurants, adequate parking, accessibility by public transportation, selling price of products and existence of movie theaters. (Note that decreasing the attractiveness to zero means that the facility is shut down).

Let J denote the set of customer locations and $j \in J$ indicate individual customer locations which is considered as aggregate population demand centers in our model. Furthermore, let E, M and S denote the set of potential location, leader and follower facilities, respectively. In our problem setting, there are n_1 demand points of new products and there are $n_2 = n - n_1$ demand points of used products which are indexed by j.

The other parameters and decision variables of the model are as follows:

h_j	Annual buying power at point j of both new and used products
c_l	Unit attractiveness cost at candidate site l
$\bar{c}_k$	Unit attractiveness cost or revenue of follower's facility at site k
f_l^f	Annualized fixed cost of opening a forward processing facility at site l
f_l^h	Annualized fixed cost of opening a hybrid processing facility at site l
f_l^r	Annualized fixed cost of opening a backward processing facility at site l
$u_l^f(u_l^h, u_l^r)$	Maximum attractiveness level for a forward (hybrid, backward) processing facility to be opened at site l
d_{lj}	Euclidean distance between candidate site l and point j
$\bar{d}_{kj}$	Euclidean distance between existing facility site k and point j
$\tilde{A}_k$	Current attractiveness level of competitor's facility at site k
$\bar{A}_k$	Maximum attractiveness level of competitor's facility at site k Decision variables
$g_l^f(g_l^h, g_l^r)$	Attractiveness of the forward (hybrid, backward) processing facility opened at site l
x_l	Binary variable which is equal to one if a facility is opened at site l, and zero if otherwise
A_k	New attractiveness level of competitor's facility at site k

A location solution vector X consists of binary decision variables $x_l^f(x_l^h, x_l^r)$ where $l \in E$ and $x_l^f = 1(x_l^h = 1, x_l^r = 1)$ indicates facility l is open for forward (hybrid, returned) processing facility, i.e. new facility is established in candidate location j.

We assume that each population center patronizes an open facility belonging to set E, M, and S proportional to the facility's attractiveness score and inversely proportional to the distance between the demand center and the facility. In other words, each demand center has a "utility" towards an open facility which is calculated as shown in the following formulation:

$$u_{ij} = \frac{A_i}{d_{ij}^2}$$

Hence the patronage probabilities of a demand center j against all open facilities are formulated as follows:

$$p_{ij} = \frac{u_{ij}}{\sum_i u_{ij}}$$

Under this patronization behavior, customers visit facilities according to probabilities p_{ij} and contribute to the overall revenues that are collected by the leader firm and its competitor follower. Consequently, the proportion p_{ij} of customers at point j (formulation is the same for both the new demand and used product) who visit a new facility at site l is expressed as:

$$p_{lj} = \frac{\frac{g_l^f + g_l^h}{d_{lj}^2}}{\frac{g_l^f + g_l^h}{d_{lj}^2} + \sum_{i=1}^{m} \frac{A_i}{\bar{d}_{ij}^2} + \sum_{k=1}^{s} \frac{A_k}{\bar{d}_{kj}^2}}$$

Now, we can formulate the following bi-level MINLP model P as follows:

$$P\colon \max \sum_{j=1}^{n_1} h_j \frac{\sum_{l\in E} \frac{g_l^f + g_l^h}{d_{lj}^2} + \sum_{i=1}^{m} \frac{A_i}{\bar{d}_{ij}^2}}{\sum_{l\in E} \frac{g_l^f + g_l^h}{d_{lj}^2} + \sum_{i=1}^{m} \frac{A_i}{\bar{d}_{ij}^2} + \sum_{k=1}^{s} \frac{A_k}{\bar{d}_{kj}^2}} + \sum_{j=n_1+1}^{n} h_j \frac{\sum_{l\in E} \frac{g_l^h + g_l^r}{\bar{d}_{lj}^2} + \sum_{i=1}^{m} \frac{A_i}{\bar{d}_{ij}^2}}{\sum_{l\in E} \frac{g_l^h + g_l^r}{d_{lj}^2} + \sum_{i=1}^{m} \frac{A_i}{\bar{d}_{ij}^2} + \sum_{k=1}^{s} \frac{A_k}{\bar{d}_{kj}^2}}$$
$$- \sum_{l\in E} f^f x_l^f + f^h x_l^h + f^r x_l^r - \sum_{l\in E} (g_l^f + g_l^h + g_l^r) c_l \tag{1}$$

$$S.T\ g_l^f \le \sum_{l \in E} u_l^f x_l^f \quad \forall l \tag{2}$$

$$g_l^h \le \sum_{l \in E} u_l^h x_l^h \quad \forall l \tag{3}$$

$$g_l^r \le \sum_{l \in E} u_l^r x_l^r \quad \forall l \tag{4}$$

$$\sum_{l \in E} x_l^f + x_l^h + x_l^r = 1 \tag{5}$$

$$x_l^f, x_l^h, x_l^r \in \{0,1\}\ ,\ g_l^f, g_l^h, g_l^r \ge 0 \tag{6}$$

$$\max \sum_{j=1}^{n_1} h_j \frac{\sum_{k=1}^{s} \frac{A_k}{d_{ij}^2}}{\sum_{l \in E} \frac{g_l^f + g_l^h}{d_{lj}^2} + \sum_{i=1}^{m} \frac{A_i}{d_{ij}^2} + \sum_{k=1}^{s} \frac{A_k}{d_{kj}^2}} + \sum_{j=n_1+1}^{n_2} h_j \frac{\sum_{k=1}^{s} \frac{A_k}{d_{ij}^2}}{\sum_{l \in E} \frac{g_l^h + g_l^r}{d_{lj}^2} + \sum_{i=1}^{m} \frac{A_i}{d_{ij}^2} + \sum_{k=1}^{s} \frac{A_k}{d_{kj}^2}}$$

$$- \sum_{k=1}^{s} \bar{c}_k (A_k - \tilde{A}_k) \tag{7}$$

$$S.T \quad A_k \le \bar{A}_k \quad \forall k \tag{8}$$

$$A_k \ge 0 \tag{9}$$

The objective function (1) of the leader firm has four components. The first one (resp. the second one) represents the revenue of new (resp. used) products which is collected by the new (resp. pervious) facility of leader firm. The third and the forth components represent the fixed cost and attractiveness cost associated with opening the new facility, respectively. Constraints (2)–(4) along with the binary location and restrictions (5) and (6) on the location variables $x_l^f(x_l^h, x_l^r)$ and non-negativity restrictions existing in (6) on attractiveness variables $g_l^f(g_l^h, g_l^r)$ ensure that if no facility is opened at site l, then the corresponding attractiveness $g_l^f(g_l^h, g_l^r)$ of the facility is zero and if a facility is opened at site i, then its attractiveness $g_l^f(g_l^h, g_l^r)$ cannot exceed the maximum level $u_l^f(u_l^h, u_l^r)$. The objective function of the competitor (7) contains the summation of three terms: the revenue collected by the competitor's facilities and the cost or revenue associated with re-adjusting the attractiveness levels. Constraints (8) and (9) ensure that the new attractiveness A_k of an existing facility at site k is between zero and an upper limit A_k.

For the solution of the above MINLP model, the GMIN-αBB algorithm was employed which will be explained in detail in Section 3. This algorithm performs a pre-processing step where all the terms in the objective function as well as in the constraints are grouped into different classes such as linear, fractional, concave, bilinear, univariate convex, and general nonconcave. Then, for each term in all classes, with the exception of the linear and concave ones, a concave over-estimator is generated.

3. Solution methodologies

Global optimization of BLPP is studied in (Gümüş & Floudas, 2005). Firstly, they propose a convex relaxation of the inner problem followed by its equivalent representation via necessary and sufficient optimality conditions. The, they introduce a BB global optimal principles presented as branch and bound framework. Their problems involve twice differentiable continuous functions. The first rigorous global optimization approach for the calculation of the flexibility test which is bi-level nonlinear optimization model is introduced by (Floudas et al., 1999). They demonstrate the applicability of their model to a heat exchanger network problem, a pump and pip problem, a reactor-cooler system, and a prototype process flow sheet model. (Pistikopoulos, Georgiadis, & Dua, 2007) introduce methods based on parametric programming to transform the bi-level problem into a family of single level optimization problems which can be solved for global optimality. The authors present computational results on several small benchmark linear-linear, linear-quadratic, quadratic-linear, and quadratic-quadratic type problems. (Gümüş & Floudas, 2005) proposes two new methods for BLPP. The first one is applicable when the outer problem includes mixed integer nonlinear and inner problem encompasses twice differentiable continuous problems; the second method is applied to problems including the inner problem featuring functions which are mixed integer nonlinear in the outer variables, and linear, polynomial or multi-linear in the inner integer variables and linear in the inner continuous variables.

In this work, the aBB global optimal approach is proposed for the efficient solution of the proposed problem.

The approach is based on the concept of feasible domain relaxation and the basic principles of the global optimization algorithm a BB. The main feature of the presented approach is that it guarantees convergence to the global optimal solution for problems that are described by general nonconvex constraints. The basic framework of the applied method is described as followed:

(1) If any of the constraints are nonconvex, the inner level of the feasibility problem is nonconvex. Therefore, the Karush–Kuhn–Tucker (KKT) optimality conditions are necessary but not sufficient to guarantee the global optimum of the inner level. Local or even suboptimal solutions may be obtained. Hence, the first step of the proposed framework involves the convexification of the feasible region defined by the inner constraints of the original problem. The basic underestimation principles of the aBB global optimization approach are utilized to underestimate the constraints nonconvex. For the convexified problem, the KKT optimality conditions are necessary and sufficient, assuming that the linear independence constraint qualification is satisfied. Under these conditions, the convexified inner level optimization problem is equivalent to its KKT optimality conditions.

(2) The convexified inner level optimization problem is replaced with its equivalent set of equations that are defined by the KKT optimality conditions; this transforms the bilevel problem into a single level optimization problem. Note that this problem is in general a nonconvex optimization problem due to the complementarity conditions. As a result, further convex underestimation may be needed which takes place according to aBB principles and the solution of the convexified single stage problem provides an upper bound of the original problem *P*.

(3) A lower bound to the problem *P* is determined through a feasible solution of the original Mixed Integer Nonlinear Programming. MINLP formulation is obtained by substituting the inner problem by the KKT optimality conditions (conditions which are only necessary) as proposed by Grossmann and Floudas.

(4) The next step after establishing an upper and a lower bound on the global solution is to refine them. This is accomplished by successfully partitioning the initial region of the variables into smaller ones. The partitioning strategy involves the successive subdivision of a hyperectangle into two sub-rectangles by halving at the middle point of the longest side of the initial rectangle (bisection). In any iteration, the lower bound of the feasibility test and the flexibility index problem is the maximum over the minima found in every sub-rectangle composing of the initial rectangle. Consequently, on-decreasing sequence of lower bounds is generated by halving the sub-rectangle that is responsible for the infimum over the minima obtained at any iteration. A non-increasing sequence of upper bounds is derived by solving the nonconvex MINLP single optimization problem obtained after the substitution of the inner problem by the KKT optimality conditions, and selecting as an upper bound the minimum over all previously determined upper bounds. If at any iteration the solution of the convexified MINLP in any sub-rectangle is found to be greater than the upper bound or the solution is not feasible, this sub-rectangle is fathomed since the global solution cannot be found inside it.

Thus, as described above, the proposed procedure for solving the problem involves the following steps:

Step 1. Set the lower bound $LB = -\infty$, upper bound $UP = \infty$ and select a tolerance ε.

Step 2. Substitute the convexified constraints instead of the inner constraints: we will show that the inner problem of P has concave objective function with convex constraints, and therefore it does not need any convexification in this step.

Proposition 1. The objective function

$$\max \sum_{j=1}^{n_1} h_j \frac{\sum_{k=1}^{s} \frac{A_k}{d_{ij}^2}}{\sum_{l\in E} \frac{g_l^f+g_l^h}{d_{lj}^2} + \sum_{i=1}^{m} \frac{A_i}{d_{ij}^2} + \sum_{k=1}^{s} \frac{A_k}{d_{kj}^2}} + \sum_{j=n_1+1}^{n_2} h_j \frac{\sum_{k=1}^{s} \frac{A_k}{d_{ij}^2}}{\sum_{l\in E} \frac{g_l^h+g_l^f}{d_{lj}^2} + \sum_{i=1}^{m} \frac{A_i}{d_{ij}^2} + \sum_{k=1}^{s} \frac{A_k}{d_{kj}^2}} - \sum_{k=1}^{s} \bar{c}_k (A_k - \tilde{A}_k)$$

is concave.

The constraints of inner problem are linear and so are convex.

Step 3. Substitute the inner optimization problem of the original problem by the KKT optimality conditions:

$$\sum_{j=1}^{n_1} h_j \frac{\frac{1}{d_{kj}^2} \sum_{k=1}^{s} \frac{A_k}{d_{kj}^2} + \frac{1}{d_{kj}^2}\left(\sum_{l\in E} \frac{g_l^f+g_l^h}{d_{lj}^2} + \sum_{i=1}^{m} \frac{A_i}{d_{ij}^2} + \sum_{k=1}^{s} \frac{A_k}{d_{kj}^2}\right)}{\left[\sum_{l\in E} \frac{g_l^f+g_l^h}{d_{lj}^2} + \sum_{i=1}^{m} \frac{A_i}{d_{ij}^2} + \sum_{k=1}^{s} \frac{A_k}{d_{kj}^2}\right]^2} +$$

$$\sum_{j=n_1+1}^{n_2} h_j \frac{\frac{1}{d_{kj}^2} \sum_{k=1}^{s} \frac{A_k}{d_{kj}^2} + \frac{1}{d_{kj}^2}\left(\sum_{l\in E} \frac{g_l^h+g_l^f}{d_{lj}^2} + \sum_{i=1}^{m} \frac{A_i}{d_{ij}^2} + \sum_{k=1}^{s} \frac{A_k}{d_{kj}^2}\right)}{\left[\sum_{l\in E} \frac{g_l^h+g_l^f}{d_{lj}^2} + \sum_{i=1}^{m} \frac{A_i}{d_{ij}^2} + \sum_{k=1}^{s} \frac{A_k}{d_{kj}^2}\right]^2} - \bar{c}_k + \lambda_{1k} - \lambda_{2k} = 0 \qquad \forall\, k$$

$$A_k - \bar{A}_k + s_{1k} = 0 \qquad \forall\, k$$

$$-A_k + s_{2k} = 0 \qquad \forall\, k$$

$$\lambda_{1k} s_{1k} = 0 \qquad \forall\, k$$

$$\lambda_{2k} s_{2k} = 0 \qquad \forall\, k$$

$$\lambda_{1k}, \lambda_{2k}, s_{1k}, s_{2k} \geq 0$$

where $\lambda_1 = (\lambda_{11}, \lambda_{12}, \ldots, \lambda_{1s})$ and $\lambda_2 = (\lambda_{21}, \lambda_{22}, \ldots, \lambda_{2s})$ are Lagrangean multiplier vectors and $s_1 = (s_{11}, s_{12}, \ldots, s_{1s})$ and $s_2 = (s_{21}, s_{22}, \ldots, s_{2s})$ are slack variables corresponding to constraint sets and , respectively. However, we can remove the nonlinearity caused by the complementarily constraints using the active set strategy suggested by Grossmann and Floudas (Grossmann & Floudas, 1987). For example, consider nonlinearity phrase, Grossmann and Floudas condition for this phrase stated as:

$$\lambda \leq Ma$$

$g(z) \leq M(1 - a)$

where $a \in \{0, 1\}$, and M is an upper bound on the slack variables s_{1k}, s_{2k} and y_{1k}, y_{2k} for $k = 1, 2, ..., s$. Consequently, the problem P transformed the following one-level problem:

$$P'\text{: } \max \sum_{j=1}^{n_1} h_j \frac{\sum_{l\in E} \frac{g_l^f + g_l^h}{d_{lj}^2} + \sum_{i=1}^{m} \frac{A_i}{\bar{d}_{ij}^2}}{\sum_{l\in E} \frac{g_l^f + g_l^h}{d_{lj}^2} + \sum_{i=1}^{m} \frac{A_i}{\bar{d}_{ij}^2} + \sum_{k=1}^{s} \frac{A_k}{\bar{d}_{kj}^2}} + \sum_{j=n_1+1}^{n} h_j \frac{\sum_{l\in E} \frac{g_l^h + g_l^r}{d_{lj}^2} + \sum_{i=1}^{m} \frac{A_i}{\bar{d}_{ij}^2}}{\sum_{l\in E} \frac{g_l^h + g_l^r}{d_{lj}^2} + \sum_{i=1}^{m} \frac{A_i}{\bar{d}_{ij}^2} + \sum_{k=1}^{s} \frac{A_k}{\bar{d}_{kj}^2}}$$

$$- \sum_{l\in E} f^f x_l^f + f^h x_l^h + f^r x_l^r - \sum_{l\in E} (g_l^f + g_l^h + g_l^r) c_l \quad (10)$$

$$S.T \quad g_l^f \leq \sum_{l\in E} u_l^f x_l^f \quad (11)$$

$$g_l^h \leq \sum_{l\in E} u_l^h x_l^h \quad (12)$$

$$g_l^r \leq \sum_{l\in E} u_l^r x_l^r \quad (13)$$

$$\sum_{l\in E} x_l^f + x_l^h + x_l^r = 1 \quad (14)$$

$$x_l^f, x_l^h, x_l^r \in \{0, 1\} \quad g_l^f, g_l^h, g_l^r \geq 0 \quad (15)$$

$$\sum_{j=1}^{n_1} h_j \frac{\frac{1}{\bar{d}_{kj}^2} \sum_{k=1}^{s} \frac{A_k}{\bar{d}_{kj}^2} + \frac{1}{\bar{d}_{kj}^2} \left(\sum_{l\in E} \frac{g_l^f + g_l^h}{d_{lj}^2} + \sum_{i=1}^{m} \frac{A_i}{\bar{d}_{ij}^2} + \sum_{k=1}^{s} \frac{A_k}{\bar{d}_{kj}^2} \right)}{\left[\sum_{l\in E} \frac{g_l^f + g_l^h}{d_{lj}^2} + \sum_{i=1}^{m} \frac{A_i}{\bar{d}_{ij}^2} + \sum_{k=1}^{s} \frac{A_k}{\bar{d}_{kj}^2} \right]^2}$$

$$+ \sum_{j=n_1+1}^{n_2} h_j \frac{\frac{1}{\bar{d}_{kj}^2} \sum_{k=1}^{s} \frac{A_k}{\bar{d}_{kj}^2} + \frac{1}{\bar{d}_{kj}^2} \left(\sum_{l\in E} \frac{g_l^h + g_l^r}{d_{lj}^2} + \sum_{i=1}^{m} \frac{A_i}{\bar{d}_{ij}^2} + \sum_{k=1}^{s} \frac{A_k}{\bar{d}_{kj}^2} \right)}{\left[\sum_{l\in E} \frac{g_l^h + g_l^r}{d_{lj}^2} + \sum_{i=1}^{m} \frac{A_i}{\bar{d}_{ij}^2} + \sum_{k=1}^{s} \frac{A_k}{\bar{d}_{kj}^2} \right]^2} - \bar{c}_k + \lambda_{1k} - \lambda_{2k} = 0 \quad (16)$$

$$A_k - \bar{A}_k + s_{1k} = 0 \qquad \forall\, k \quad (17)$$

$$-A_k + s_{2k} = 0 \qquad \forall\, k \quad (18)$$

$$\lambda_{1k} - Ma_{1k} \leq 0 \qquad \forall\, k \quad (19)$$

$$\lambda_{2k} - Ma_{2k} \leq 0 \qquad \forall\, k \quad (20)$$

$$s_{1k} - M(1 - a_{1k}) \leq 0 \qquad \forall\, k \quad (21)$$

$$s_{2k} - M(1 - a_{2k}) \leq 0 \qquad \forall\, k \quad (22)$$

$$\lambda_{1k}, \lambda_{2k}, s_{1k}, s_{2k}, A_k \geq 0, \; a_{1k}, a_{2k} \in \{0, 1\} \quad (23)$$

Step 4. The resulting formulation P' is an MINLP model. For its solution, the GMIN-αBB algorithm was employed. To do this, both upper bound and lower bound of the problem P were obtained. First, it was necessary to get rid of nonconvex terms. To do so, new variables v_{1j},v_{2j} for $j = 1, 2, ..., n_1$ and new variables v_{3j}, v_{4j} for $j = n_1 + 1, 2, ..., n$ were defined as follows:

$$v_{1j} = \frac{\sum_{l\in E} \frac{g_l^f + g_l^h}{d_{lj}^2} + \sum_{i=1}^{m} \frac{A_i}{\bar{d}_{ij}^2}}{v_{2j}} \tag{24}$$

$$\text{And } v_{2j} = \sum_{l\in E} \frac{g_l^f + g_l^h}{d_{lj}^2} + \sum_{i=1}^{m} \frac{A_i}{\bar{d}_{ij}^2} + \sum_{k=1}^{s} \frac{A_k}{\bar{d}_{kj}^2} \tag{25}$$

$$v_{3j} = \frac{\sum_{l\in E} \frac{g_l^h + g_l^r}{d_{lj}^2} + \sum_{i=1}^{m} \frac{A_i}{\bar{d}_{ij}^2}}{v_{4j}} \tag{26}$$

$$\text{And } v_{4j} = \sum_{l\in E} \frac{g_l^h + g_l^r}{d_{lj}^2} + \sum_{i=1}^{m} \frac{A_i}{\bar{d}_{ij}^2} + \sum_{k=1}^{s} \frac{A_k}{\bar{d}_{kj}^2} \tag{27}$$

$$v_{5j} = \frac{\frac{1}{\bar{d}_{kj}^2} \sum_{k=1}^{s} \frac{A_k}{\bar{d}_{kj}^2}}{v_{2j}} \tag{28}$$

$$\text{And } v_{6j} = \frac{\frac{1}{\bar{d}_{kj}^2} \sum_{k=1}^{s} \frac{A_k}{\bar{d}_{kj}^2}}{v_{4j}} \tag{29}$$

Therefore, by using these new variables P' is written as below:

$$P'': \max \sum_{j=1}^{n_1} h_j v_{1j} + \sum_{j=n_1+1}^{n} h_j v_{3j} - \sum_{l\in E} f^f x_l^f + f^h x_l^h + f^r x_l^r \tag{30}$$

$$S.T \quad g_l^f \le \sum_{l\in E} u_l^f x_l^f \qquad \forall\, l \tag{31}$$

$$g_l^h \le \sum_{l\in E} u_l^h x_l^h \qquad \forall\, l \tag{32}$$

$$g_l^r \le \sum_{l\in E} u_l^r x_l^r \qquad \forall\, l \tag{33}$$

$$\sum_{l\in E} x_l^f + x_l^h + x_l^r = 1 \tag{34}$$

$$x_l^f, x_l^h, x_l^r \in \{0,1\}\ ,\ g_l^f, g_l^h, g_l^r \ge 0 \tag{35}$$

$$\sum_{j=1}^{n_1} h_j \frac{1}{\bar{d}_{kj}^2}\left(\frac{v_{5j}}{v_{2j}} + \frac{v_{1j}}{v_{2j}}\right) + \sum_{j=n_1+1}^{n_2} h_j \frac{1}{\bar{d}_{kj}^2}\left(\frac{v_{6j}}{v_{4j}} + \frac{v_{3j}}{v_{4j}}\right) - \bar{c}_k + \lambda_{1k} - \lambda_{2k} = 0 \qquad \forall\, k \tag{36}$$

$$A_k - \bar{A}_k + s_{1k} = 0 \qquad \forall\, k \tag{37}$$

$$-A_k + s_{2k} = 0 \quad \forall\, k \tag{38}$$

$$\lambda_{1k} - Ma_{1k} \le 0 \quad \forall\, k \tag{39}$$

$$\lambda_{2k} - Ma_{2k} \le 0 \quad \forall\, k \tag{40}$$

$$\lambda_{2k} - Ma_{2k} \le 0 \quad \forall\, k \tag{41}$$

$$\lambda_{2k} - Ma_{2k} \le 0 \quad \forall\, k \tag{42}$$

$$s_{1k} - M(1 - a_{1k}) \le 0 \quad \forall\, k \tag{43}$$

$$s_{2k} - M(1 - a_{2k}) \le 0 \quad \forall\, k \tag{44}$$

$$-v_{1j}v_{2j} + \sum_{l \in E} \frac{g_l^f + g_l^h}{d_{lj}^2} + \sum_{i=1}^{m} \frac{A_i}{\bar{d}_{ij}^2} = 0 \quad \forall\, j \tag{45}$$

$$-v_{3j}v_{4j} - \sum_{l \in E} \frac{g_l^h + g_l^r}{d_{lj}^2} + \sum_{i=1}^{m} \frac{A_i}{\bar{d}_{ij}^2} = 0 \quad \forall\, j \tag{46}$$

$$-v_{2j} + \sum_{l \in E} \frac{g_l^f + g_l^h}{d_{lj}^2} + \sum_{i=1}^{m} \frac{A_i}{\bar{d}_{ij}^2} + \sum_{k=1}^{s} \frac{A_k}{\bar{d}_{kj}^2} = 0 \quad \forall\, j \tag{47}$$

$$v_{4j} - \sum_{l \in E} \frac{g_l^h + g_l^r}{d_{lj}^2} + \sum_{i=1}^{m} \frac{A_i}{\bar{d}_{ij}^2} + \sum_{k=1}^{s} \frac{A_k}{\bar{d}_{kj}^2} = 0 \quad \forall\, j \tag{48}$$

$$-v_{5j}v_{2j} + \frac{1}{\bar{d}_{kj}^2} \sum_{k=1}^{s} \frac{A_k}{\bar{d}_{kj}^2} = 0 \quad \forall\, j \tag{49}$$

$$-v_{6j}v_{4j} + \frac{1}{\bar{d}_{kj}^2} \sum_{k=1}^{s} \frac{A_k}{\bar{d}_{kj}^2} = 0 \quad \forall\, j \tag{50}$$

$$\lambda_{1k}, \lambda_{2k}, s_{1k}, s_{2k}, v_{1j}, v_{2j}, v_{3j}, v_{4j}, v_{5j}, v_{6j}, A_k \ge 0,\ a_{1k}, a_{2k} \in \{0, 1\} \tag{51}$$

In the above formulation, nonconvex terms are eliminated, but the cost of this work is introducing new bilinear and fractional terms into the formulation. It helps to avoid the computationally intensive calculations used in the convexification procedure of the terms. To obtain the upper bound, the relaxed version of P'' needed to be solved. Firstly, it was necessary to convexify the nonconvex in it except for the linear and convex ones. The terms for which an over-estimator should be constructed are the fractional and bi-linear terms. To do this, the procedure given by (Floudas, 2000) was used:

Let $v_{1j}v_{2j} = w_{1j}$, $v_{3j}v_{4j} = w_{2j}$, $\frac{v_{1j}}{v_{2j}} = w_{3j}$, $\frac{v_{3j}}{v_{4j}} = w_{4j}$, $\frac{v_{5j}}{v_{2j}} = w_{5j}$, $\frac{v_{6j}}{v_{4j}} = w_{6j}$, $v_{5j}v_{2j} = w_{7j}$, and $v_{6j}v_{2j} = w_{8j}$. Then, to overestimate the fractional terms the following linear constraints are added to P'':

$$\frac{v_{1j}^L}{v_{2j}} + \frac{v_{1j}}{v_{2j}^U} - \frac{v_{1j}^L}{v_{2j}^U} - w_{3j} \le 0 \quad \forall\, j \tag{52}$$

$$\frac{v_{3j}^L}{v_{4j}} + \frac{v_{3j}}{v_{4j}^U} - \frac{v_{3j}^L}{v_{4j}^U} - w_{4j} \leq 0 \qquad \forall j \tag{53}$$

$$\frac{v_{1j}^U}{v_{2j}} + \frac{v_{1j}}{v_{2j}^L} - \frac{v_{1j}^U}{v_{2j}^L} - w_{3j} \leq 0 \qquad \forall j \tag{54}$$

$$\frac{v_{3j}^U}{v_{4j}} + \frac{v_{3j}}{v_{4j}^L} - \frac{v_{3j}^U}{v_{4j}^L} - w_{4j} \leq 0 \qquad \forall j \tag{55}$$

$$\frac{v_{5j}^L}{v_{2j}} + \frac{v_{5j}}{v_{2j}^U} - \frac{v_{5j}^L}{v_{2j}^U} - w_{5j} \leq 0 \qquad \forall j \tag{56}$$

$$\frac{v_{6j}^L}{v_{4j}} + \frac{v_{6j}}{v_{4j}^U} - \frac{v_{6j}^L}{v_{4j}^U} - w_{6j} \leq 0 \qquad \forall j \tag{57}$$

$$\frac{v_{5j}^U}{v_{2j}} + \frac{v_{5j}}{v_{2j}^L} - \frac{v_{5j}^U}{v_{2j}^L} - w_{5j} \leq 0 \qquad \forall j \tag{58}$$

$$\frac{v_{6j}^U}{v_{4j}} + \frac{v_{6j}}{v_{4j}^L} - \frac{v_{6j}^U}{v_{4j}^L} - w_{6j} \leq 0 \qquad \forall j \tag{59}$$

And the bilinear terms are overestimated using the following linear constraints:

$$v_{1j}^L v_{2j} + v_{1j} v_{2j}^L - v_{1j}^L v_{2j}^L - w_{1j} \leq 0 \qquad \forall j \tag{60}$$

$$v_{1j}^U v_{2j} + v_{1j} v_{2j}^U - v_{1j}^U v_{2j}^U - w_{1j} \leq 0 \qquad \forall j \tag{61}$$

$$-v_{1j}^L v_{2j} - v_{1j} v_{2j}^L + v_{1j}^L v_{2j}^L + w_{1j} \leq 0 \qquad \forall j \tag{62}$$

$$-v_{1j}^U v_{2j} - v_{1j} v_{2j}^U + v_{1j}^U v_{2j}^U + w_{1j} \leq 0 \qquad \forall j \tag{63}$$

$$v_{3j}^L v_{4j} + v_{3j} v_{4j}^L - v_{3j}^L v_{4j}^L - w_{2j} \leq 0 \qquad \forall j \tag{64}$$

$$v_{3j}^U v_{4j} + v_{3j} v_{4j}^U - v_{3j}^U v_{4j}^U - w_{2j} \leq 0 \qquad \forall j \tag{65}$$

$$-v_{3j}^L v_{4j} - v_{3j} v_{4j}^L + v_{3j}^L v_{4j}^L + w_{2j} \leq 0 \qquad \forall j \tag{66}$$

$$-v_{3j}^U v_{4j} - v_{3j} v_{4j}^U + v_{3j}^U v_{4j}^U + w_{2j} \leq 0 \qquad \forall j \tag{67}$$

$$v_{5j}^L v_{2j} + v_{5j} v_{2j}^L - v_{5j}^L v_{2j}^L - w_{7j} \leq 0 \qquad \forall j \tag{68}$$

$$v_{5j}^U v_{2j} + v_{5j} v_{2j}^U - v_{5j}^U v_{2j}^U - w_{7j} \le 0 \quad \forall j \tag{69}$$

$$-v_{5j}^L v_{2j} - v_{5j} v_{2j}^L + v_{5j}^L v_{2j}^L + w_{7j} \le 0 \quad \forall j \tag{70}$$

$$-v_{5j}^U v_{2j} - v_{5j} v_{2j}^U + v_{5j}^U v_{2j}^U + w_{7j} \le 0 \quad \forall j \tag{71}$$

$$v_{6j}^L v_{4j} + v_{6j} v_{4j}^L - v_{6j}^L v_{4j}^L - w_{8j} \le 0 \quad \forall j \tag{72}$$

$$v_{6j}^U v_{4j} + v_{6j} v_{4j}^U - v_{6j}^U v_{4j}^U - w_{8j} \le 0 \quad \forall j \tag{73}$$

$$-v_{6j}^L v_{4j} - v_{6j} v_{4j}^L + v_{6j}^L v_{4j}^L + w_{8j} \le 0 \quad \forall j \tag{74}$$

$$-v_{6j}^U v_{4j} - v_{6j} v_{4j}^U + v_{6j}^U v_{4j}^U + w_{8j} \le 0 \quad \forall j \tag{75}$$

where v_{tj}^L and v_{tj}^U are the lower and upper bounds of the on v_{tj} , $t = 1, \ldots 6$, respectively. Therefore, the upper bound of the problem P'' will be obtained by solving the following convexified problem P''' at a given node of the B&B tree when binary variables are relaxed.

$$P''': \max \sum_{j=1}^{n_1} h_j v_{1j} + \sum_{j=n_1+1}^{n} h_j v_{3j} - \sum_{l \in E} f^f x_l^f + f^h x_l^h + f^r x_l^r - \sum_{l \in E} (g_l^f + g_l^h + g_l^r) c_l$$

$$\text{S.T} \quad \sum_{j=1}^{n_1} h_j \frac{1}{d_{kj}^2} \left(w_{5j} + w_{3j} \right) + \sum_{j=n_1+1}^{n_2} h_j \frac{1}{d_{kj}^2} \left(w_{6j} + w_{4j} \right) - \bar{c}_k + \lambda_{1k} - \lambda_{2k} = 0 \quad \forall k \tag{76}$$

$$\lambda_{1k}, \lambda_{2k}, s_{1k}, s_{2k}, v_{tj}, w_{lj}, \ t = 1, ..6, l = 1, .., 8, A_k \ge 0, \ 0 \le a_{1k}, a_{2k} \le 1 \tag{77}$$

And Equations (31–35) and (37–50) and (52–75)

In addition, at the root node of the tree, a local optimal solution is found for the original NLP, which gives a lower bound. In fact a lower bound is obtained by finding a local optimal solution to the problem using, for example, a commercial solver CPLEX that can handle a nonlinear programming problem. To guarantee ε-convergence from the global maximum, i.e. the difference between the lower and upper bounds is within ε of the optimal objective value, the feasible region of the problem is divided at each node such that the rectangles defined by the lower and upper limits on the decision variables are partitioned into smaller ones. This partitioning constitutes the branching mechanism in the B&B tree only.

Algorithm

(1) Set the lower bound $LB = -\infty$, Obtain the relaxed problem P' by relaxing all integer variables. While there are active nodes in the tree, repeat the following steps:

(2) Select an active node according to some branching rule. (Branching at a node is performed by considering the solution at that node and selecting the relaxed binary variable whose value is the closest to 0.5.)

(3) Apply the αBB algorithm and obtain an upper bound at the active node from the solution of P'.

(4) If a feasible solution is obtained, let $LB = UB$, where LB indicates a lower bound to the problem. Update Z_{LB} by $Z_{LB} = \max\{Z_{LB}, LB\}$, prune that node and backtrack. Otherwise, if an infeasible solution is obtained or $UP \leq Z_{LB}$, prune that node and backtrack.

Report the solution with the objective value Z_{LB} as the optimal solution.

4. Computational analysis

The performance of the presented approach was verified experimentally. It was implemented in MATLAB software. All tests were conducted for a fixed number of iterations in random test instances for assessment. The instances were classified based on the number of potential facilities and number of customers. The following policy was applied for generating test instances (for each data-set, five different data instances were created):

- The number of demand points belonging to the {5, 10, 20, 30} where every point encompasses both used and new products
- The number of candidate sites s belonging to {2, 5, 7}
- The number of existing facility of competitors belonging to {2, 3}
- Points representing the facilities and clients are uniformly distributed in [0,100] × [0,100]
- The annualized buying power h_j of customers at point j of new product are uniformly distributed in [10, 10,000] with normal probability distributions in N (50.05, 16. 5^2) and the annualized buying power h_j of customers at point j of used product are uniformly distributed in [5, 5,000] and have normal probability distributions in N (40, 10.5^2)
- The attractiveness cost c_i of candidate facility for the leader at point i are real-valued quantities and uniformly distributed in [0.5,5] with normal probability distributions in N ($5,1.5^2$)
- The attractiveness cost $\tilde{c}_k$ of facility of follower at point k are real-valued quantities and are uniformly distributed in [0.5,5] with normal probability distributions in N (5, 1.5^2)
- The current attractiveness cost $\tilde{A}_k$ of facility at point k are real-valued quantities and are uniformly distributed in [10, 1,000] with normal probability distributions in N (500, 140^2)
- The fixed costs f_i^f, f_i^h and f_i^r are set as 800c_i, 1,100c_i and 550c_i respectively
- Upper bounds u_l^f, u_l^h and u_l^r are taken as 7,500c_i, 8,000c_i, and 7,000c_i respectively, and upper bound $\bar{A}_k$ is assigned a value of $7,500\tilde{c}_k$.

As can be seen, GMIN-αBB is computationally expensive especially for very small values of the convergence parameter ε. To improve the running time efficiency of GMIN-αBB, instead of terminating the αBB algorithm when the difference $Z_{UB} - Z_{LB} \leq \varepsilon\%$, the iterations are stopped as soon as the improvement in the gap $Z_{UB} - Z_{LB}$ between two non-successive iterations becomes less than a user-specified threshold value δ, namely the exploration of BB tree terminated when

$$\frac{\left(Z_{UB}^{t-k} - Z_{LB}^{t-k}\right) - \left(Z_{UB}^{t} - Z_{LB}^{t}\right)}{Z_{UB}^{t-k} - Z_{LB}^{t-k}} \leq \delta,$$

where Z_{UB}^{t} and Z_{UB}^{t-k} are the upper bounds at iterations t and $t - k$, respectively, while Z_{LB}^{t} and Z_{LB}^{t-k} are the best lower bounds at iterations t and $t-k$, respectively. k is a user-specified parameter that represents the number of iterations during which the improvement in the gap between Z_{UB} and Z_{LB} is measured. In our examples, $k = 4$ and $\delta = 0.1$.

We present the results on the test instances in the following Tables. In these tables, Z_L^{best} indicates the best lower bound of the leader obtained by the GMIN-αBB algorithm run with the above-mentioned approximation strategy, and Z_F^{best}, stands for the profit realized by the follower as a reaction to the leader.

Table 1. computational results obtained by GMIN-αBB

Instance (n,m,r)	Z_L^{best}	Z_F^{best}	Computation time solution
(5,2,2)	162,183.6007	235,145.6705	12.929
	36,349.04157	157,060.0119	31.402
	66,631.85411	58,711.14141	38.603
	89,943.47257	223,194.9707	38.526
	138,772.4512	56,199.83209	11.546
(5,2,3)	89,739.0741	75,874.64574	9.728
	173,197.0197	202,176.3623	7.789
	127,315.7748	173,560.5099	35.743
	163,068.2135	285,158.7187	18.609
	154,481.1468	225,034.012	7.734
(5,5,2)	203,396.9806	144,857.8099	23.755
	105,957.5555	253,223.8754	32.685
	86,379.93404	68,352.56194	5.314
	40,739.87958	172,223.5862	5.56
	201,139.1605	214,540.4041	7.532
(5,5,3)	62,911.88963	90,561.81037	22.072
	113,893.9911	247,933.2591	28.99
	85,188.16331	52,256.00463	38.594
	84,427.9808	273,325.108	15.525
	175,078.0963	301,299.9646	45.921
(5,7,2)	138,233.9526	221,840.5653	19.901
	233,885.4458	44,592.87104	34.749
	263,980.0847	182,909.0165	53.742
	122,832.3421	172,596.3891	48.382
	241,437.2665	269,588.1328	56.482
(5,7,3)	69,482.35423	53,840.55126	52.555
	193,436.4351	211,785.1145	67.448
	218,915.7749	175,336.0339	28.899
	153,821.2247	214,380.0306	17.932
	134,490.9615	181,012.1815	17.941
(10,2,2)	70,604.33593	271,916.4984	78.562
	188,830.8969	291,307.906	72.487
	52,431.17585	264,983.3772	99.124
	223,174.397	184,677.3347	57.186
	171,770.5513	55,232.11059	78.522
(10,2,3)	114,472.1819	129,969.3082	60.394
	59,810.44828	179,560.6885	59.202
	96,562.32776	130,097.7135	101.514
	75,264.63391	275,796.4832	103.686
	137,495.253	209,843.7231	97.645
(10,5,2)	330,410.1511	222,858.3721	121.385
	281,186.4398	317,812.4356	90.415
	122,673.6612	339,240.4703	98.017
	212,524.2916	282,221.2561	124.526

(Continued)

Table 1. (Continued)

Instance (n,m,r)	z_L^{best}	z_F^{best}	Computation time solution
	108,363.6277	111,240.4921	52.729
(10,5,2)	335,998.9269	141,323.166	124.512
	299,940.1545	226,142.0719	62.797
	284,357.9325	134,539.5229	95.809
	117,829.0574	141,430.8419	92.808
	106,278.132	51,260.48358	137.385
(10,7,2)	109,500.8225	276,289.3792	118.054
	296,213.3181	128,327.8353	87.64
	363,374.5414	125,776.3877	150.382
	375,706.9231	227,135.0875	157.824
	397,676.5038	251,334.3833	76.778
(10,7,3)	372,090.7241	168,049.5399	136.328
	189,520.9862	295,089.6073	103.129
	67,882.32438	300,067.2645	92.484
	186,791.8758	477,048.4731	41.554
	342,006.7619	281,694.029	102.742
(20,2,2)	199,654.8371	359,331.5461	48.299
	62,380.77262	236,306.7942	77.227
	95,416.56812	81,357.14524	96.299
	120,847.4246	340,503.0967	96.097
	174,115.4013	77,400.55178	34.636
(20,2,3)	118,277.0718	114,184.2124	165.227
	269,544.5983	93,956.01716	72.66
	186,384.8419	385,585.4354	54.061
	251,186.137	206,843.1408	80.703
	235,622.0786	93,386.83148	103.082
(20,5,2)	256,457.0542	271,809.5933	13.227
	521,261.3496	363,142.3941	38.208
	69,508.14009	83,214.97575	54.142
	323,328.3122	85,730.53802	43.77
	426,734.0701	105,891.3373	68.188
(20,5,3)	294,766.428	319,680.5018	126.679
	513,666.5635	376,284.4593	229.163
	241,483.9162	454,857.6198	261.407
	548,986.0524	266,111.7024	110.178
	587,898.447	514,811.3983	237.254
(20,7,2)	42,709.65389	315,852.3485	247.766
	370,151.9575	485,110.7658	284.165
	345,738.3904	437,345.0982	191.173
	575,351.498	509,527.9404	163.559
	86,137.15274	467,642.1405	142.926
(20,7,3)	556,747.484	546,487.4743	389.231
	448,144.1009	342,408.2299	105.065

(Continued)

Table 1. (Continued)

Instance (n,m,r)	Z_L^{best}	Z_F^{best}	Computation time solution
	564,479.2747	284,345.5649	460.78
	465,056.7041	284,392.409	353.689
	199,654.8371	359,331.5461	128.299
(30,2,2)	197,218.2698	400,247.3211	341.568
	468,154.1387	431,408.0801	316.507
	155,571.4447	389,106.2362	426.387
	546,857.9932	260,059.5321	253.391
	429,057.5134	52,049.28754	341.404
(30,2,3)	513,975.5685	262,227.7069	208.015
	343,157.6509	293,709.8753	201.347
	458,007.2743	262,309.2226	492.373
	391,451.981	354,803.3854	424.872
	585,922.6657	312,934.4984	412.929
(30,5,2)	824,098.3337	182,133.6065	231.402
	759,456.5808	288,156.2301	238.603
	531,479.4225	130,667.8153	438.526
	479,196.0078	348,556.6453	211.546
	402,414.2866	270,371.3102	29.728
(30,5,3)	1018,464.901	245,159.8741	47.789
	907,927.4439	333,519.9054	135.743
	551,968.9233	353,459.8821	148.609
	753,738.76	300,400.3355	467.734
	519,834.1113	141,293.2357	203.755
(30,7,2)	726,401.7239	311,873.8549	405.314
	682,214.9942	178,286.8042	215.56
	209,986.6841	188,336.6445	527.532
	177,231.5601	56,838.20522	422.072
	360,750.0297	473,714.6627	728.99
(30,7,3)	707,342.0875	223,085.925	838.594
	832,012.6217	218,764.0853	415.525
	854,905.0668	390,453.3115	45.921
	895,686.9233	431,443.9555	19.901
	162,183.6007	235,145.6705	12.929

The computational results in Table 1 show that the GMIN-αBB method solves the proposed model very efficiently and the model is extremely efficient in finding integer solutions. Most of the time, GMIN-αBB method was not invoked to obtain integer solutions, which shows that our model is integer- friendly. It is worth noting that our method for obtaining a lower bound often generated an optimal solution or a high-quality approximate solution with only one or two locations different from the optimal location. Furthermore, for showing the affectedness of attractiveness adjustment of the follower facilities, the advantages of the leader and the disadvantages of the competitor were firstly quantified when the leader takes into consideration the reaction of the competitor (*fixed* A_k). In the second stage, the aim was to quantify the benefit of the leader and the benefit of the competitor when the leader takes into consideration the reaction of the competitor but when the upper bound of the attractiveness is changed to half ($\bar{A}_k/2$). For each instance, the percentage of loss from the competitor was computed.

Table 2. The loss of the competitor, in the case of (fixed $\bar{A}_k$) and ($\bar{A}_k/2$)

Instance (n,m,r)	Loss (%) (fixed A_k)	Loss (%) ($\bar{A}_k/2$)	Instance (n, m, r)	Loss (%) (fixed A_k)	Loss (%) ($\bar{A}_k/2$)
(5,2,2)	−71.2528	−26.7198	(20,2,2)	−65.0711	−24.4017
	−54.6343	−20.4879		−5.15566	0
	−43.8187	−16.432		−52.3502	−19.6313
	3.29196	1.23449		−4.4203	−1.65761
	−72.4594	−27.1723		−18.5769	−6.96633
(5,2,3)	−55.606	−20.8522	(20,2,3)	−40.0674	−15.0253
	0	0		−21.0946	−7.91047
	−23.7305	−8.89895		−78.0695	−29.2761
	−38.513	−14.4424		−49.0011	−18.3754
	−6.34794	0		−24.699	−9.26214
(5,5,2)	−5.82299	−2.18362	(20,5,2)	−53.3279	−19.998
	−18.7321	−7.02455		4.55657	3.70872
	−56.6354	−21.2383		−2.64937	−0.99351
	−2.28611	−0.85729		−67.5641	−25.3365
	−60.41	−22.6538		−15.3931	−5.77243
(5,5,3)	−8.75085	−3.28157	(20,5,3)	−27.1221	−10.1708
	−12.6961	−4.76103		−21.8111	−8.17915
	−15.6241	−5.85904		−14.8479	−5.56798
	−56.4283	−21.1606		−68.186	−25.5698
	−63.1049	−23.6643		−7.77452	−2.91545
(5,7,2)	−36.635	−13.7381	(20,7,2)	−62.7091	−23.5159
	−58.6022	−21.9758		−4.28833	0
	−25.311	−9.49163		−2.02326	−0.75872
	−28.8598	−10.8224		−7.85695	−2.94636
	2.16348	0.8113		−77.3864	−29.0199
(5,7,3)	−23.4194	−8.78227	(20,7,3)	−65.1301	−24.4238
	−9.48925	0		−50.1543	−18.8079
	−54.4435	−20.4163		−41.2046	−15.4517
	−15.1416	−5.67811		−14.8676	−5.57536
	−41.2847	−15.4818		−70.3426	−26.3785
(10,2,2)	−32.813	−12.3049	(30,2,2)	−69.0811	−25.9054
	−11.2048	−4.20178		−71.9083	−26.9656
	0	0		−2.18569	−0.81963
	−11.7362	−4.40109		−70.5922	−26.4721
	−47.7947	−17.923		−57.2363	−21.4636
(10,2,3)	−67.9163	−25.4686	(30,2,3)	−36.1381	−13.5518
	−41.6369	−15.6138		−5.20567	0
	−55.1919	−20.697		−44.5298	−16.6987
	−20.8229	−7.80859		−40.7943	−15.2979
	−72.8108	−27.304		−3.7339	−1.40021
(10,5,2)	−51.3014	−19.238	(30,5,2)	−73.7731	−27.6649

(Continued)

Table 2. (Continued)

Instance (n,m,r)	Loss (%) (fixed A_k)	Loss (%) ($\bar{A}_k/2$)	Instance (n, m, r)	Loss (%) (fixed A_k)	Loss (%) ($\bar{A}_k/2$)
	−78.657	−29.4964		−73.1438	−27.4289
	−13.7076	−5.14034		−66.9033	−25.0888
	−37.6759	−14.1285		−64.4928	−24.1848
	−65.2035	−24.4513		−71.6976	−26.8866
(10,5,2)	−20.0422	−7.51584	(30,5,3)	−37.2121	−13.9546
	−66.7327	−25.0248		−63.3512	−23.7567
	−23.6039	−8.85146		−76.108	−28.5405
	−54.6415	−20.4906		−68.0937	−25.5351
	−21.5778	−8.09167		−48.574	−18.2152
(10,7,2)	−1.36702	−0.51263	(30,7,2)	−77.2526	−28.9697
	−38.0556	−14.2708		−61.5623	−23.0859
	−24.5988	−9.22454		−69.5192	−26.0697
	−1.97686	−0.74132		−39.0315	−14.6368
	−62.3955	−23.3983		−13.8508	−5.19405
(10,7,3)	−60.4663	−22.6749	(30,7,3)	−63.5692	−23.8384
	−27.6192	−10.3572		−32.8455	−12.3171
	−5.1332	0		−60.7151	−22.7682
	−73.6578	−27.6217		−2.09586	0
	−1.82969	−0.68613		−66.4812	−24.9304

The conclusion drawn from the results is that the competitor almost takes disadvantage of the leader's effort when the attractiveness of competitor's facility is decreased or fixed. In fact, with A_k decreasing, the objective value increases while the market share captured by Firm B shrinks, and the optimal locations tend to be nodes which are closer to the warehouse/plant since the transportation cost becomes the main cost (Table 2).

5. Conclusion

In this paper, we considered a relatively new discrete CFL problem in a closed chain with two competitors, which at first attempts to add new facility to its facilities so as to maximize own profits. It is assumed that there is a second competitor in the market which reacts to the opening of the new facility by adjusting the new attractiveness levels of its facilities with the objective of maximizing its own profits. As the market includes forward and reverse logistics network, we considered three kinds of facilities in this market: forward, backward, and hybrid processing facilities. A programming model is proposed which is a bi-level model and solves the problem outlined in this paper by using a global optimization method called GMIN-aBB after converting the bi-level model into an equivalent one-level MINLP model. This conversion is possible since the objective function of the competitor, which is the follower in the bi-level programming formulation, is concave in terms of the attractiveness variables when the locations and the attractiveness levels of the leader firm are fixed. We applied this method on the random generated examples and the computation results show the efficiency of the proposed method. For future research, one can consider an extension of the proposed bi-level model where the competitor reacts to the market entrant firm in the case of adjusting the attractiveness levels and also opening new facilities and/or closing existing ones. Furthermore, a combination of available algorithms such as branch and bound and multi-parametric could be helpful.

Acknowledgments
The author is grateful to the referees for their constructive suggestions for improving this paper.

Funding
This research was supported by Kurdistan University [grant number 3732194681].

Author details
Arsalan Rahmani[1]
E-mail: arsalan.rah@gmail.com
[1] Department of Mathematics, University of Kurdistan, Pasdaran Boulevard, P. O. Box 416, Sanandaj, Iran.

References

Cooper, L. G., & Nakanishi, M. (1974). Parameter estimate for multiplicative interactive choice model: Least squares approach. *Journal of Marketing Research, 11*, 303–311.

Dempe, S., & Gadhi, N. (2007). Necessary optimality conditions for bilevel set optimization problems. *Journal of Global Optimization, 39*, 529–542. http://dx.doi.org/10.1007/s10898-007-9154-0

Drezner, T. (1994). Locating a single new facility among existing unequally attractive facilities. *Journal of Regional Science, 34*, 237–252. http://dx.doi.org/10.1111/jors.1994.34.issue-2

Fernández, J., Pelegrín, B., Plastria, F., & Tóth, B. (2007). Solving a Huff-like competitive location and design model for profit maximization in the plane. *European Journal of Operational Research, 179*, 1274–1287. http://dx.doi.org/10.1016/j.ejor.2006.02.005

Floudas, C. A. (2000). Deterministic global optimization: Theory algorithms and applications. *Journal of Global Optimization, 18*, 103–105.

Floudas, C. A., Pardalos, P. M., Adjiman, C. S., Esposito, W. R., Gümüş, Z., Harding, S.T, & Schweiger, C. A. (1999). *Handbook of test problems for local and global optimization*. Kluwer Academic. http://dx.doi.org/10.1007/978-1-4757-3040-1

Grossmann, I. E., & Floudas, C. A. (1987). Active constraint strategy for flexibility analysis in chemical processes. *Computers and Chemical Engineering, 11*, 675–693. http://dx.doi.org/10.1016/0098-1354(87)87011-4

Gümüş, Z. H., & Floudas, C. A. (2005). Global optimization of mixed-integer bilevel programming problems. *Computational Management Science, 2*, 181–212. http://dx.doi.org/10.1007/s10287-005-0025-1

Hadgson, M. J. (1981). A location—allocation model maximizing consumers' welfare. *Regional Studies, 15*, 493–506. http://dx.doi.org/10.1080/09595238100185441

Hakimi, S. L. (1983). On locating new facilities in a competitive environment. *European Journal of Operational Research, 12*, 29–35. http://dx.doi.org/10.1016/0377-2217(83)90180-7

Huff, D. L. (1964). Defining and estimating a trade area. *Journal of Marketing, 28*, 34–38. http://dx.doi.org/10.2307/1249154

Huff, D. L. (1966). A programmed solution for approximating an optimum retail location. *Land Economics, 42*, 293–303. http://dx.doi.org/10.2307/3145346

Jane, A. K., & Mahajan, V. (1979). *Evaluating the competitive environment in retailing using multiplicative competitive intreaction model, research in marketing, 2*, 217–235.

Nakanishi, M., & Cooper, L. G. (1974). Parameter estimation for a multiplicative competitive interaction model-least squares approach. *Journal of Marketing Research, 11*, 303–311. http://dx.doi.org/10.2307/3151146

Pistikopoulos, E. N., Georgiadis, M. C., & Dua, V. (2007). *Multi-parametric programming*, Weinheim: Wiley-VCH, 1. http://dx.doi.org/10.1002/9783527631216

Revelle, C. (1986). The maximum capture or "sphere of influence" location problem: Hotelling revisited on a network. *Journal of Regional Science, 26*, 343–358. http://dx.doi.org/10.1111/jors.1986.26.issue-2

7

Two-dimensional steady-state temperature distribution of a thin circular plate due to uniform internal energy generation

Kishor R. Gaikwad[1]*
*Corresponding author: Kishor R. Gaikwad, Department of Mathematics, Nanded Education Society's, Science College, Nanded 431605, Maharashtra, India
E-mail: drkr.gaikwad@yahoo.in
Reviewing editor: Xiao-Jun Yang, China unviersity of Mining and Technology, China

Abstract: This article is concerned with the determination of temperature, displacement, and thermal stresses in a thin circular plate due to uniform internal energy generation within it. The fixed circular edge ($r = a$) is kept at zero temperature and the upper ($z = h$) and lower ($z = 0$) surfaces are thermally insulated. The governing heat conduction equation has been solved using finite Hankel transform technique. The results are obtained in a series form in terms of Bessel's functions. The results for temperature, displacement, and stresses have been computed numerically and illustrated graphically.

Subjects: Applied Mathematics; Applied Mechanics; Mathematics Statistics; Science

Keywords: thermal stresses; thin circular plate; steady state; internal energy generation

AMS Subject classifications: 35B07; 35G30; 35K05; 44A10

1. Introduction

During the second half of the twentieth century, non-isothermal problems of the theory of elasticity became increasingly important. This is due mainly to their many applications in diverse fields. First, the high velocity of modern aircrafts give rise to an aerodynamic heating, which produce intense thermal stresses, reducing the strength of aircrafts structure. Secondly, in the nuclear field, the extremely high temperature and temperature gradients originating inside nuclear reactor influence their design and operations (Nowinski, 1978).

ABOUT THE AUTHOR

The author Kishor R. Gaikwad is an assistant professor in Post Graduate Department of Mathematics, NES, Science College, Nanded. He has 8 years teaching experience in different institutions. He has received PhD (2012) in area Thermoelasticity from Dr. Babasaheb Ambedkar Marathwada University, Aurangabad, M.S., India. He is actively involved in the field of Thermoelasticity particularly Boundary Value Problem of Heat Conduction, Partial Differential Equation, Integral Transform, Mathematical Modeling. The author has published more than 20 research publications in reputed National/ International Journals. The author has sanctioned one minor research project under the scheme of University Grant Commission, New Delhi, India.

PUBLIC INTEREST STATEMENT

Studying the two-dimensional steady-state temperature distribution of a thin circular plate due to uniform internal energy generation. The results obtained here are more useful in engineering problems, particularly in aerospace engineering for stations of a missile body not influenced by nose tapering. The missile skill material is assumed to have physical properties independent of temperature, so that the temperature *T(r, z)* is a function of radius and thickness only.

Nowacki (1957) determined the steady-state thermal stresses in a circular plate subjected to an axisymmetric temperature distribution on the upper surface with zero temperature on the lower surface and with the circular edge thermally insulated. The problem of the thermal deflection of an axisymmetrically heated circular plate in the case of fixed and simply supported edges have been considered by Boley and Weiner (1960). Roy Choudhary (1972) studied quasi-static thermal stresses in a thin circular plate due to transient temperature applied along the circumference of a circle on the upper face with the lower face at zero temperature and a fixed circular edge thermally insulated.

Furukawa, Noda, and Ashida (1991) studied a problem of finding the temperature and thermal stresses in an infinitely long cylinder within a generalized thermoelasticity with one relaxation time. Ishihara, Tanigawa, Kawamura, and Noda (1997) studied a heat conduction problem to determined the temperature distribution and discuss the thermoelastic deformation in thin circular plates subjected to partially distributed and axisymmetric heat supply on the outer curved surface. Gaikwad and Ghadle (2010) solved the quasi-static thermal stresses of an infinitely long circular cylinder having constant initial temperature under steady-state field. Parihar and Patil (2011) determined the transient heat conduction problem and analyzed thermal stresses in thin circular plate. Gaikwad and Ghadle (2012) solved non-homogeneous heat conduction problem and obtained thermal deflection due to internal heat generation in a thin hollow circular disk. Recently, Gaikwad (2013) analyzed thermoelastic deformation of a thin hollow circular disk due to partially distributed heat supply.

This article deals with the determination of temperature, displacement, and thermal stresses in a thin circular plate due to uniform internal energy generation within it. The fixed circular edge ($r = a$) is kept at zero temperature and the upper ($z = h$) and lower ($z = 0$) surfaces are thermally insulated. The governing heat conduction equation has been solved using finite Hankel transform technique. The results are obtained in a series form in terms of Bessel's functions. The results for temperature, displacement, and stresses have been computed numerically and illustrated graphically.

To the author knowledge, no literature on steady-state temperature distribution and thermal stresses of a thin circular plate due to internal heat generation has been published. The results presented here should prove useful in engineering problem particularly in the determination of the state of strain in thin circular plate.

2. Formulation of the problem

Consider two-dimensional thin circular plate under a steady-state temperature field of radius r and thickness h occupying space D: $0 \le r \le a,\ 0 \le z \le h$, as shown in Figure 1. The fixed circular edge ($r = a$) is kept at zero temperature and the upper ($z = h$) and lower ($z = 0$) surfaces are thermally insulated, while the plate is also subjected to uniform internal energy generation g_0 (W/m ^{3}). Under these realistic prescribed conditions, temperature, displacement, and thermal stresses in a thin circular plate due to uniform internal heat generation are required to be determined.

The mathematical formulation of this problem, using the general cylindrical heat equation is given as

$$\frac{\partial^2 T}{\partial r^2} + \frac{1}{r}\frac{\partial T}{\partial r} + \frac{\partial^2 T}{\partial z^2} + \frac{g_0}{k} = 0 \qquad \text{in} \quad 0 \le r \le a,\ 0 \le z \le h \tag{1}$$

with the boundary conditions,

$$T = 0 \qquad \text{at } r = a \tag{2}$$

$$\frac{\partial T}{\partial z} = 0 \qquad \text{at } z = 0 \tag{3}$$

$$\frac{\partial T}{\partial z} = 0 \qquad \text{at } z = h, \tag{4}$$

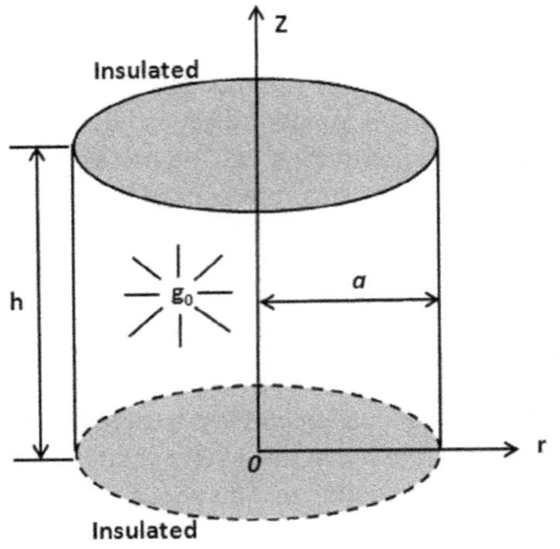

Figure 1. Geometry of the problem.

where k is the thermal conductivity of the material of the circular plate.

Following Roy Choudhary (1972), we assume that a circular plate of small thickness h is in a plane state of stress. In fact "the smaller the thickness of the circular plate compared to its diameter, the nearer to a plane state of stress is the actual state". Then the displacements equations of thermoelasticity have the form

$$U_{i,kk} + \left(\frac{1+\nu}{1-\nu}\right)e_{,i} = 2\left(\frac{1+\nu}{1-\nu}\right)a_t T_{,i}$$
$$e = U_{k,k}; \quad k, i = 1, 2,$$

where U_i is the displacements component, e is the dilatation, T is the temperature, and ν and a_t are, respectively, the Poisson ratio and linear coefficients of thermal expansions of the circular plate material.

Introducing $U_i = \phi_{,i} \quad i = 1, 2$, we have

$$\nabla_1^2\phi = (1+\nu)a_t T,$$
$$\nabla_1^2 = \frac{\partial^2}{\partial x_1^2} + \frac{\partial^2}{\partial x_2^2}$$
$$\sigma_{ij} = 2\mu(\phi_{,ij} - \delta_{ij}\phi_{,kk}) \qquad i, j, k = 1, 2,$$

where μ is Lamé constant and δ_{ij} is the well-known Kronecker symbol.

In the axisymmetric case,

$$\phi = \phi(r, z) \qquad T = (r, z)$$

and the differential equation governin.g the displacements potential function $\phi(r, z)$ is given by

$$\frac{\partial^2\phi}{\partial r^2} + \frac{1}{r}\frac{\partial\phi}{\partial r} = (1+\nu)a_t T \tag{5}$$

The stress components σ_{rr} and $\sigma_{\theta\theta}$ of the circular plate are given by,

$$\sigma_{rr} = -\frac{2\mu}{r}\frac{\partial\phi}{\partial r} \tag{6}$$

$$\sigma_{\theta\theta} = -2\mu\frac{\partial^2\phi}{\partial r^2} \tag{7}$$

with $\phi = 0$ at $r = a$.

Also, in the planar state of stress within the circular plate,

$$\sigma_{rz} = \sigma_{zz} = \sigma_{\theta z} = 0 \tag{8}$$

Equations (1–8) constitute the mathematical formulation of the problem under consideration.

3. Solution of the heat conduction equation

We first reduce Poissons Equation (1) to the Laplace equation by defining a new dependent variable as described next:

A new dependent variable $\theta(r, z)$ is defined as

$$T(r, z) = \theta(r, z) + P(r, z) \tag{9}$$

where the $P(r, z)$ function in the cylindrical co-ordinate system is

$$P(r, z) = -\frac{g_0}{k}\frac{z^2}{4} \tag{10}$$

Then Equation (9) becomes

$$T(r, z) = \theta(r, z) - \frac{g_0}{k}\frac{z^2}{4} \tag{11}$$

Substituting Equation (11) into Equations (1–4), one obtains the Laplace equation with one inhomogeneous boundary condition,

$$\frac{\partial^2\theta}{\partial r^2} + \frac{1}{r}\frac{\partial\theta}{\partial r} + \frac{\partial^2\theta}{\partial z^2} = 0 \tag{12}$$

with the boundary conditions,

$$\theta = 0 \quad \text{at } r = a \tag{13}$$

$$\frac{\partial\theta}{\partial z} = 0 \quad \text{at } z = 0 \tag{14}$$

$$\frac{\partial\theta}{\partial z} = \frac{g_0 z}{2k} \quad \text{at } z = h \tag{15}$$

To obtain the expression of the function $\theta(r, z)$, we develop the finite Hankel transform and its inverse transform over the variable r in the range $0 \le r \le a$ defined in Sneddon (1972) as

$$H[\theta(r, z)] = \bar{\theta}(\alpha_m, z) = \int_{r'=0}^{a} r' K_0(\alpha_m, r')\theta(r', z)dr' \tag{16}$$

$$H^{-1}[\bar{\theta}(\alpha_m, z)] = \theta(r, z) = \sum_{m=1}^{\infty} K_0(\alpha_m, r)\bar{\theta}(\alpha_m, z) \tag{17}$$

where

$$K_0(\alpha_m, r) = \frac{\sqrt{2}}{a}\frac{I_0(\alpha_m r)}{I_0'(\alpha_m a)}$$

and $\alpha_1, \ \alpha_2, \ \alpha_3, \ldots$ are the positive root of transcendental equation

$$I_0(\alpha_m a) = 0 \tag{18}$$

This transform satisfies the relations,

$$H\left[\frac{\partial^2\theta}{\partial r^2} + \frac{1}{r}\frac{\partial\theta}{\partial r}\right] = -\alpha_m^2\overline{\theta}(\alpha_m, z)$$

$$H\left[\frac{\partial^2\theta}{\partial z^2}\right] = \frac{d^2\overline{\theta}}{dz^2}$$

Applying the finite Hankel transform defined in Equation (16) to (12) and using the conditions (13–15), one obtains

$$\frac{d^2\overline{\theta}}{dz^2} - \alpha_m^2\overline{\theta} = 0 \tag{19}$$

with

$$\frac{\partial\overline{\theta}}{\partial z} = \frac{\sqrt{2}}{a}\frac{g_0 h}{2k}\frac{1}{I_0'(\alpha_m a)}\int_{r'=0}^{a} r' I_0(\alpha_m r')dr' \qquad \text{at } z = h \tag{20}$$

Solution of the differential Equation (19) is obtained as

$$\overline{\theta}(\alpha_m, z) = \frac{\sqrt{2}}{a}\frac{g_0 h}{2k}\frac{1}{I_0'(\alpha_m a)}\frac{\cos(\alpha_m z)}{\alpha_m \sin(\alpha_m h)}\int_{r'=0}^{a} r' I_0(\alpha_m r')dr' \tag{21}$$

On applying the inverse Hankel transform defined in Equation (17), one obtains

$$\theta(r, z) = \frac{g_0 h}{a^2 k}\sum_{m=1}^{\infty}\frac{I_0(\alpha_m r)}{I_0'^2(\alpha_m a)}\frac{\cos(\alpha_m z)}{\alpha_m \sin(\alpha_m h)}\int_{r'=0}^{a} r' I_0(\alpha_m r')dr' \tag{22}$$

Substituting Equation (22) into Equation (9), one obtains the expression of the temperature distribution function as

$$T(r, z) = \frac{g_0 h}{a^2 k}\sum_{m=1}^{\infty}\frac{I_0(\alpha_m r)}{I_0'^2(\alpha_m a)}\frac{\cos(\alpha_m z)}{\alpha_m \sin(\alpha_m h)}\int_{r'=0}^{a} r' I_0(\alpha_m r')dr' - \frac{g_0 z^2}{4k} \tag{23}$$

4. Displacement potential and thermal stresses

Using Equation (23) in (5), one obtains

$$\frac{\partial^2\phi}{\partial r^2} + \frac{1}{r}\frac{\partial\phi}{\partial r} = (1+\nu)a_t\left[\frac{g_0 h}{a^2 k}\sum_{m=1}^{\infty}\frac{I_0(\alpha_m r)}{I_0'^2(\alpha_m a)}\frac{\cos(\alpha_m z)}{\alpha_m \sin(\alpha_m h)}\int_{r'=0}^{a} r' I_0(\alpha_m r')dr' - \frac{g_0 z^2}{4k}\right] \tag{24}$$

Now suitable form of ϕ satisfying (24) is given by

$$\phi = (1+\nu)a_t\left[\frac{g_0 h}{a^2 k}\sum_{m=1}^{\infty}\frac{1}{\alpha_m^3}\frac{I_0(\alpha_m r)}{I_0'^2(\alpha_m a)}\frac{\cos(\alpha_m z)}{\sin(\alpha_m h)}\int_{r'=0}^{a} r' I_0(\alpha_m r')dr' - \frac{g_0 z^2 r^2}{16k}\right] \tag{25}$$

Using Equation (25) in Equations (6) and (7), one obtains the expressions of thermal stresses as

$$\sigma_{rr} = -2(1+\nu)a_t\mu\left[\frac{g_0 h}{a^2 k}\sum_{m=1}^{\infty}\frac{1}{\alpha_m^2}\frac{1}{r}\frac{I_1(\alpha_m r)}{I_0'^2(\alpha_m a)}\frac{\cos(\alpha_m z)}{\sin(\alpha_m h)}\int_{r'=0}^{a} r' I_0(\alpha_m r')dr' - \frac{g_0 z^2}{8k}\right] \tag{26}$$

$$\sigma_{\theta\theta} = 2(1+\nu)a_t\mu\left[\frac{g_0 h}{a^2 k}\sum_{m=1}^{\infty}\frac{1}{\alpha_m}\frac{1}{I_0'^2(\alpha_m a)}\left(\frac{I_1(\alpha_m r)}{\alpha_m r} - I_0(\alpha_m r)\right)\frac{\cos(\alpha_m z)}{\sin(\alpha_m h)}\right.$$
$$\left.\times\int_{r'=0}^{a} r' I_0(\alpha_m r')dr' - \frac{g_0 z^2}{8k}\right] \tag{27}$$

5. Numerical results and discussion

5.1. Dimensions

The constants associated with the numerical calculation are taken as

Radius of a circular plate $a = 1$ m,
Thickness of a circular plate $h = 0.1$ m,

5.2. Material properties

The numerical calculation has been carried out for a aluminum (pure) circular plate with the material properties as,

Thermal diffusivity $\alpha = 84.18\text{m}^2/\text{s}$
Thermal conductivity $k = 204$ W/mK
Density $\rho = 2707\text{kg/m}^3$
Specific heat capacity $c_p = 896\text{J/kg K}$
Poisson ratio $\nu = 0.35$
Lamé constant $\mu = 26.67$ GPa
Coefficients of linear thermal expansion $a_t = 22.2 \times 10^{-6}\frac{1}{K}$
Young's modulus elasticity of the material of the plate $E = 70$ GPa.
The rate of internal energy generation is $g_0 = 1 \times 10^6\text{W/m}^3$

5.3. Roots of the transcendental equation

The first five positive root of the transcendental equation $I_0(\alpha_m a) = 0$ as defined in Ozisik (1968) are $\alpha_1 = 1.0025,\ \alpha_2 = 1.3262,\ \alpha_3 = 2.4463,\ \alpha_4 = 5.2945,\ \alpha_5 = 7.3663.$

For convenience, we set

$$A = (1+\nu)\,a_t, \qquad B = -2(1+\nu)\mu a_t.$$

Also noticed that

$$\frac{\partial}{\partial r}(I_0(\alpha_m r)) = \alpha_m I_1(\alpha_m r)$$
$$\frac{\partial^2}{\partial r^2}(I_0(\alpha_m r)) = -\alpha_m^2\left[\frac{I_1(\alpha_m r)}{\alpha_m r} - I_0(\alpha_m r)\right].$$

The numerical calculation has been carried out with the help of computational software MATLAB-2007 and the graphs are plotted with the help of MATLAB tools.

From Figure 2, it is observed that, near the centerline ($r \sim 0$), the temperature is increasing primarily by the internal energy generation and decreasing toward the outer surface in radial direction.

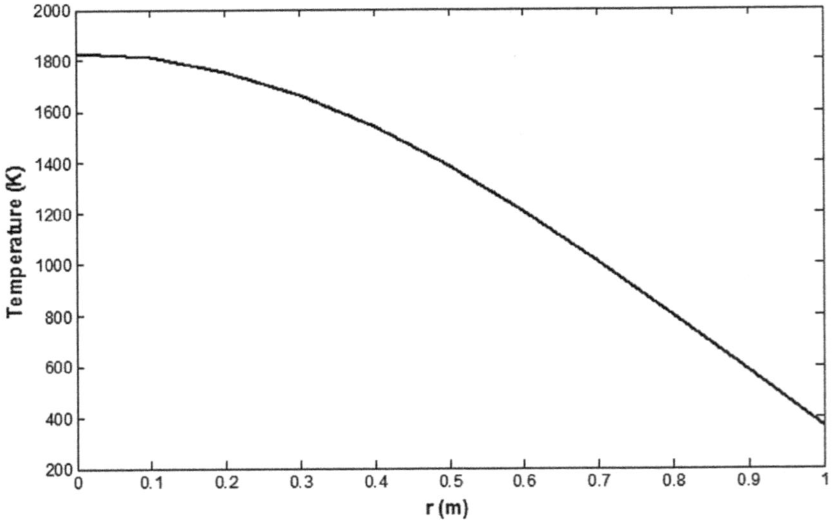

Figure 2. The temperature distribution function T in the radial direction.

From Figure 3, it is observed that, due to the internal energy generation at a constant rate g_0 of a thin circular plate, the displacement function ϕ/A is maximum at the center and decreasing toward the outer surface in radial direction. It is proportional to the temperature.

From Figure 4, it can be observed that, due to internal heat generation at a constant rate g_0, the radial stress decreases from center $r = 0$ of a thin circular plate to the outer circular boundary $r = 1$ in radial direction.

From Figure 5, it is seen that, the angular stress function $\sigma_{\theta\theta}$ is maximum at center of a thin circular plate and decreasing toward the outer surface in radial direction.

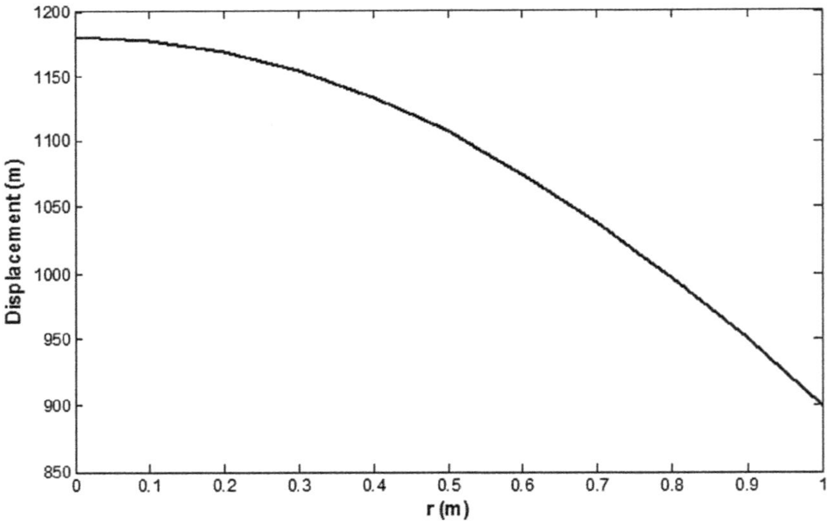

Figure 3. The displacement potential function ϕ/A in the radial direction.

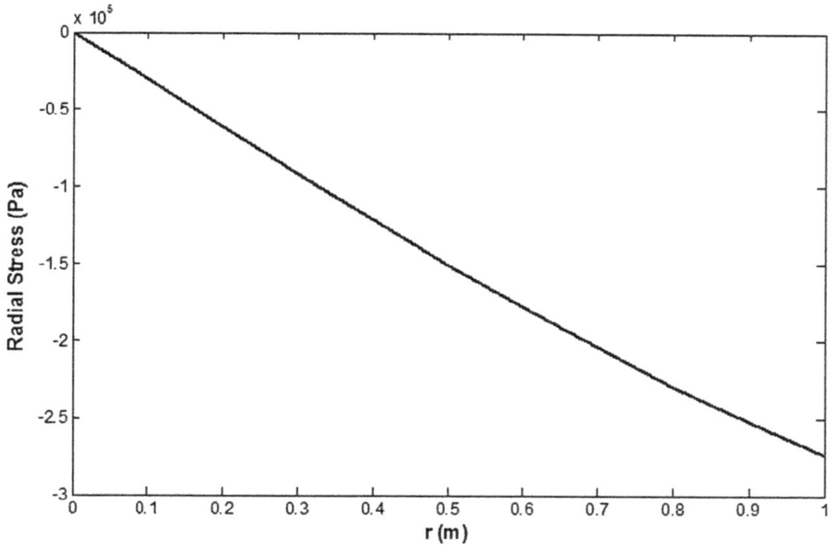

Figure 4. The radial stress function σ_{rr}/B in the radial direction.

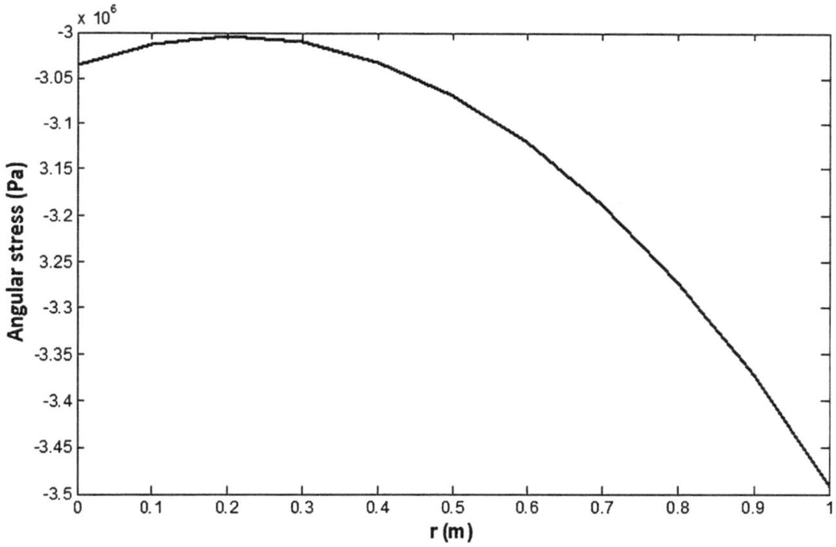

Figure 5. The angular stress function $\sigma_{\theta\theta}/B$ in the radial direction.

6. Concluding remarks

In this paper, we analyzed the steady-state thermoelastic problem and determined the expressions of temperature, displacement, and thermal stresses of a thin circular plate due to uniform internal energy generation at a constant rate $g_0 = 1 \times 10^6 \text{W/m}^3$. The present method is based on the direct method, using the finite Hankel transform and using their inversion. As a special case, a mathematical model is constructed for aluminum (pure) a thin circular plate, with the material properties specified as above and examined the thermoelastic behaviors in the steady-state field for temperature change, displacement, thermal stresses in radial direction.

We conclude that, the displacement and the stress components occur near heat region. Due to the uniform internal energy generation of a thin circular plate at a constant rate $g_0 = 1 \times 10^6$ W/m^3, the radial stress and axial stress develops the tensile stresses in radial direction. Also, it can be observed from the figure, temperature and displacement, the direction of heat flow and direction of body displacement are the same and they are proportionate. In the plane state of stress, the stress components σ_{rz}, σ_{zz}, $\sigma_{\theta z}$ are zero. Also any particular case of special interest can be derived by assigning suitable values to the parameters and functions in the expressions (23–27).

Acknowledgements
The author offers his grateful thanks to Editor and the referee for their kind help and active guidance in the preparation of this revised paper.

Funding
The author is thankful to University Grant Commission, New Delhi, India, for providing partial financial assistance under Minor Research Project Scheme [file number 47-892/14(WRO)].

Author details
Kishor R. Gaikwad[1]
E-mail: drkr.gaikwad@yahoo.in
[1] Department of Mathematics Nanded Education Society's, Science College, Nanded, 431605 Maharashtra, India.

References

Boley, B. A., & Weiner, J. H. (1960). *Theory of thermal stresses.* New York, NY: Wiley.

Furukawa, T., Noda, N., & Ashida, F. (1991). Generalized Thermoelasticity for an infinite solid cylinder. *JSME International Journal Series I, 34*, 281–286.

Gaikwad, K. R. (2013). Analysis of thermoelastic deformation of a thin hollow circular disk due to partially distributed heat supply. *Journal of Thermal Stresses, 36*, 207–224.

Gaikwad, K. R., & Ghadle, K. P. (2010). Quasi-static thermoelastic problem of an infinitely long circular cylinder. *Journal of Korean Society for Industrial and Applied Mathematics, 14*, 141–149.

Gaikwad, K. R., & Ghadle, K. P. (2012). Nonhomogeneous heat conduction problem and its thermal deflection due to internal heat generation in a thin hollow circular disk. *Journal of Thermal Stresses, 35*, 485–498.

Ishihara, M., Tanigawa, Y., Kawamura, R., & Noda, N. (1997). Theoretical analysis of thermoelastic deformation of a circular plate due to a partially distributed heat supply. *Journal of Thermal Stresses, 20*, 203–225.

Nowacki, W. (1957). The state of stresses in a thick circular plate due to temperature field. *Bulletin of the Polish Academy of Sciences Technical Sciences, 5*, 227.

Nowinski, J. L. (1978). *Theory of thermoelasticity with application* (p. 407). Alphen Aan Den Rijn: Sijthof Noordhoff.

Ozisik, N. M. (1968). *Boundary value problem of heat conduction* (pp. 135–148). Scranton, PA: International Textbook.

Parihar, K. S., & Patil, S. S. (2011). Transient heat conduction and analysis of thermal stresses in thin circular plate. *Journal of Thermal Stresses, 34*, 335–351.

Roy Choudhary, S. K. (1972). A note of quasi-static stress in a thin circular plate due to transient temperature applied along the circumference of a circle over the upper face. *Bulletin de l'Acadèmie Polonaise des Sciences, Sèrie des Sciences Mathèmatiques, Astronomiques et Physiques, 20*, 20–21.

Sneddon, I. N. (1972). *The use of integral transform* (pp. 235–238). New York, NY: McGraw Hill.

8

Integral inequalities of Hermite–Hadamard type for logarithmically *h*-preinvex functions

Muhammad Aslam Noor[1], Khalida Inayat Noor[1], Muhammad Uzair Awan[1] and Feng Qi[2,3*]

*Corresponding author: Feng Qi, Department of Mathematics, College of Science, Tianjin Polytechnic University, Tianjin City 300387, China; Institute of Mathematics, Henan Polytechnic University, Jiaozuo City, Henan Province 454010, China

E-mails: qifeng618@gmail.com; qifeng618@hotmail.com; qifeng618@ qq.com

Reviewing editor: Yong Hong Wu, Curtin University of Technology, Australia

Abstract: In the paper, the authors introduce the notion "logarithmically *h*-preinvex functions", reveal that the class of *h*-preinvex functions include several new and known classes of preinvex functions, and establish several integral inequalities of Hermite–Hadamard type.

Subjects: Advanced Mathematics; Analysis - Mathematics; Functional Analysis; Mathematical Analysis; Mathematics & Statistics; Real Functions; Science; Special Functions

Keywords: Hermite–Hadamard type inequality; logarithmically *h*-preinvex function; convex functions

AMS subject classifications: 26B25; 26B35; 26B99; 26D15

1. Introduction

Due to extensive applications of convex functions in different fields of pure and applied sciences, many researchers have paid much attention to study and investigate the theory of convex functions. As a result, the concepts of classical convex functions have been extended and generalized in several directions using various innovative approaches (see, e.g. Bai, Qi, & Xi, 2013; Breckner, 1978; Cristescu & Lupsa, 2002; Dragomir, Pečarić, & Persson, 1995; Godunova & Levin, 1985; Jiang, Niu, & Qi, 2014; Noor, Awan, & Noor, 2013; Noor, Noor, & Awan, 2014; Varošanec, 2007; Wang & Qi, 2014; Wang, Wang, & Qi, 2013; Wang, Xi, & Qi, 2014; Weir & Mond, 1988).

ABOUT THE AUTHOR

Feng Qi is a full professor in mathematics at Tianjin Polytechnic University in China. He received his PhD degree from University of Science and Technology of China. He was the founder and the former head of School of Mathematics and Informatics at Henan Polytechnic University. He was a visiting professors at Victoria University in Australia and University of Hong Kong in China. He was a part-time professor at Henan University, Henan Normal University, and Inner Mongolia University for Nationalities in China. He is being editor of over 15 international journals. He visited Copenhagen University in Denmark and several universities such as Dongguk University in South Korea. He published over 400 research papers in over 130 reputed journals. His research areas include the analytic combinatorics, analytic number theory, special functions, mathematical inequalities, mathematical means, differential geometry, and so on. For more detailed information, please click \url http://qifeng618.wordpress.com and related links therein.

PUBLIC INTEREST STATEMENT

The Hermite–Hadamard type inequalities for convex functions and sequences are a milestone of the theory of convex analysis. The concept of convexity for functions has been generalized and extended in many directions and in diverse forms. In the paper, the authors introduce a new notion "logarithmically *h*-preinvex functions", reveal that the class of *h*-preinvex functions include the logarithmically *s*-preinvex functions, logarithmically *P*-preinvex functions, and logarithmically *Q*-preinvex functions, and establish several integral inequalities of Hermite–Hadamard type for these convex functions.

Motivated by this ongoing research, we now introduce a new class of preinvex functions, which are called logarithmically h-preinvex functions, and derive several new integral inequalities of Hermite-Hadamard type for logarithmically h-preinvex functions.

2. Definitions and a lemma

Let K be a nonempty closed set in $\mathbb{R}^n$, let $f : K \to \mathbb{R}$ be a continuous function, and let $\eta(\cdot,\cdot) : K \times K \to \mathbb{R}^n$ be a continuous bi-function.

Definition 2.1 (Weir & Mond, 1988) A set K is said to be invex with respect to $\eta(\cdot,\cdot)$, if

$$a + t\eta(b, a) \in K$$

for $a, b \in K$ and $t \in [0, 1]$. The invex set K is also called an η-connected set.

Remark 2.1 (Antczak, 2005) The above Definition 2.1 has a geometric interpretation. This definition essentially says that there is a path starting from a point a which is contained in K. The point b may not be one of the end points of the path. This observation plays an important role in our analysis. If b is an end point of the path for every pair of points $a, b \in K$, then $\eta(b, a) = b - a$ and, consequently, invexity reduces to convexity. Thus, it is true that every convex set is also an invex set with respect to $\eta(b, a) = b - a$, but not conversely (see Mohan & Neogy, 1995; Weir & Mond, 1988 and related references therein). For the sake of simplicity, we always assume that $K = [a, a + \eta(b, a)]$, unless otherwise specified.

Definition 2.2 (Weir & Mond, 1988) A function f is said to be preinvex with respect to an arbitrary bi-function $\eta(\cdot,\cdot)$, if

$$f(a + t\eta(b, a)) \le (1 - t)f(a) + tf(b) \quad (2.1)$$

is valid for $a, b \in K$ and $t \in [0, 1]$.

A function f is said to be preconcave if and only if its negative $-f$ is preinvex. For different aspects and applications of the preinvex functions in variational inequalities (see Antczak, 2005; Barani, Ghazanfari, & Dragomir, 2012; Farajzadeh, Noor, & Noor, 2009; Jiang, Niu, Hua, & Qi, 2012; Matloka, 2013; Mishra & Noor, 2005; Mohan & Neogy, 1995; Noor, 2005, 2007a, 2007b, 1994; Noor, Qi, & Awan, 2013; Sarikaya, Alp, & Bozkurt, 2013; Sarikaya, Saglam, & Yildrim, 2008; Wang & Qi, 2014; Wang et al., 2013, 2014; Weir & Mond, 1988; Yang, Yang, & Teo, 2003).

For $\eta(b, a) = b - a$ in Equation 2.1, the preinvex function becomes a convex function in the classical sense.

Definition 2.3 (Noor, Noor, Awan, & Li, 2015) Let $h : J \to \mathbb{R}$, where $(0, 1) \subseteq J$ and $h \not\equiv 0$, be an interval in $\mathbb{R}$ and let K be an invex set with respect to $\eta(\cdot,\cdot)$. A nonnegative function $f : K \to \mathbb{R}$ is called h-preinvex with respect to $\eta(\cdot,\cdot)$, if

$$f(a + t\eta(b, a)) \le h(1 - t)f(a) + h(t)f(b) \quad (2.2)$$

holds for $a, b \in K$ and $t \in (0, 1)$.

In Noor et al. (2015), it was showed that the class of h-preinvex functions generalizes several other classes of convex functions. For example, if we take $h(t) = t, h(t) = \frac{1}{t}, h(t) = t^s$, and $h(t) = 1$ in (2.2), then the h-preinvex function reduces to the preinvex function in Weir and Mond (1988), the Q-preinvex function, the s-preinvex function, and the P-preinvex function, respectively. If we take $\eta(b, a) = b - a$, then the definition of h-preinvex functions reduces to the definition of h-convex functions, which was introduced in Varošanec (2007). Noor (2007a) showed that a function f is preinvex if and only if

$$f\left(\frac{2a+\eta(b,a)}{2}\right) \leq \frac{1}{\eta(b,a)}\int_a^{a+\eta(b,a)} f(x)\mathrm{d}x \leq \frac{f(a)+f(b)}{2} \tag{2.3}$$

The double inequality Equation 2.3 is known as the Hermite–Hadamard–Noor inequality for prepinvex functions. If $\eta(b,a) = b - a$, then the double inequality Equation 2.3 reduces to the classical Hermite–Hadamard inequality for convex functions. For recent developments and applications (see Sarikaya et al., 2013).

Definition 2.4 A function $f: K \to (0,\infty)$ is said to be logarithmically h-preinvex with respect to $\eta(\cdot,\cdot)$, if

$$f(a+t\eta(b,a)) \leq [f(a)]^{h(1-t)}[f(b)]^{h(t)}$$

for $a, b \in I$ and $t \in (0,1)$.

Remark 2.2 From Definition 2.4, we may obtain

$$\begin{aligned}\ln f(a+t\eta(b,a)) &\leq \ln\{[f(a)]^{h(1-t)}[f(b)]^{h(t)}\}\\ &= \ln[f(a)]^{h(1-t)} + \ln[f(b)]^{h(t)}\\ &= h(1-t)\ln f(a) + h(t)\ln f(b)\end{aligned}$$

Remark 2.3 If $h(t) = t^s$, then the definition of logarithmically h-prinvex function reduces to the definition of logarithmically s-preinvex function.

Definition 2.5 A function $f: K \to (0,\infty)$ is said to be logarithmically s-preinvex, where $s \in (0,1]$, with respect to $\eta(\cdot,\cdot)$, if

$$f(a+t\eta(b,a)) \leq [f(a)]^{(1-t)^s}[f(b)]^{t^s}$$

for $a, b \in I$ and $t \in [0,1]$.

Remark 2.4 If $h(t) = 1$, then the definition of logarithmically h-preinvex function reduces to the definition of logarithmically P-preinvex function.

Definition 2.6 A function $f: K \to (0,\infty)$ is said to be logarithmically P-preinvex with respect to $\eta(\cdot,\cdot)$, if

$$f(a+t\eta(b,a)) \leq [f(a)][f(b)]$$

for $a, b \in I$ and $t \in [0,1]$.

Remark 2.5 If $h(t) = \frac{1}{t}$, then the definition of logarithmically h-preinvex function reduces to the definition of logarithmically Q-preinvex function.

Definition 2.7 A function $f: K \to (0,\infty)$ is said to be logarithmically Q-preinvex with respect to $\eta(\cdot,\cdot)$, if

$$f(a+t\eta(b,a)) \leq [f(a)]^{1/(1-t)}[f(b)]^{1/t}$$

for $a, b \in I$ and $t \in (0,1)$.

To prove some results in this paper, we need the following well-known Condition C introduced by Mohan and Neogy.

Condition C (Mohan & Neogy, 1995) Let $K \subset \mathbb{R}$ be an invex set with respect to the bi-function $\eta(\cdot,\cdot)$. Then for any $a, b \in K$ and $t \in [0,1]$, we have

$$\eta(b, b + t\eta(a,b)) = -t\eta(a,b) \quad \text{and} \quad \eta(a, b + t\eta(a,b)) = (1-t)\eta(a,b)$$

From Condition C, it follows that

$$\eta(b + t_2\eta(a,b), b + t_1\eta(a,b)) = (t_2 - t_1)\eta(a,b)$$

for every $a, b \in K$ and $t_1, t_2 \in [0,1]$.

It is worth mentioning that Condition C plays a crucial and significant role in the development of the variational-like inequalities and optimization problems (see Farajzadeh et al., 2009; Mohan & Neogy, 1995; Noor, 1994; Noor et al., 2013 and related references therein).

The following lemma is also necessary for us.

LEMMA 2.1 (Barani et al., 2012) *Let $f : K \to (0,\infty)$ be a differentiable mapping on $[a, a+\eta(b,a)] \subseteq K$ with $\eta(b,a) > 0$. If $f' \in L_1[a, a+\eta(b,a)]$, then*

$$\frac{1}{\eta(b,a)}\int_a^{a+\eta(b,a)} f(x)dx - \frac{f(a) + f(a+\eta(b,a))}{2} = \frac{\eta(b,a)}{2}\int_0^1 (1-2t)f'(a+t\eta(b,a))dt$$

3. Main results

We now start out to establish several new integral inequalities of Hermite–Hadamard type for logarithmically h-preinvex functions.

THEOREM 3.1 *Let f be a logarithmically h-preinvex function such that $h(\frac{1}{2}) \neq 0$. Also suppose that Condition C holds for η, then, for $\eta(b,a) > 0$, we have*

$$\ln f\left(\frac{2a+\eta(b,a)}{2}\right)^{1/2h(1/2)} \leq \frac{1}{\eta(b,a)}\int_a^{a+\eta(b,a)} \ln f(x)dx \leq [\ln f(a) + \ln f(b)]\int_0^1 h(t)dt$$

Consequently,

$$f\left(\frac{2a+\eta(b,a)}{2}\right)^{1/2h(1/2)} \leq \exp\left[\frac{1}{\eta(b,a)}\int_a^{a+\eta(b,a)} \ln f(x)dx\right] \leq [f(a)f(b)]^{\int_0^1 h(t)dt}$$

Proof Since f is logarithmically h-preinvex, using Condition C, we have

$$\begin{aligned} f\left(\frac{2a+\eta(b,a)}{2}\right) &= f\left(a + (1-t)\eta(b,a) + \frac{\eta(a+t\eta(b,a), a+(1-t)\eta(b,a))}{2}\right) \\ &\leq [f(a+t\eta(b,a))]^{h(1/2)}[f(a+(1-t)\eta(b,a))]^{h(1/2)} \\ &= \{[f(a+t\eta(b,a))][f(a+(1-t)\eta(b,a))]\}^{h(1/2)} \end{aligned}$$

Taking the logarithm on both sides of the above inequality yields

$$\begin{aligned} \ln f\left(\frac{2a+\eta(b,a)}{2}\right) &\leq \ln[f(a+t\eta(b,a))f(a+(1-t)\eta(b,a))]^{h(1/2)} \\ &= h\left(\frac{1}{2}\right)\ln[f(a+t\eta(b,a))f(a+(1-t)\eta(b,a))] \end{aligned}$$

which implies that

$$\frac{1}{h(1/2)}\ln f\left(\frac{2a+\eta(b,a)}{2}\right) \le \ln[f(a+t\eta(b,a))f(a+(1-t)\eta(b,a))]$$
$$= \ln f(a+t\eta(b,a)) + \ln f(a+(1-t)\eta(b,a))$$

Integrating on both sides of the above inequality with respect to $t \in [0,1]$ gives

$$\frac{1}{h(1/2)}\ln f\left(\frac{2a+\eta(b,a)}{2}\right) = \frac{1}{\eta(b,a)}\int_a^{a+\eta(b,a)} \ln f(x)dx$$
$$+\frac{1}{\eta(b,a)}\int_a^{a+\eta(b,a)} \ln f(x)dx = \frac{2}{\eta(b,a)}\int_a^{a+\eta(b,a)} \ln f(x)dx$$

which means that

$$\frac{1}{2h(1/2)}\ln f\left(\frac{2a+\eta(b,a)}{2}\right) \le \frac{1}{\eta(b,a)}\int_a^{a+\eta(b,a)} \ln f(x)dx \tag{3.1}$$

Integrating on both sides of

$\ln f(a+t\eta(b,a)) \le h(1-t)\ln f(a) + \ln h(t) f(b)$

with respect to $t \in [0,1]$ shows

$$\frac{1}{\eta(b,a)}\int_a^{a+\eta(b,a)} \ln f(x)dx \le [\ln f(a) + \ln f(b)]\int_0^1 h(t)dt \tag{3.2}$$

Combining Equations 3.1 and 3.2 reveals that

$$\ln f\left(\frac{2a+\eta(b,a)}{2}\right)^{1/2h(1/2)} \le \frac{1}{\eta(b,a)}\int_a^{a+\eta(b,a)} \ln f(x)dx$$
$$\le [\ln f(a) + \ln f(b)]\int_0^1 h(t)dt$$

which is equivalent to

$$f\left(\frac{2a+\eta(b,a)}{2}\right)^{1/2h(1/2)} \le \exp\left[\frac{1}{\eta(b,a)}\int_a^{a+\eta(b,a)} \ln f(x)dx\right]$$
$$\le \exp^{[\ln f(a)+\ln f(b)]\int_0^1 h(t)dt}$$
$$= \exp^{\ln[f(a)f(b)]\int_0^1 h(t)dt}$$
$$= [f(a)f(b)]^{\int_0^1 h(t)dt}$$

The proof of Theorem 3.1 is complete.

COROLLARY 3.1 *Let f be a logarithmically s-preinvex function. Also suppose that Condition C holds for η, then, for $\eta(b,a) > 0$, we have*

$$\ln f\left(\frac{2a+\eta(b,a)}{2}\right)^{2^{s-1}} \le \frac{1}{\eta(b,a)}\int_a^{a+\eta(b,a)} \ln f(x)dx \le \frac{\ln f(a)+\ln f(b)}{s+1}$$

Consequently,

$$f\left(\frac{2a+\eta(b,a)}{2}\right)^{2^{s-1}} \le \exp\left[\frac{1}{\eta(b,a)}\int_a^{a+\eta(b,a)} \ln f(x)dx\right] \le [f(a)f(b)]^{1/(s+1)}$$

Proof This follows from taking $h(t) = t^s$ for $s \in (0,1]$ in Theorem 3.1.

COROLLARY 3.2 *Let f be a logarithmically P-preinvex function. Also suppose that Condition C holds for η, then, for $\eta(b,a) > 0$, we have*

$$\ln f\left(\frac{2a+\eta(b,a)}{2}\right) \le \frac{2}{\eta(b,a)}\int_a^{a+\eta(b,a)} \ln f(x)dx \le 2[\ln f(a) + \ln f(b)]$$

Consequently,

$$f\left(\frac{2a+\eta(b,a)}{2}\right) \le \exp\left[\frac{2}{\eta(b,a)}\int_a^{a+\eta(b,a)} \ln f(x)dx\right] \le [f(a)f(b)]^2$$

Proof This follows from letting $h(t) = 1$ in Theorem 3.1

COROLLARY 3.3 *Let f be a logarithmically Q-preinvex function. Also suppose that Condition C holds for η, then, for $\eta(b,a) > 0$, we have*

$$\frac{1}{4}\ln f\left(\frac{2a+\eta(b,a)}{2}\right) \le \frac{1}{\eta(b,a)}\int_a^{a+\eta(b,a)} \ln f(x)dx$$

Consequently,

$$f\left(\frac{2a+\eta(b,a)}{2}\right)^{1/4} \le \exp\left[\frac{1}{\eta(b,a)}\int_a^{a+\eta(b,a)} \ln f(x)dx\right]$$

Proof This follows from setting $h(t) = \frac{1}{t}$ in Theorem 3.1.

Remark 3.1 When $\eta(b,a) = b - a$, the above results reduce to ones for classical logarithmically h-convex functions, logarithmic s-convex functions, logarithmic P-convex functions, and logarithmic Q-convex functions, respectively (see Noor et al., 2013).

THEOREM 3.2 *Let $f, g: K \to \mathbb{R}$ be logarithmically h-preinvex functions and $a, a+\eta(b,a) \in K$ with $\eta(b,a) > 0$. Then*

$$\frac{1}{\eta(b,a)}\int_a^{a+\eta(b,a)} f(x)g(x)dx$$
$$\le \alpha\int_0^1 \left\{[f(a)]^{h(1-t)/\alpha}[f(b)]^{h(t)/\alpha}\right\}dt + \beta\int_0^1 \left\{[g(a)]^{h(1-t)/\beta}[g(b)]^{h(t)/\beta}\right\}dt$$

Proof Using Young's inequality $ab \le \alpha a^{1/\alpha} + \beta b^{1/\beta}$ for $\alpha, \beta > 0$ and $\alpha + \beta = 1$ produces

$$\frac{1}{\eta(b,a)}\int_a^{a+\eta(b,a)} f(x)g(x)dx = \int_0^1 f(a+t\eta(b,a))g(a+t\eta(b,a))dt$$
$$\le \int_0^1 \left\{\alpha[f(a+t\eta(b,a))]^{1/\alpha} + \beta[g(a+t\eta(b,a))]^{1/\beta}\right\}dt$$
$$\le \int_0^1 \left\{\alpha\left([f(a)]^{h(1-t)}[f(b)]^{h(t)}\right)^{1/\alpha} + \beta\left([g(a)]^{h(1-t)}[g(b)]^{h(t)}\right)^{1/\beta}\right\}dt$$
$$= \alpha\int_0^1 \left([f(a)]^{h(1-t)/\alpha}[f(b)]^{h(t)/\alpha}\right)dt + \beta\int_0^1 \left([g(a)]^{h(1-t)/\beta}[g(b)]^{h(t)/\beta}\right)dt$$

The proof of Theorem 3.2 is complete.

THEOREM 3.3 *Let $f: K \to (0,\infty)$ be a differentiable function such that $f' \in L_1[a, a+\eta(b,a)]$. If $|f'|^q$ is logarithmically h-preinvex on K for $q > 1$, $h(t) + h(1-t) = 1$ and $\eta(b,a) > 0$, then*

$$\left|\frac{1}{\eta(b,a)}\int_a^{a+\eta(b,a)} f(x)\mathrm{d}x - \frac{f(a)+f(a+\eta(b,a))}{2}\right| \le \frac{(b-a)|f'(a)|}{2^{(2q-1)/q}}\left[\int_0^1 |1-2t|\left|\frac{f'(b)}{f'(a)}\right|^{qh(t)}\mathrm{d}t\right]^{1/q}$$

Proof Using Lemma 2.1, the well-known power mean inequality, and the condition that $|f'|^q$ is logarithmically h-preinvex gives

$$\begin{aligned}
&\left|\frac{1}{\eta(b,a)}\int_a^{a+\eta(b,a)} f(x)\mathrm{d}x - \frac{f(a)+f(a+\eta(b,a))}{2}\right| \\
&= \left|\frac{\eta(b,a)}{2}\left[\int_0^1 (1-2t)f'(a+t\eta(b,a))\mathrm{d}t\right]\right| \\
&\le \frac{\eta(b,a)}{2}\left[\int_0^1 |1-2t|\mathrm{d}t\right]^{1-1/q}\left[\int_0^1 |1-2t||f'(a+t\eta(b,a))|^q\mathrm{d}t\right]^{1/q} \\
&\le \frac{\eta(b,a)}{2}\left(\frac{1}{2}\right)^{1-1/q}\left[\int_0^1 |1-2t||f'(a)|^{qh(1-t)}|f'(b)|^{q(h(t))}\mathrm{d}t\right]^{1/q} \\
&= \frac{\eta(b,a)|f'(a)|}{2^{(2q-1)/q}}\left[\int_0^1 |1-2t|\left|\frac{f'(b)}{f'(a)}\right|^{qh(t)}\mathrm{d}t\right]^{1/q}
\end{aligned}$$

This completes the proof of Theorem 3.3

Remark 3.2 For different suitable choices of h, we can obtain corresponding results for logarithmically preinvex functions, logarithmically s-preinvex functions, and logarithmically P-preinvex functions.

COROLLARY 3.4 *Let $f: K \to \mathbb{R}$ be a differentiable function such that $f' \in L_1[a, a+\eta(b,a)]$. If $|f'|$ is logarithmically h-preinvex on K, then, for $\eta(b,a) > 0$, we have*

$$\left|\frac{1}{\eta(b,a)}\int_a^{a+\eta(b,a)} f(x)\mathrm{d}x - \frac{f(a)+f(a+\eta(b,a))}{2}\right| \le \frac{\eta(b,a)|f'(a)|}{2}\left[\int_0^1 |1-2t|\left|\frac{f'(b)}{f'(a)}\right|^{h(t)}\mathrm{d}t\right]$$

Proof This is a direct consequence of Theorem 3.3 for $q = 1$.

THEOREM 3.4 *Let $f: K \to (0,\infty)$ be a differentiable function such that $f' \in L[a, a+\eta(b,a)]$. If $|f'|^q$ is logarithmically h-preinvex on K for $q > 1$ such that $\frac{1}{p}+\frac{1}{q} = 1$ and if $h(t)+h(1-t) = 1$, then, for $\eta(b,a) > 0$, we have*

$$\left|\frac{1}{\eta(b,a)}\int_a^{a+\eta(b,a)} f(x)\mathrm{d}x - \frac{f(a)+f(a+\eta(b,a))}{2}\right| \le \frac{\eta(b,a)|f'(a)|}{2}\left(\frac{1}{p+1}\right)^{1/p}\left[\int_0^1 \left|\frac{f'(b)}{f'(a)}\right|^{qh(t)}\mathrm{d}t\right]^{1/q}$$

Proof This directly follows from the proof of Theorem 3.3.

Acknowledgements
The authors are grateful to Dr S.M. Junaid Zaidi, Rector, COMSATS Institute of Information Technology, Pakistan for providing excellent research facilities.

Funding
The author Feng Qi was partially supported by the Natural Science Foundation of Shaanxi Province of China [grant number 2014JQ1006] and by the National Natural Science Foundation of China [grant number 11361038].

Author details
Muhammad Aslam Noor[1]
E-mail: noormaslam@hotmail.com
Khalida Inayat Noor[1]
E-mail: khalidanoor@hotmail.com

Muhammad Uzair Awan[1]
E-mail: awan.uzair@gmail.com
Feng Qi[2,3]
E-mails: qifeng618@gmail.com; qifeng618@hotmail.com; qifeng618@qq.com
ORCID ID: http://orcid.org/0000-0001-6239-2968
[1] Department of Mathematics, COMSATS Institute of Information Technology, Islamabad, Pakistan.
[2] Department of Mathematics, College of Science, Tianjin Polytechnic University, Tianjin City 300387, China.
[3] Institute of Mathematics, Henan Polytechnic University, Jiaozuo City, Henan Province 454010, China.

References

Antczak, T. (2005). Mean value in invexity analysis. *Nonlinear Analysis: Theory, Methods & Applications, 60*, 1473–1484. doi:10.1016/j.na.2004.11.005

Bai, R.-F., Qi, F., & Xi, B.-Y. (2013). Hermite–Hadamard type inequalities for the *m*- and (α, *m*)-logarithmically convex functions. *Filomat, 27*, 1–7. doi:10.2298/FIL1301001B

Barani, A., Ghazanfari, A. G., & Dragomir, S. S. (2012). Hermite–Hadamard inequality for functions whose derivatives absolute values are preinvex. *Journal of Inequalities and Applications, 2012*, 247, 9 pp. doi:10.1186/1029-242X-2012-247

Breckner, W. W. (1978). Stetigkeitsaussagen für eine Klasse verallgemeinerter konvexer Funktionen in topologischen linearen Räumen [Continuity statements for a class of generalized convex functions in topological vector spaces]. *Publications of the Institute of Mathematics (Beograd) (New Series) (Beograd) (N.S.), 23*, 13–20 (German).

Cristescu, G., & Lupsa, L. (2002). *Non-connected convexities and applications, applied optimization* (Vol. 68). Dordrecht: Kluwer.

Dragomir, S. S., Pečarić, J., & Persson, L. E. (1995). Some inequalities of Hadamard type. *Soochow Journal of Mathematics, 21*, 335–341.

Farajzadeh, A., Noor, M. A., & Noor, K. I. (2009). Vector nonsmooth variational-like inequalities and optimization problems. *Nonlinear Analysis, 71*, 3471–3476. doi:10.1016/j.na.2009.02.011

Godunova, E. K., & Levin, V. I. (1985). Inequalities for functions of a broad class that contains convex, monotone and some other forms of functions. *Numerical mathematics and mathematical physics, 166*, 138–142 (Moskov. Gos. Ped. Inst., Moscow Russian).

Jiang, W.-D., Niu, D.-W., Hua, Y., & Qi, F. (2012). Generalizations of Hermite–Hadamard inequality to *n*-time differentiable functions which are *s*-convex in the second sense. *Analysis (Munich), 32*, 209–220. doi:10.1524/anly.2012.1161

Jiang, W.-D., Niu, D.-W., & Qi, F. (2014). Some inequalities of Hermite-Hadamard type for r-φ-preinvex functions. *Tamkang Journal of Mathematics, 45*, 31–38. doi:10.5556/j.tkjm.45.2014.1261

Matloka, M. (2013). On some Hadamard-type inequalities for (h_1, h_2)-preinvex functions on the co-ordinates. *Journal of Inequalities and Applications, 2013*, 227, 12 pp. doi:10.1186/1029-242X-2013-227

Mishra, S. K., & Noor, M. A. (2005). On vector variational-like inequality problems. *Journal of Mathematical Analysis and Applications, 311*, 69–75. doi:10.1016/j.jmaa.2005.01.070

Mohan, S. R., & Neogy, S. K. (1995). On invex sets and preinvex functions. *Journal of Mathematical Analysis and Applications, 189*, 901–908. doi:10.1006/jmaa.1995.1057

Noor, M. A. (1994). Variational-like inequalities. *Optimization, 30*, 323–330. doi:10.1080/02331939408843995

Noor, M. A. (2005). Invex equilibrium problems. *Journal of Mathematical Analysis and Applications, 302*, 463–475. doi:10.1016/j.jmaa.2004.08.014

Noor, M. A. (2007a). Hermite–Hadamard integral inequalities for log-preinvex functions. *Journal of Mathematical Analysis and Approximation Theory Theory, 2*, 126–131.

Noor, M. A. (2007b). On Hadamard integral inequalities involving two log-preinvex functions. *Journal of Inequalities and Applications, 8*, 6 pp. Art. 75. Retrieved from http://www.emis.de/journals/JIPAM/article883.html

Noor, M. A., Awan, M. U., & Noor, K. I. (2013). On some inequalities for relative semi-convex functions. *Journal of Inequalities and Applications, 2013*, 332, 16 pp. doi:10.1186/1029-242X-2013-332

Noor, M. A., Noor, K. I., & Awan, M. U. (2014). Hermite–Hadamard inequalities for relative semi-convex functions and applications. *Filomat, 28*, 221–230. doi:10.2298/FIL1402221N

Noor, M. A., Noor, K. I., Awan, M. U., & Li, J. (2015). On Hermite–Hadamard Inequalities for *h*-preinvex functions. *Filomat, 28*, 1463–1474. doi:10.2298/FIL1407463N

Noor, M. A., Qi, F., & Awan, M. U. (2013). Some Hermite–Hadamard type inequalities for log -*h*-convex functions. *Analysis (Berlin), 33*, 367–375. doi:10.1524/anly.2013.1223

Sarikaya, M. Z., Alp, N., & Bozkurt, H. (2013). On Hermite–Hadamard type integral inequalities for preinvex and log-preinvex functions. *Contemporary Analysis and Applied Mathmatics, 1*, 237–252.

Sarikaya, M. Z., Saglam, A., & Yildrim, H. (2008). On some Hadamard-type inequalities for *h*-convex functions. *Journal of Mathematical Inequalities, 2*, 335–341. doi:10.7153/jmi-02-30

Varošanec, S. (2007). On *h*-convexity. *Journal of Mathematical Analysis and Applications, 326*, 303–311. doi:10.1016/j.jmaa.2006.02.086

Wang, S.-H., & Qi, F. (2014). Hermite–Hadamard type inequalities for *n*-times differentiable and preinvex functions. *Journal of Inequalities and Applications, 2014*, 49, 9 pp. doi:10.1186/1029-242X-2014-49

Wang, Y., Wang, S.-H., & Qi, F. (2013). Simpson type integral inequalities in which the power of the absolute value of the first derivative of the integrand is *s*-preinvex. *Facta Universitatis, Series: Mathematics and Informatics, 28*, 151–159.

Wang, Y., Xi, B.-Y., & Qi, F. (2014). Hermite–Hadamard type integral inequalities when the power of the absolute value of the first derivative of the integrand is preinvex. *Matematiche (Catania), 69*, 89–96. doi:10.4418/2014.69.1.6

Weir, T., & Mond, B. (1988). Pre-invex functions in multiple objective optimization. *Journal of Mathematical Analysis and Applications, 136*, 29–38. doi:10.1016/0022-247X(88)90113-8

Yang, X. M., Yang, X. Q., & Teo, K. L. (2003). Generalized invexity and generalized invariant monotonicity. *Journal of Optimization Theory and Applications, 117*, 607–625. doi:10.1023/A:1023953823177

9

Two dimensional deformation in microstretch thermoelastic half space with microtemperatures and internal heat source

Praveen Ailawalia[1]*, Sunil Kumar Sachdeva[2,3] and Devinder Pathania[4]

*Corresponding author: Praveen Ailawalia, Department of Applied Sciences and Humanities, M.M. University, Sadopur, Ambala City, Haryana, India

E-mail: praveen_2117@rediffmail.com

Reviewing editor: Timothy Marchant, University of Wollongong, Australia

Abstract: The purpose of this paper is to study the two dimensional deformation due to internal heat source in a microstretch thermoelastic solid with microtemperatures (MTSM). A mechanical force is applied along the interface of fluid half space and microstretch thermoelastic half space. The normal mode analysis has been applied to obtain the exact expressions for component of normal displacement, microtemperature, normal force stress, microstress tensor, heat flux moment tensor, and couple stress for MTSM. The effect of internal heat source, micropolarity, and microstretch on the above components has been depicted graphically.

Subjects: Applied Mathematics; Mathematics & Statistics; Science

Keywords: thermoelasticity; microstretch; microtemperature; heat source; fluid half space

1. Introduction

The dynamical interaction between the thermal and mechanical has great practical applications in modern aeronautics, astronautics, nuclear reactors, and high-energy particle accelerators. Classical elasticity is not adequate to model the behavior of materials possessing internal structure. Furthermore, the micropolar elastic model is more realistic than the purely elastic theory for studying the response of materials to external stimuli. Eringen and Suhubi (1964a, 1964b) developed a

ABOUT THE AUTHOR

The author, Praveen Ailawalia working as a professor at M M University, Sadopur, Ambala, Haryana (India) is actively involved in the field of thermoelasticity, micropolar elasticity. He has 17 years of teaching experience in different Universities and institutions. The author has more than 70 research publications in international journals of repute. He has guided four PhD students and four students are currently working with him. The author has discussed deformation in thermoelastic medium and micropolar elastic medium in many of his research papers. The research problem discussed in the paper helps in analyzing the behavior of a medium with temperature changes if the medium undergoes deformation due to an internal heat source. The results obtained in the paper may be applied to various geological problems which involves sources acting in the medium. The problem can be further discussed in case of mechanical sources applied on the free surface of the medium or along the interface of two different mediums.

PUBLIC INTEREST STATEMENT

Studying the two dimensional deformation due to internal heat source in a microstretch thermoelastic solid with microtemperatures is very useful in the study of earthquake engineering, seismology, and volcanic eruptions. It helps us to study the effect of a heat source in the medium and the deformation caused in the medium due to the heat source.

nonlinear theory of microelastic solids. Later Eringen (1965, 1966a, 1996b) developed a theory for the special class of microelastic materials and called it the "linear theory of micropolar elasticity". Under this theory, solids can undergo macro-deformations and microrotations. Eringen (1990) developed a theory of thermo-microstretch elastic solids in which he included microstructural expansions and contractions. The material points of microstretch solids can stretch and contract independently of their translations and rotations. Microstretch continuum is a model for Bravais lattice with a basis on the atomic level and a two-phase dipolar solid with a core on the macroscopic level. For example, composite materials reinforced with chopped elastic fibers, porous media whose pores are filled with gas or inviscid liquid, other elastic inclusions and "solid-liquid" crystals, etc., should be characterizable by microstretch solids. Eringen (1968) developed a theory of microstretch elastic solid in which he included microstructural expansions and contractions, Singh and Kumar (1998) studied wave propagation in a generalized thermo-microstretch elastic solid, Kumar and Rupender (2008) studied the reflection at free surface of magneto-thermo-microstretch elastic solid, Tomar and Khurana (2009) discussed reflection and transmission of elastic waves from a plane interface between two thermo-microstretch solid half-spaces. Marin (2010) discussed Lagrange identity method for microstretch thermoelastic materials, Othman and Lotfy (2010) studied the plane waves of generalized thermo-microstretch elastic half space under three theories, Kumar and Partap (2009) presented the analysis of free vibrations for Rayleigh–Lamb waves in a microstretch thermoelastic plate with two relaxation times, Othman, Lotfy, and Farouk (2010) studied generalized thermo-microstretch elastic medium with temperature-dependent properties for different theories, Kumar and Kansal (2011) studied fundamental solution in the theory of thermo-microstretch elastic diffusive solids, Othman and Lotfy (2011) studied the effect of rotation on plane waves in generalized thermo-microstretch elastic solid with one relaxation time, Kumar, Sharma, and Sharma (2011) discussed the generalized thermoelastic waves in microstretch plates loaded with fluid of varying temperature. Abbas and Othman (2012) studied the plane waves in generalized thermo-microstretch elastic solid with thermal relaxation using finite element method, Kumar and Rupender (2009) discussed the propagation of plane waves at imperfect boundary of elastic and electro-microstretch generalized thermoelastic solids.

Grot (1969) discussed a theory of thermodynamics of elastic bodies with microstructure whose microelements possess microtemperatures. Říha (1976) studied heat conduction in materials with microtemperatures. Iesan and Quintanilla (2000) studied a theory of thermoelasticity with microtemperatures. Iesan (2001) proposed the theory of micromorphic elastic solids with microtemperatures. Exponential stability in thermoelasticity with microtemperatures was studied by Casas and Quintanilla (2005). Scalia and Svanadze (2006) gave the solutions of the theory of thermoelasticity with microtemperatures. Magaña and Quintanilla (2006) discussed the time decay of solutions in one-dimensional theories of porous materials. Aouadi (2008) discussed some theorems in the isotropic theory of microstretch thermoelasticity with microtemperatures. Ieşan and Quintanilla (2009) discussed thermoelastic bodies with inner structure and microtemperatures. Scalia, Svanadze, and Tracinà (2010) studied basic theorems in the equilibrium theory of thermoelasticity with microtemperatures. Quintanilla (2011) discussed the growth and continuous dependence in thermoelasticity with microtemperatures. Steeb, Singh, and Tomar (2013) studied time harmonic waves in thermoelastic material with microtemperatures. Chiriţă, Ciarletta, and D'Apice (2013) studied the theory of thermoelasticity with microtemperatures. Singh, Kumar, and Kumar (2014) discussed a problem in microstretch thermoelastic diffusive medium. Kumar and Kaur (2014) studied the reflection and refraction of plane waves at the interface of an elastic solid and microstretch thermoelastic solid with microtemperatures (MTSM).

In the present problem, the authors have discussed deformation due to internal heat source and a mechanical force which is applied along the interface of fluid half space and microstretch thermoelastic half space with microtemperatures. The normal mode analysis has been applied to obtain the exact expressions for component of normal displacement, microtemperature, normal force stress, microstress tensor, heat flux moment tensor, and couple stress for MTSM. The effect of internal heat source, micropolarity, and microstretch on the above components has been depicted graphically.

The behavior of a thermo-microstretch isotropic material with microtemperatures without body forces, body couples, stretch force, heat sources, and first heat source moment is governed by the following equations given by Eringen (1990) and Ieşan (2007) as,

$$t_{ji,j} = \rho \ddot{u}_i \tag{1}$$

$$m_{ij,i} + \varepsilon_{ijk} t_{jk} - \mu_1 \varepsilon_{ijr} w_{r,j} = \rho J \ddot{\varphi}_j \tag{2}$$

$$h_{i,i} - s = \frac{\rho j_0}{2} \ddot{\phi}^* \tag{3}$$

The constitutive relations are,

$$t_{ij} = \lambda u_{r,r} \delta_{ij} + \mu(u_{i,j} + u_{j,i}) + K(u_{j,i} - \varepsilon_{ijr}\varphi_r) - \nu T \delta_{ij} + \lambda_0 \phi^* \delta_{ij} \tag{4}$$

$$m_{ij} = \alpha \varphi_{r,r} \delta_{ij} + \beta \varphi_{i,j} + \gamma \varphi_{j,i} + b_0 \varepsilon_{mji} \phi^*_{,m} \tag{5}$$

$$\lambda_i^* = \alpha_0 \phi^*_{,i} + b_0 \varepsilon_{ijm} \varphi_{j,m} \tag{6}$$

$$q_{ij} = -k_4 w_{r,r} \delta_{ij} - k_5 w_{i,j} - k_6 w_{j,i} \tag{7}$$

$$h_i = \alpha_0 \phi^*_{,i} - \mu_2 w_i \tag{8}$$

$$s = \lambda_0 e_{rr} - \nu_1 T + \lambda_1 \varphi^*; \quad i,j,m = 1,2,3 \tag{9}$$

using Equations 4–9 in Equations 1–3, we get the equations,

$$(\mu + K) u_{i,jj} + (\lambda + \mu) u_{j,ij} - K \varepsilon_{ijr} \varphi_{r,j} + \lambda_0 \phi^*_{,i} - \nu T_{,i} = \rho \ddot{u}_i \tag{10}$$

$$\gamma \varphi_{i,jj} + K \varepsilon_{ijr} u_{r,j} - 2K \varphi_i - \mu_1 \varepsilon_{ijr} w_{r,j} = \rho J \ddot{\varphi}_i \tag{11}$$

$$\alpha_0 \phi^*_{,ii} + \nu_1 T - \lambda_1 \phi^* - \lambda_0 u_{i,i} - \mu_2 w_{i,i} = \rho \frac{j_0}{2} \ddot{\phi}^* \tag{12}$$

$$K^* T_{,ii} - \rho c^* \dot{T} - \nu_1 T_0 \dot{\phi}^* - \nu T_0 \dot{u}_{i,i} + k_1 w_{i,i} = Q_1 \tag{13}$$

$$k_6 w_{i,jj} + (k_4 + k_5) w_{j,ij} + \mu_1 \varepsilon_{ijr} \dot{\varphi}_{r,j} - \mu_2 \dot{\phi}^*_{,i} - b \dot{w}_i - k_2 w_i - k_3 T_{,i} = 0 \tag{14}$$

where

$\nu = (3\lambda + 2\mu + K)\alpha_{t_1}$, $\nu_1 = (3\lambda + 2\mu + K)\alpha_{t_2}$, α_{t_1}, α_{t_2} are the coefficents of linear thermal expansion, λ and μ are Lame's constants, K, α, β, γ are the micropolar constants of the solid, α_0, λ_0, λ_1 are the stretch constants, and j_0, μ_1, μ_2, k_1, k_2, k_3, k_4, k_5, k_6 are the constitutive coefficients. t_{ij} is the component of stress tensor, m_{ij} is the coupled stress tensor, λ_i^* is the microstress tensor, q_{ij} is the first heat flux moment tensor, $\vec{u} = (u_i)$ is the displacement vector, $\vec{\varphi} = (\varphi_i)$ is the microrotation vector, $\vec{w} = (w_i)$ is the microtemperature vector and ϕ^* is the scalar microstretch, ρ is the density, J is the microinertia, c^* is the specific heat at constant strain, Q_1 is the internal heat source, K^* is the thermal conductivity, and T is the thermodynamic temperature above reference temperature T_0.

The equations of motion and stress components in fluid (Ewing, Jardetzky, & Press, 1957) are:

$$\lambda^f u^f_{i,ij} = \rho^f \ddot{u}^f_i, \tag{15}$$

$$t_{ij}^{f} = \lambda^{f} u_{r,r}^{f} \delta_{ij} \tag{16}$$

where $\vec{u}^{f} = (u_i^{f})$ is the displacement vector, λ^{f} is the fluid constant, and ρ^{f} is the density of fluid.

We consider a normal force of magnitude F_1 acting along the interface of microstretch thermoelastic medium with microtemperatures (medium I) occupying the region $0 \le z \le \infty$ and a non-viscous fluid (medium II) in the region $-\infty \le z \le 0$ is shown in Figure 1.

A homogeneous isotropic, microstretch thermoelastic solid half space with microtemperatures is considered. We have restricted our analysis to the plane strain parallel to xz plane with displacement vector $u_i = (u_1, 0, u_3)$, microtemperature vector $w_i = (w_1, 0, w_3)$, and microrotation vector $\varphi_i = (0, \varphi_2, 0)$.

For convenience, the following non-dimensional variables are used:

$$x' = \tfrac{1}{L}x,\ z' = \tfrac{1}{L}z,\ u_i' = \tfrac{1}{L}u_i,\ u_i^{f'} = \tfrac{1}{L}u_i^{f},\ w_i' = Lw_i,\ t' = \tfrac{c_1}{L}t,\ t_{ij}' = \tfrac{t_{ij}}{\nu T_0},\ t_{ij}^{f'} = \tfrac{t_{ij}^{f}}{\nu T_0},\ \varphi_i' = \varphi_i,\ \phi^{*'} = \phi^{*},$$
$$m_{ij}' = \tfrac{m_{ij}}{L\nu T_0}, q_{ij}' = \tfrac{q_{ij}}{Lc_1 \nu T_0}, \lambda_i^{*'} = \tfrac{\lambda_i^{*}}{L\nu T_0}, T' = \tfrac{T}{T_0}, F_1' = \tfrac{F_1}{\nu T_0}, Q_1' = \tfrac{Q_1}{Q_0}.$$

where $L = \left(\frac{b}{\rho c^{*} T_0}\right)^{\frac{1}{2}}, c_1^2 = \frac{\lambda + 2\mu + K}{\rho}$.

Assuming the scalar potential functions $\psi_1(x, z, t)$, $\psi_2(x, z, t)$, $\psi_3(x, z, t)$, and $\psi_4(x, z, t)$ defined by the relation in non-dimensional form as,

$$u_1 = \frac{\partial \psi_1}{\partial x} - \frac{\partial \psi_2}{\partial z};\quad u_3 = \frac{\partial \psi_1}{\partial z} + \frac{\partial \psi_2}{\partial x};\quad w_1 = \frac{\partial \psi_3}{\partial x} - \frac{\partial \psi_4}{\partial z};\quad w_3 = \frac{\partial \psi_3}{\partial z} + \frac{\partial \psi_4}{\partial x}. \tag{17}$$

using above non-dimensional variables and relation given by Equation 17, Equations 10–14 reduce to (after dropping superscripts),

$$\left\{(A_1 + 1)\nabla^2 - A_2 \frac{\partial^2}{\partial t^2}\right\}\psi_1 + A_3 \phi^{*} - A_4 T = 0 \tag{18}$$

$$\left(\nabla^2 - A_2 \frac{\partial^2}{\partial t^2}\right)\psi_2 + A_5 \varphi_2 = 0 \tag{19}$$

$$\left(\nabla^2 - 2A_6 - A_7 \frac{\partial^2}{\partial t^2}\right)\varphi_2 - A_6 \nabla^2 \psi_2 + A_8 \nabla^2 \psi_4 = 0 \tag{20}$$

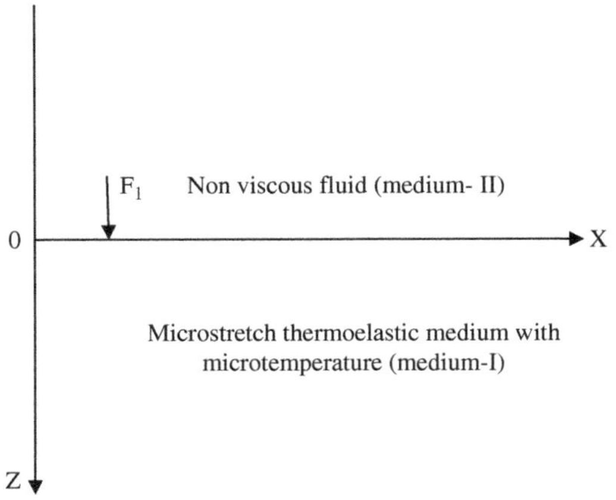

Figure 1. Geometry of the problem.

$$\left(\nabla^2 - A_9 - A_{10}\frac{\partial^2}{\partial t^2}\right)\phi^* - A_{11}\nabla^2\psi_1 - A_{12}\nabla^2\psi_3 + A_{13}T = 0 \tag{21}$$

$$\left(\nabla^2 - A_{14}\frac{\partial}{\partial t}\right)T - A_{15}\frac{\partial\phi^*}{\partial t} - A_{16}\nabla^2\psi_1 + A_{17}\nabla^2\psi_3 = YQ_1 \tag{22}$$

$$\left(\nabla^2(1 + A_{18}) - A_{19} - A_{20}\frac{\partial}{\partial t}\right)\psi_3 - A_{21}\frac{\partial\phi^*}{\partial t} - A_{22}T = 0 \tag{23}$$

$$\left(\nabla^2 - A_{19} - A_{20}\frac{\partial}{\partial t}\right)\psi_4 + A_{23}\frac{\partial\varphi_2}{\partial t} = 0 \tag{24}$$

where

$A_1 = \frac{\lambda+\mu}{\mu+K}$, $A_2 = \frac{\rho c_1^2}{\mu+K}$, $A_3 = \frac{\lambda_0}{\mu+K}$, $A_4 = \frac{\nu T_0}{\mu+K}$, $A_5 = \frac{K}{\mu+K}$, $A_6 = \frac{KL^2}{\gamma}$, $A_7 = \frac{\rho J c_1^2}{\gamma}$, $A_8 = \frac{\mu_1}{\gamma}$, $A_9 = \frac{\lambda_1 L^2}{\alpha_0}$, $A_{10} = \frac{\rho j_0 c_1^2}{2\alpha_0}$, $A_{11} = \frac{\lambda_0 L^2}{\alpha_0}$, $A_{12} = \frac{\mu_2}{\alpha_0}$, $A_{13} = \frac{\nu_1 T_0 L^2}{\alpha_0}$, $A_{14} = \frac{\rho c^* c_1 L}{K^*}$, $A_{15} = \frac{\nu_1 c_1 L}{K^*}$, $A_{16} = \frac{\nu c_1 L}{K^*}$, $A_{17} = \frac{k_1}{K^* T_0}$, $A_{18} = \frac{k_4+k_5}{k_6}$, $A_{19} = \frac{k_2 L^2}{k_6}$, $A_{20} = \frac{b c_1 L}{k_6}$, $A_{21} = \frac{\mu_2 c_1 L}{k_6}$, $A_{22} = \frac{k_3 T_0 L^2}{k_6}$, $A_{23} = \frac{\mu_1 c_1 L}{k_6}$, $Y = \frac{L^2}{K^*}Q_0$.

2. Analytic solution

The solution of the considered physical variable can be decomposed in terms of normal mode and can be considered in the following form,

$$(\psi_i, \phi^*, T, \varphi_2, t_{ij}, q_{ij}, u_i^f, t_{ij}^f, m_{ij}, \lambda_i^*, Q_1)(x, z, t) = (\bar{\psi}_i, \bar{\phi}^*, \bar{T}, \bar{\varphi}_2, \bar{t}_{ij}, \bar{q}_{ij}, \bar{u}_i^f, \bar{t}_{ij}^f, \bar{m}_{ij}, \bar{\lambda}_i^*, \bar{Q}_1)(z)e^{\omega t + iax}$$

where ω is the complex frequency, a is the wave number in x-direction, and $\bar{\psi}_i(z), \bar{\phi}^*(z), \bar{T}(z), \bar{\varphi}_2(z), \bar{\sigma}_{ij}(z), \bar{q}_{ij}(z), \bar{u}_i^f, \bar{t}_{ij}^f, \bar{m}_{ij}(z), \bar{\lambda}_i^*(z), \bar{Q}_1(z)$ are the amplitudes of field quantities.

Using normal mode in Equations 18–24, we get,

$$(D^2 - B_8)\bar{\psi}_1 + B_2\bar{\phi}^* - B_3\bar{T} = 0 \tag{25}$$

$$(D^2 - B_9)\bar{\psi}_2 + A_5\bar{\varphi}_2 = 0 \tag{26}$$

$$(D^2 - B_{10})\bar{\varphi}_2 - A_6(D^2 - a^2)\bar{\psi}_2 + A_8(D^2 - a^2)\bar{\psi}_4 = 0 \tag{27}$$

$$(D^2 - B_{11})\bar{\phi}^* - A_{11}(D^2 - a^2)\bar{\psi}_1 - A_{12}(D^2 - a^2)\bar{\psi}_3 + A_{13}\bar{T} = 0 \tag{28}$$

$$(D^2 - B_{12})\bar{T} - A_{15}\omega\bar{\phi}^* - A_{16}(D^2 - a^2)\bar{\psi}_1 + A_{17}(D^2 - a^2)\bar{\psi}_3 = Y\bar{Q}_1 \tag{29}$$

$$(D^2 - B_{13})\bar{\psi}_3 - B_6\bar{\phi}^* - B_7\bar{T} = 0 \tag{30}$$

$$(D^2 - B_{14})\bar{\psi}_4 + A_{23}\omega\bar{\varphi}_2 = 0 \tag{31}$$

where

$B_1 = \frac{A_2}{A_1+1}$, $B_2 = \frac{A_3}{A_1+1}$, $B_3 = \frac{A_4}{A_1+1}$, $B_4 = \frac{A_{19}}{A_{18}+1}$, $B_5 = \frac{A_{20}}{A_{18}+1}$, $B_6 = \frac{A_{21}\omega}{A_{18}+1}$, $B_7 = \frac{A_{22}}{A_{18}+1}$, $B_8 = a^2 + B_1\omega^2$, $B_9 = a^2 + A_2\omega^2$, $B_{10} = a^2 + 2A_6 + A_7\omega^2$, $B_{11} = a^2 + A_9 + A_{10}\omega^2$, $B_{12} = a^2 + A_{14}\omega$, $B_{13} = a^2 + B_4 + B_5\omega$, $B_{14} = a^2 + A_{19} + A_{20}\omega$.

and constitutive relations (4–7) become,

$$\bar{t}_{xx} = \left(A_{25}D^2 - a^2A_{24}\right)\bar{\psi}_1 + ia\left(A_{25} - A_{24}\right)D\bar{\psi}_2 - \bar{T} + A_{26}\bar{\phi}^* \tag{32}$$

$$\bar{t}_{xz} = ia\left(A_{27} + A_{28}\right)D\bar{\psi}_1 - \left(A_{27}D^2 + a^2A_{28}\right)\bar{\psi}_2 - K\bar{\varphi}_2 \tag{33}$$

$$\bar{t}_{zz} = \left(A_{24}D^2 - a^2A_{25}\right)\bar{\psi}_1 + ia\left(A_{24} - A_{25}\right)D\bar{\psi}_2 - \bar{T} + A_{26}\bar{\phi}^* \tag{34}$$

$$\bar{q}_{xx} = \left(A_{29}a^2 - A_{30}D^2\right)\bar{\psi}_3 + ia\left(A_{29} - A_{30}\right)D\bar{\psi}_4 \tag{35}$$

$$\bar{q}_{xz} = -ia\left(A_{31} + A_{32}\right)D\bar{\psi}_3 + \left(A_{31}D^2 + a^2A_{32}\right)\bar{\psi}_4 \tag{36}$$

$$\bar{q}_{zz} = \left(A_{30}a^2 - A_{29}D^2\right)\bar{\psi}_3 + ia\left(A_{30} - A_{29}\right)D\bar{\psi}_4 \tag{37}$$

$$\bar{m}_{yz} = A_{33}D\bar{\phi}_2 - iaA_{34}\bar{\phi}^* \tag{38}$$

$$\bar{\lambda}_3^* = A_{35}D\bar{\phi}^* - iaA_{34}\bar{\varphi}_2 \tag{39}$$

where

$A_{24} = \frac{\lambda+2\mu+K}{\nu T_0}$, $A_{25} = \frac{\lambda}{\nu T_0}$, $A_{26} = \frac{\lambda_0}{\nu T_0}$, $A_{27} = \frac{\mu}{\nu T_0}$, $A_{28} = \frac{\mu+K}{\nu T_0}$, $A_{29} = \frac{k_4+k_5+k_6}{L^3c_1\nu T_0}$, $A_{30} = \frac{k_4}{L^3c_1\nu T_0}$, $A_{31} = \frac{k_5}{L^3c_1\nu T_0}$, $A_{32} = \frac{k_6}{L^3c_1\nu T_0}$, $A_{33} = \frac{\beta}{L^2\nu T_0}$, $A_{34} = \frac{b_0}{L^2\nu T_0}$, $A_{35} = \frac{\alpha_0}{L^2\nu T_0}$.

Eliminating $\bar{\phi}^*(z), \bar{\psi}_3(z), \bar{T}(z)$ from Equations 25, 28–30, we get the following eight-order differential equation for $\bar{\psi}_1(z)$ as,

$$(D^8 + PD^6 + QD^4 + RD^2 + S)\bar{\psi}_1(z) = B_{15}\bar{Q}_1 \tag{40}$$

Eliminating $\bar{\psi}_4(z)$ and $\bar{\varphi}_2(z)$ from Equations 26–27 and 31, we get the following sixth-order differential equation for $\bar{\psi}_2(z)$ as,

$$(D^6 + ED^4 + FD^2 + G)\bar{\psi}_2(z) = 0 \tag{41}$$

where

$$P = \left[-(B_{12} + B_{13}) + B_7A_{17} - B_8 - B_{11} - A_{12}B_6 + B_2A_{11} - B_3A_{16}\right]$$

$$\begin{aligned} Q = \Big[&B_{12}B_{13} - B_7A_{17}a^2 + (B_8 + B_{11})(B_{12} + B_{13}) - B_7A_{17}(B_8 + B_{11}) + B_8B_{11} + A_{13}A_{15}\omega - A_{13}A_{17}B_6 \\ &- A_{12}A_{15}B_7\omega + A_{12}B_{12}B_6 + A_{12}(a^2 + B_8)B_6 + B_2B_7(A_{11}A_{17} + A_{12}A_{16}) \\ &- B_2A_{11}(B_{12} + B_{13}) - B_2A_{13}A_{16} - a^2B_2A_{11} + (A_{11}A_{17} + A_{12}A_{16})B_3B_6 - B_3A_{11}A_{15}\omega + B_3B_{11} \\ &+ A_{16}(a^2 + B_{13})B_3\Big] \end{aligned}$$

$$\begin{aligned} R = \Big[&-B_{12}B_{13}(B_8 + B_{11}) - B_7A_{17}a^2(B_8 + B_{11}) - B_8B_{11}(B_{12} + B_{13}) + B_7A_{17}B_8B_{11} - A_{13}A_{15}B_{13}\omega \\ &+ a^2A_{13}A_{17}B_6 - B_8A_{13}A_{15}\omega + B_8B_6A_{13}A_{17} + A_{12}(a^2 + B_8)A_{15}B_7\omega - A_{12}(a^2 + B_8)B_6B_{12} \\ &- A_{12}B_8B_6a^2 - 2a^2B_2B_7(A_{11}A_{17} + A_{12}A_{16}) + B_2A_{11}B_{12}B_{13} + B_2A_{13}A_{16}B_{13} + a^2B_2A_{11}(B_{12} + B_{13}) \\ &+ a^2A_{13}A_{16}B_2 - 2a^2B_3B_6(A_{11}A_{17} + A_{12}A_{16}) + B_3(B_{13} + a^2)A_{11}A_{15}\omega - B_3(B_{13} + a^2)A_{16}B_{11} \\ &- a^2B_3A_{16}B_{13}\Big] \end{aligned}$$

$$S = \Big[B_{12}B_{13}B_8B_{11} - B_7A_{17}a^2B_8B_{11} + B_8B_{13}A_{13}A_{15}\omega - B_6B_8A_{13}A_{17}a^2 - B_7B_8A_{12}A_{15}a^2\omega + B_6B_8B_{12}A_{12}a^2 + a^4B_2B_7(A_{11}A_{17} + A_{12}A_{16}) - a^2B_2A_{11}B_{12}B_{13} - a^2B_2A_{13}A_{16}B_{13} + a^4B_3B_6(A_{11}A_{17} + A_{12}A_{16}) - B_3B_{13}A_{11}A_{15}a^2\omega + B_3B_{13}B_{11}A_{16}a^2\Big]$$

$$E = [-(B_{10} + B_{14}) - A_8A_{23}\omega - B_9 + A_5A_6]$$

$$F = \Big[B_{10}B_{14} + A_8A_{23}\omega a^2 + B_9(B_{10} + B_{14}) + B_9A_8A_{23}\omega - A_5A_6(B_{14} + a^2)\Big]$$

$$G = \Big[-B_9B_{10}B_{14} - B_9A_8A_{23}\omega a^2 + a^2A_5A_6B_{14}\Big]$$

$$B_{15} = YA_{12}B_3\Big[-A_3B_{13}B_2 + B_7A_{12}B_2a^2 - a^2A_{12}B_6B_3 - B_3B_{13}B_{11}\Big]$$

In a similar manner, we can show that $\bar{\phi}^*(z), \bar{\psi}_3(z)$, and $\bar{T}(z)$ satisfy the equation,

$$(D^8 + PD^6 + QD^4 + RD^2 + S)(\bar{\phi}^*(z), \bar{\psi}_3(z), \bar{T}(z)) = B_{15}\bar{Q}_1 \quad (42)$$

which can be factorized as follows,

$$(D^2 - r_1^2)(D^2 - r_2^2)(D^2 - r_3^2)(D^2 - r_4^2)\bar{\psi}_1(z) = B_{15}\bar{Q}_1 \quad (43)$$

where r_n^2; (n = 1, 2, 3, 4) are the roots of Equation 42.

and $\bar{\psi}_4(z)$ and $\bar{\varphi}_2(z)$ satisfy the equation,

$$(D^6 + ED^4 + FD^2 + G)(\bar{\psi}_4(z), \bar{\varphi}_2(z)) = 0 \quad (44)$$

which can be factorized as follows,

$$(D^2 - h_1^2)(D^2 - h_2^2)(D^2 - h_3^2)\bar{\psi}_2(z) = 0 \quad (45)$$

where h_n^2; (n = 1, 2, 3) are the roots of Equation 44.

The series solution of Equation 42 has the form,

$$\bar{\psi}_1(z) = \sum_{n=1}^{4}[M_n(a,\omega)e^{-r_nz}] + N \quad (46)$$

$$\bar{\phi}^*(z) = \sum_{n=1}^{4}[M'_n(a,\omega)e^{-r_nz}] + N_1 \quad (47)$$

$$\bar{T}(z) = \sum_{n=1}^{4}[M''_n(a,\omega)e^{-r_nz}] + N_2 \quad (48)$$

$$\bar{\psi}_3(z) = \sum_{n=1}^{4}[M'''_n(a,\omega)e^{-r_nz}] + N_3 \quad (49)$$

The series solution of Equation 44 has the form,

$$\bar{\psi}_2(z) = \sum_{n=1}^{3}[L_n(a,\omega)e^{-h_n z}] \tag{50}$$

$$\bar{\varphi}_2(z) = \sum_{n=1}^{3}[L'_n(a,\omega)e^{-h_n z}] \tag{51}$$

$$\bar{\psi}_4(z) = \sum_{n=1}^{3}[L''_n(a,\omega)e^{-h_n z}] \tag{52}$$

where $M_n(a,\omega)$, $M'_n(a,\omega)$, $M''_n(a,\omega)$, $M'''_n(a,\omega)$, and $L_n(a,\omega)$, $L'_n(a,\omega)$, $L''_n(a,\omega)$ are the specific function depending upon a and ω.

Using Equations 46–49 in Equations 25, 28–30, we get the following relations,

$$M'_n(a,\omega) = H_{1n}M_n(a,\omega) \tag{53}$$

$$M''_n(a,\omega) = H_{2n}M_n(a,\omega) \tag{54}$$

$$M'''_n(a,\omega) = H_{3n}M_n(a,\omega) \tag{55}$$

similarly, using Equations 50–52 in Equations 26–27 and 31, we get the following relations,

$$L'_n(a,\omega) = R_{1n}L_n(a,\omega) \tag{56}$$

$$L''_n(a,\omega) = R_{2n}L_n(a,\omega) \tag{57}$$

Thus we have,

$$\bar{\phi}^*(z) = \sum_{n=1}^{4}[H_{1n}M_n(a,\omega)e^{-r_n z}] + N_1 \tag{58}$$

$$\bar{T}(z) = \sum_{n=1}^{4}[H_{2n}M_n(a,\omega)e^{-r_n z}] + N_2 \tag{59}$$

$$\bar{\psi}_3(z) = \sum_{n=1}^{4}[H_{3n}M_n(a,\omega)e^{-r_n z}] + N_3 \tag{60}$$

$$\bar{\varphi}_2(z) = \sum_{n=1}^{3}[R_{1n}L_n(a,\omega)e^{-h_n z}] \tag{61}$$

$$\bar{\psi}_4(z) = \sum_{n=1}^{3}[R_{2n}L_n(a,\omega)e^{-h_n z}] \tag{62}$$

$$\bar{t}_{xx}(z) = \sum_{n=1}^{4}[H_{4n}M_n(a,\omega)e^{-r_n z}] + \sum_{n=1}^{3}[R_{3n}L_n(a,\omega)e^{-h_n z}] - N_4 \tag{63}$$

$$\bar{t}_{xz}(z) = \sum_{n=1}^{4}[H_{5n}M_n(a,\omega)e^{-r_n z}] + \sum_{n=1}^{3}[R_{4n}L_n(a,\omega)e^{-h_n z}] \tag{64}$$

$$\bar{t}_{zz}(z) = \sum_{n=1}^{4}[H_{6n}M_n(a,\omega)e^{-r_n z}] - \sum_{n=1}^{3}[R_{3n}L_n(a,\omega)e^{-h_n z}] - N_5 \tag{65}$$

$$\bar{q}_{xx}(z) = \sum_{n=1}^{4}[H_{7n}M_n(a,\omega)e^{-r_n z}] + \sum_{n=1}^{3}[R_{5n}L_n(a,\omega)e^{-h_n z}] + N_6 \tag{66}$$

$$\bar{q}_{xz}(z) = \sum_{n=1}^{4}[H_{8n}M_n(a,\omega)e^{-r_n z}] + \sum_{n=1}^{3}[R_{6n}L_n(a,\omega)e^{-h_n z}] \tag{67}$$

$$\bar{q}_{zz}(z) = \sum_{n=1}^{4}[H_{9n}M_n(a,\omega)e^{-r_n z}] - \sum_{n=1}^{3}[R_{5n}L_n(a,\omega)e^{-h_n z}] + N_7 \tag{68}$$

$$\bar{m}_{yz}(z) = \sum_{n=1}^{4}[H_{10n}M_n(a,\omega)e^{-r_n z}] + \sum_{n=1}^{3}[R_{7n}L_n(a,\omega)e^{-h_n z}] + N_8 \tag{69}$$

$$\bar{\lambda}_3^*(z) = \sum_{n=1}^{4}[H_{11n}M_n(a,\omega)e^{-r_n z}] + \sum_{n=1}^{3}[R_{8n}L_n(a,\omega)e^{-h_n z}] \tag{70}$$

where

$$N = \tfrac{B_{15}}{S}\bar{Q}, N_1 = \tfrac{A_{12}YB_3\bar{Q}_1 - [(A_{12}B_{12} - A_{13}A_{17})B_8 + a^2B_3(A_{11}A_{17} + A_{12}A_{16})]N}{B_2(A_{17}A_{13} - A_{12}B_{12}) - B_3(A_{17}B_{11} + A_{12}A_{15}\omega)},$$

$$N_2 = \tfrac{-B_8N + B_2N_1}{B_3}, N_3 = \tfrac{-(B_6N_1 + B_7N_2)}{B_{13}}, N_4 = (A_{24}a^2N + N_2 - A_{26}N_1),$$

$$N_5 = (A_{25}a^2N + N_2 - A_{26}N_1), N_6 = (A_{24}a^2N_3), N_7 = (A_{30}a^2N_3), N_8 = (-iaA_{34}N_1),$$

$$H_{1n} = -\tfrac{[A_{12}r_n^4 - B_{16}r_n^2 + B_{17}]}{[(A_{12}B_2 + A_{17}B_3)r_n^2 - A_{12}B_{12}B_2 + A_{13}A_{17}B_2 - A_{17}B_3B_{11} - A_{12}A_{15}B_3]},$$

$$H_{2n} = \tfrac{(r_n^2 - B_8 + B_2H_{1n})}{B_3},\ H_{3n} = \tfrac{(B_6H_{1n} + B_7H_{2n})}{(r_n^2 - B_{13})},\ H_{4n} = (A_{25}r_n^2 - a^2A_{24}) - H_{2n} + A_{26}H_{1n},$$

$$H_{5n} = -iar_n(A_{27} + A_{28}), H_{6n} = (A_{24}r_n^2 - a^2A_{25}) - H_{2n} + A_{26}H_{1n}, H_{7n} = (a^2A_{29} - A_{30}r_n^2)H_{3n},$$

$$H_{8n} = ia(A_{31} + A_{32})r_nH_{3n}, H_{9n} = (a^2A_{30} - A_{29}r_n^2)H_{3n}, H_{10n} = -iaA_{34}H_{1n}, H_{11n} = -A_{35}r_nH_{1n},$$

$$R_{1n} = \tfrac{(B_9 - h_n^2)}{(A_5)}, R_{2n} = -\tfrac{(A_{23}\omega R_{1n})}{(h_n^2 - B_{14})}, R_{3n} = ia(A_{24} - A_{25})h_n, R_{4n} = -(A_{27}h_n^2 + a^2A_{28} + KR_{1n}),$$

$$R_{5n} = -ia(A_{29} - A_{30})h_nR_{2n}, R_{6n} = (h_n^2A_{31} + A_{32}a^2)R_{2n}, R_{7n} = -A_{33}h_nR_{1n}, R_{8n} = -iaA_{34}R_{1n},$$

$$B_{16} = (A_{12}\{B_{12} + B_8\} + A_{13}A_{17} - \{A_{11}A_{17} + A_{12}A_{16}\}B_3),$$

$$B_{17} = \{A_{12}B_{12} - A_{13}A_{17}\}B_8 + a^2B_3(A_{11}A_{17} + A_{12}A_{16}).$$

similarly for medium II (i.e. fluid half space), the solutions are of the form,

$$\bar{u}_1^f(z) = M_5(b,\omega)e^{-r_5 z} \tag{71}$$

$$\bar{u}_3^f(z) = M_5'(b,\omega)e^{-r_5 z} \tag{72}$$

where $M_5(a, \omega)$ and $M_5'(a,\omega)$ are the specific functions depending upon a and ω and r_5 is the root of characteristic equation,

$$(D^2 - a^2 + l\omega^2)\bar{u}_1^f(z) = 0 \tag{73}$$

where $l = \frac{\rho^f c_1^2}{\lambda^f}$ and $r_5 = \sqrt{a^2 - l\omega^2}$

Thus we have,

$$\bar{u}_3^f(z) = HM_5(b,\omega)e^{-r_5 z} \tag{74}$$

$$\bar{t}_{zz}^f(z) = IM_5(b,\omega)e^{-r_5 z} \tag{75}$$

$$\bar{t}_{xz}^f(z) = 0 \tag{76}$$

where $H = \frac{r_5^2 - l\omega^2}{iar_5}$ and $I = \frac{(\lambda^f)(iaH - r_5)}{\rho c_1^2}$.

3. Applications

In this section, we determine the parameters M_n; $(n = 1, 2, 3, 4, 5)$ and L_n;$(n = 1, 2, 3)$. In the physical problem, we should suppress the positive exponentials that are unbounded at infinity. Constants M_1, M_2, M_3, M_4, M_5 and L_1, L_2, L_3 have to be selected such that boundary conditions at the surface $z = 0$ take the form,

$$t_{zz} = t_{zz}^f - F_1 e^{\omega t + iax},\ t_{xz} = t_{xz}^f,\ m_{yz} = 0,\ \lambda_3^* = 0,\ q_{zz} = 0,\ q_{xz} = 0,\ \frac{\partial T}{\partial z} = 0,\ \frac{\partial u_3}{\partial t} = \frac{\partial u_3^f}{\partial t} \tag{77}$$

where F_1 is the magnitude of mechanical force.

Using the expressions of $t_{zz}, t_{xz}, t_{zz}^f, t_{zx}^f, m_{yz}, \lambda_3^*, q_{zz}, q_{xz}, T, u_3, u_3^f$ into above boundary conditions (77), give the following equations satisfied by the parameters,

$$\sum_{n=1}^{4}[H_{6n}M_n] - \sum_{n=1}^{3}[R_{3n}L_n] - IM_5 = N_5 - F_1$$

$$\sum_{n=1}^{4}[H_{5n}M_n] + \sum_{n=1}^{3}[R_{4n}L_n] = 0$$

$$\sum_{n=1}^{4}[H_{10n}M_n] + \sum_{n=1}^{3}[R_{7n}L_n] = -N_8$$

$$\sum_{n=1}^{4}[H_{11n}M_n] + \sum_{n=1}^{3}[R_{8n}L_n] = 0$$

$$\sum_{n=1}^{4}[H_{9n}M_n] - \sum_{n=1}^{3}[R_{5n}L_n] = -N_7$$

$$\sum_{n=1}^{4}[H_{8n}M_n] + \sum_{n=1}^{3}[R_{6n}L_n] = 0$$

$$\sum_{n=1}^{4}[H_{2n}r_nM_n] = 0$$

$$\sum_{n=1}^{4}[-r_nM_n] + ia\sum_{n=1}^{3}L_n - HM_5 = 0$$

After solving these non-homogeneous system of equations, we get the values of constants M_1, M_2, M_3, M_4, M_5, L_1, L_2, L_3 and hence obtain the component of normal displacement, microtemperature, normal force stress, microstress tensor, heat flux moment tensor, and couple stress at the interface of fluid half space and MTSM.

4. Special case

(1) If we neglect micropolarity effect i.e. $\alpha = \beta = \gamma = b_0 = \mu = K = J = 0$, we obtain the results for microstretch thermoelastic solid with microtemperatures without microrotational effect (TSMWM).

(2) If we neglect microstretch effect i.e. $\alpha_0 = \lambda_0 = \lambda_1 = v_1 = b_0 = \mu_2 = J_0 = 0$, we obtain the results for thermoelastic solid with microtemperatures without microstretch effect (TSMWS).

(3) If we neglect both micropolarity effect and microstretch effect i.e. $\alpha = \beta = \gamma = 0$, $\mu = K = J = \alpha_0 = \lambda_0 = \lambda_1 = v_1 = b_0 = \mu_2 = J_0 = 0$, we obtain the results for thermoelastic solid with microtemperatures (TSM).

5. Numerical results and discussions

In order to illustrate the theoretical results obtained in the preceding section, we present some numerical results for the physical constants,

The values of micropolar constants are (Eringen, 1984):

$\lambda = 9.4 \times 10^{10}$ N/m^2, $\mu = 4.0 \times 10^{10}$ N/m^2, $\rho = 1.74 \times 10^3$ kg/m^3, $K = 10^{10}$ Nm^{-2}, $\gamma = 7.79 \times 10^{-10}$ N, $J = 0$ $.0000002 \times 10^{-14}$ m^2, $\beta = 0.32 \times 10^{10}$ N/m^2 K, $b_0 = 0.0098 \times 10^{10}$ N.

The values of thermal parameters are (Dhaliwal & Singh, 1980):

$c^* = 0.104 \times 10^4$ Nm/kg/K, $T_0 = 298$ K, $K^* = 1.7 \times 10^2$ Ns^{-1} K^{-1}, $\alpha_{t_1} = 0.05$K^{-1}, $\alpha_{t_2} = 0.05$ K^{-1}, $\tau_1 = 0.613 \times 10^3$ s.

The values of microstretch parameters are (Kumar & Kaur, 2014):

$j_0 = 0.000019 \times 10^{-13}$ m^2, $\lambda_0 = 0.21 \times 10^{11}$N/m^2, $\lambda_1 = 0.007 \times 10^{12}$N/m^2, $\alpha_0 = 0.008 \times 10^{-7}$ N, $b = 0.15 \times 10^{-10}$ N.

The values of microtemperature parameters are (Kumar & Kaur, 2014):

$k_1 = 0.0035$ Ns^{-1}, $k_2 = 0.045$ Ns^{-1}, $k_3 = 0.055$ NK^{-1}s^{-1}, $k_4 = 0.065Ns^{-1}$ m^2, $k_5 = 0.076$ Ns^{-1} m^2, $k_6 = 0.09$ 6 Ns^{-1} m^2, $\mu_1 = 0.0085$ N, $\mu_2 = 0.0095$ N.

The physical constants for water are given by Ewing et al. (1957):

$\lambda^f = 2.14 \times 10^9 \text{N/m}^2, \rho^f = 10^3 \text{ kg/m}^3.$

The computations are carried out for the value of non-dimensional time $t = 0.2$ in the range $0 \leq x \leq 10$ and on the surface $z = 1.3$. The numerical values for normal displacement, microtemperature, normal force stress, microstress tensor, heat flux moment tensor, and couple stress are shown in Figures 2–7 for mechanical force with magnitude.

$F_1 = 1.0, Q_0 = 1, \omega = \omega_0 + \iota\xi, \omega_0 = -0.3, \xi = 0.1, Q_1 = 10$ and $a = 0.9$ for

(a) MTSM by solid line with centered symbol ♦.

(b) TSMWM by solid line with centered symbol ■.

(c) TSMWS by dashed line with centered symbol ▲.

(d) TSM by dashed line with centered symbol ×.

6. Discussion

The variation of normal displacement for MTSM, TSMWM, and TSMWS is similar in nature. These values decrease sharply in the entire range. The values of normal displacement for TSM increase in the range $0 \leq x \leq 2.3$ and then the values approach zero with a straight curve. The variations of microtemperature for MTSM and TSMWS are opposite in nature which shows that microstructure has significant effect on microtemperature. The values of microtemperature for TSM are very less and lie in a very short range. These variations of normal displacement and microtemperature are shown in Figures 2 and 3, respectively.

Figure 4 depicts that the variations of normal force stress are opposite in nature for both MTSM and TSMWM. This concludes that micropolarity effect is more prominent in the study of normal force stress. The variations of normal force stress for TSMWS and TSM are similar in nature. The values are

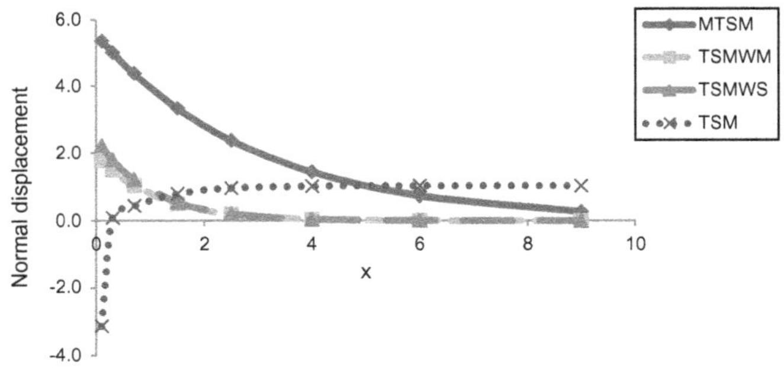

Figure 2. Variation of normal displacement with horizontal distance.

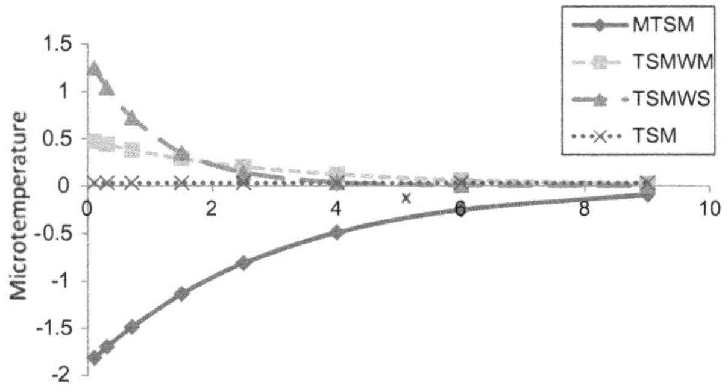

Figure 3. Variation of microtemperature with horizontal distance.

also quite close to each other. The values for these medium (TSMWS and TSM) decrease sharply and then follow a straight curve to converge. With difference in magnitude, the variation of microstress tensor for MTSM and TSMWM is similar in nature. These values decrease uniformly and then approach to zero with increase in horizontal distance. The variation of microstress tensor is shown in Figure 5.

Figure 6 shows that the variations of heat flux moment tensor are similar in nature for all mediums. There is difference in magnitude among all the solids which proves the effect of micropolarity and microstress in the medium. It is again observed that the values of heat flux moment tensor for TSM are very less and hence as compared to other medium, the variation lies in a very short range.

In the absence of stretch effect, the variation of couple stress is effected to a great extent as visible in Figure 7. The values increase in the range $0 \le x \le 4.0$ and then show a constant behavior. The variations are sharper for TSMWS in comparison to MTSM.

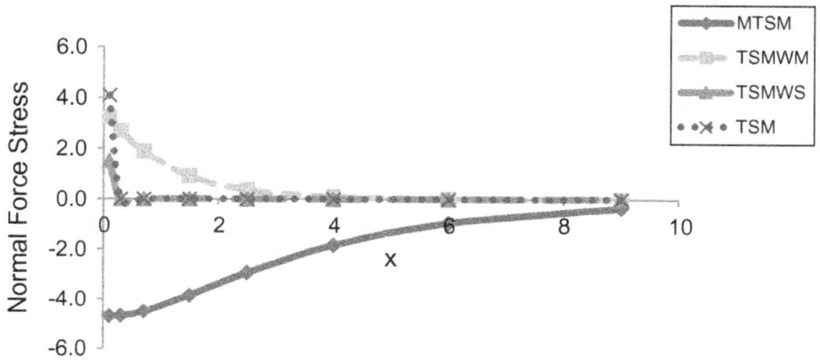

Figure 4. Variation of normal force stress with horizontal distance.

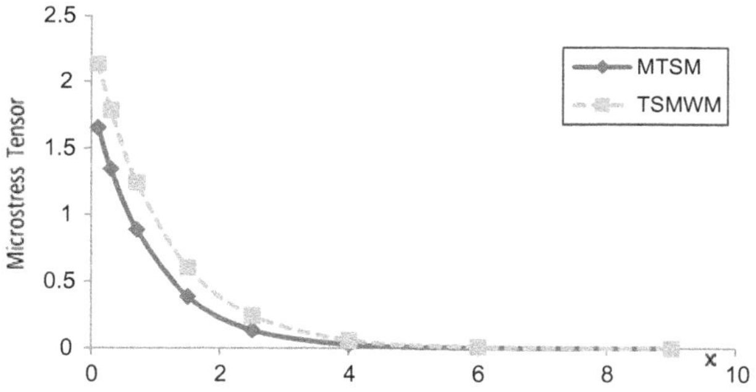

Figure 5. Variation of microstress tensor with horizontal distance.

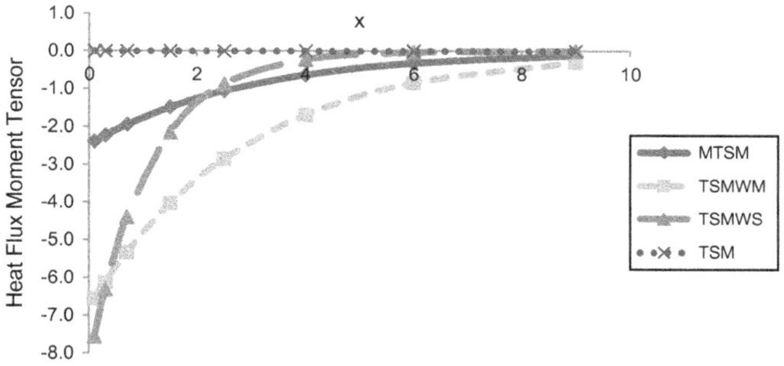

Figure 6. Variation of heat flux moment tensor with horizontal distance.

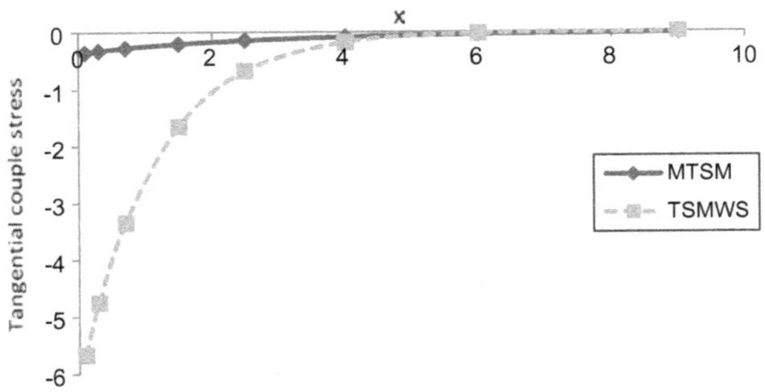

Figure 7. Variation of tangential coule stress with horizontal distance.

7. Conclusion

Both micropolarity and stretch effect have a significant effect on the normal displacement, microtemperature, normal force stress, microstress tensor, heat flux moment tensor, and tangential couple stress. The values of all the quantities for a generalized TSM are less in magnitude as compared to the medium with micropolarity and stretch effect. Micropolarity does not show appreciable effect on microstress tensor but microstretch has a significant effect on couple stress. Such type of problems is very useful in the study of earthquake engineering, seismology, and volcanic eruptions. It helps us to study the effect of a heat source in the medium and the deformation caused in the medium due to the heat source.

Funding
The authors received no direct funding for this research.

Author details
Praveen Ailawalia[1]
E-mail: praveen_2117@rediffmail.com
Sunil Kumar Sachdeva[2,3]
E-mail: sunilsachdeva.daviet@gmail.com
Devinder Pathania[4]
E-mail: despathania@yahoo.com
ORCID ID: http://orcid.org/0000-0002-3324-9633
[1] Department of Applied Sciences and Humanities, M.M. University, Sadopur, Ambala City, Haryana, India.
[2] Department of Applied Sciences, D.A.V. Institute of Engineering and Technology, Kabir Nagar, Jalandhar, Punjab, India.
[3] Punjab Technical University, Jalandhar, Punjab, India.
[4] Department of Applied Sciences, Guru Nanak Engineering College, Ludhiana, Punjab, India.

References

Abbas, I. A., & Othman, M. I. A. (2012). Plane waves in generalized thermo-microstretch elastic solid with thermal relaxation using finite element method. *International Journal of Thermophysics, 33*, 2407–2423. http://dx.doi.org/10.1007/s10765-012-1340-8

Aouadi, M. (2008). Some theorems in the isotropic theory of microstretch thermoelasticity with microtemperatures. *Journal of Thermal Stresses, 31*, 649–662. http://dx.doi.org/10.1080/01495730801981772

Casas, P. S., & Quintanilla, R. (2005). Exponential stability in thermoelasticity with microtemperatures. *International Journal of Engineering Science, 43*, 33–47. http://dx.doi.org/10.1016/j.ijengsci.2004.09.004

Chiriţă, S., Ciarletta, M., & D'Apice, C. (2013). On the theory of thermoelasticity with microtemperatures. *Journal of Mathematical Analysis and Applications, 397*, 349–361. http://dx.doi.org/10.1016/j.jmaa.2012.07.061

Dhaliwal, R. S., & Singh, A. (1980). *Dynamic coupled thermoelasticity*. New Delhi: Hindustan Publication Corporation.

Eringen, A. C. (1965). *Linear theory of micropolar elasticity* (ONR Technical Report No. 29). School of Aeronautics, Aeronautics and Engineering Science, Purdue University, West Lafayette, IN.

Eringen, A. C. (1966). A unified theory of thermomechanical materials. *International Journal of Engineering Science, 4*, 179–202. http://dx.doi.org/10.1016/0020-7225(66)90022-X

Eringen, A. C. (1968). *Micropolar elastic solids with stretch.* In Mustafa Inan Anisina, Ari. Kitapevi Matbaassi, Istanbul (pp. 1–18).

Eringen, A. C. (1984). Plane waves in nonlocal micropolar elasticity. *International Journal of Engineering Science, 22*, 1113–1121. http://dx.doi.org/10.1016/0020-7225(84)90112-5

Eringen, A. C. (1990). Theory of thermo-microstretch elastic solids. *International Journal of Engineering Science, 28*, 1291–1301. http://dx.doi.org/10.1016/0020-7225(90)90076-U

Eringen, A. C. (1996). Linear theory of micropolar elasticity. *Journal of Mathematics and Mechanics, 15*, 909–923.

Eringen, A. C., & Suhubi, E. S. (1964). Nonlinear theory of simple micro-elastic solids I. *International Journal of Engineering Science, 2*, 189–203. http://dx.doi.org/10.1016/0020-7225(64)90004-7

Eringen, A. C., & Suhubi, E. S. (1964). Nonlinear theory of simple micro-elastic solids II. *International Journal of Engineering Science, 2*, 389–404.

Ewing, W. M., Jardetzky, W. S., & Press, F. (1957). *Elastic waves in layered media*. New York, NY: McGraw Hill.

Grot, R. A. (1969). Thermodynamics of a continuum with microstructure. *International Journal of Engineering Science, 7*, 801–814. http://dx.doi.org/10.1016/0020-7225(69)90062-7

Iesan, D. (2001). On a theory of micromorphic elastic solids with microtemperatures. *Journal of Thermal Stresses, 24*, 737–752.

Iesan, D., & Quintanilla, R. (2000). On a theory of thermoelasticity with microtemperatures. *Journal of Thermal Stresses, 23*, 199–215.

Ieşan, D. (2007). Thermoelasticity of bodies with microstructure and microtemperatures. *International Journal of Solids and Structures, 44*, 8648–8662. http://dx.doi.org/10.1016/j.ijsolstr.2007.06.027

Ieşan, D., & Quintanilla, R. (2009). On thermoelastic bodies with inner structure and microtemperatures. *Journal of Mathematical Analysis and Applications, 354*, 12–23. http://dx.doi.org/10.1016/j.jmaa.2008.12.017

Kumar, R., & Kansal, T. (2011). Fundamental solution in the theory of thermomicrostretch elastic diffusive solids. *ISRN Applied Mathematics, 2011*, 1–15. http://dx.doi.org/10.5402/2011/764632

Kumar, R., & Kaur, M. (2014). Reflection and refraction of plane waves at the interface of an elastic solid and microstretch thermoelastic solid with microtemperatures. *Archive of Applied Mechanics, 84*, 571–590. http://dx.doi.org/10.1007/s00419-014-0818-1

Kumar, R., & Partap, G. (2009). Analysis of free vibrations for Rayleigh–Lamb waves in a microstretch thermoelastic plate with two relaxation times. *Journal of Engineering Physics and Thermophysics, 82*, 35–46.

Kumar, R., & Rupender, R. (2008). Reflection at free surface of magneto-thermo-microstretch elastic solid. *Bulletin of the Polish Academy of Sciences, 56*, 263–271.

Kumar, R., & Rupender, R. (2009). Propagation of plane waves at the imperfect boundary of elastic and electro-microstretch generalized thermoelastic solids. *Applied Mathematics and Mechanics, 30*, 1445–1454. http://dx.doi.org/10.1007/s10483-009-1110-6

Kumar, S., Sharma, J. N., & Sharma, Y. D. (2011). Generalized thermoelastic waves in microstretch plates loaded with fluid of varying temperature. *International Journal of Applied Mechanics, 3*, 563–586. http://dx.doi.org/10.1142/S1758825111001135

Magaña, A., & Quintanilla, R. (2006). On the time decay of solutions in one-dimensional theories of porous materials. *International Journal of Solids and Structures, 43*, 3414–3427. http://dx.doi.org/10.1016/j.ijsolstr.2005.06.077

Marin, M. (2010). Lagrange identity method for microstretch thermoelastic materials. *Journal of Mathematical Analysis and Applications, 363*, 275–286. http://dx.doi.org/10.1016/j.jmaa.2009.08.045

Othman, M. I. A., & Lotfy, Kh. (2010). On the plane waves of generalized thermo-microstretch elastic half space under three theories. *International Communications in Heat and Mass Transfer, 37*, 192–200. http://dx.doi.org/10.1016/j.icheatmasstransfer.2009.09.017

Othman, M. I. A., & Lotfy, Kh. (2011). Effect of rotation on plane waves in generalized thermo- microstretch elastic solid with one relaxation time. *Multidiscipline Modeling in Materials and Structures, 7*, 43–62. http://dx.doi.org/10.1108/15736101111141430

Othman, M. I. A., Lotfy, Kh., & Farouk, R. M. (2010). Generalized thermo-microstretch elastic medium with temperature dependent properties for different theories. *Engineering Analysis with Boundary Elements, 34*, 229–237. http://dx.doi.org/10.1016/j.enganabound.2009.10.003

Quintanilla, R. (2011). On growth and continuous dependence in thermoelasticity with microtemperatures. *Journal of Thermal Stresses, 34*, 911–922. http://dx.doi.org/10.1080/01495739.2011.586278

Říha, P. (1976). On the microcontinuum model of heat conduction in materials with inner structure. *International Journal of Engineering Science, 14*, 529–535. http://dx.doi.org/10.1016/0020-7225(76)90017-3

Scalia, A., & Svanadze, M. (2006). On the representations of solutions of the theory of thermoelasticity with microtemperatures. *Journal of Thermal Stresses, 29*, 849–863. http://dx.doi.org/10.1080/01495730600705448

Scalia, A., Svanadze, M., & Tracinà, R. (2010). Basic theorems in the equilibrium theory of thermoelasticity with microtemperatures. *Journal of Thermal Stresses*, 721–753. http://dx.doi.org/10.1080/01495739.2010.482348

Singh, B., & Kumar, R. (1998). Wave propagation in a generalized thermo-microstretch elastic solid. *International Journal of Engineering Science, 36*, 891–912. http://dx.doi.org/10.1016/S0020-7225(97)00099-2

Singh, D., Kumar, A., & Kumar, R. (2014). A problem in microstretch thermoelastic diffusive medium. *International Journal of Mathematical, Computational, Physical, Electrical and Computer Engineering, 8*, 24–27.

Steeb, H., Singh, J., & Tomar, S. K. (2013). Time harmonic waves in thermoelastic material with microtemperatures. *Mechanics Research Communications, 48*, 8–18. http://dx.doi.org/10.1016/j.mechrescom.2012.11.006

Tomar, S. K., & Khurana, A. (2009). Reflection and transmission of elastic waves from a plane interface between two thermo-microstretch solid half-spaces. *International Journal of Applied Mathematics and Mechanics, 5*, 48–68.

10

Effects of two temperatures and thermal phase-lags in a thick circular plate with axisymmetric heat supply

Rajneesh Kumar[1], Nidhi Sharma[2] and Parveen Lata[3]*

*Corresponding author: Parveen Lata, Department of Basic and Applied Sciences, Punjabi University, Patiala, Punjab, India

E-mail: parveenlata@pbi.ac.in

Reviewing editor: Xiao-Jun Yang, China University of Mining and Technology, China

Abstract: The present investigation is concerned with thermomechanical interactions for the dual-phase-lag in a homogeneous isotropic thick circular plate in the light of two-temperature thermoelasticity theory. The upper and lower surfaces of the thick plate are traction free and subjected to an axisymmetric heat supply. The solution is found by using Laplace and Hankel transform technique and a direct approach without the use of potential functions. The analytical expressions of displacement components, stresses, conductive temperature, temperature change and cubic dilatation are computed in transformed domain. Numerical inversion technique has been applied to obtain the results in the physical domain. Numerically simulated results are depicted graphically. The effects of thermal phase-lags and two temperatures are shown on the various components. Some particular cases are also deduced from the present investigation.

Subjects: Earth Sciences; Engineering & Technology; Geology – Earth Sciences; Materials Science; Science; Technology

Keywords: two-temperature; two phase lags; isotropic; thick circular plate; Laplace transform; Hankel transform

1. Introduction

The use of thermal phase-lags in the heat conduction equation gives a more realistic model of thermoelastic media as it allows a delayed response to the relative heat flux vector. The result of the problem is useful in the two-dimensional problem of dynamic response due to various thermal and mechanical sources which has various geophysical and industrial applications.

Classical Fourier heat conduction law implies an infinitely fast propagation of a thermal signal which is violated in ultrafast heat conduction system due to its very small dimensions and short

ABOUT THE AUTHOR

The corresponding author Parveen Lata is an assistant professor in Mathematics, Department of Basic and Applied Sciences, Punjabi University Patiala and is pursuing the research work with collaboration of Rajneesh Kumar and Nidhi Sharma. The primary field of research is thermoelasticity and further extended to the various theories of thermoelasticity which is the most demanding area due to its wide applications in the field of aeronautics, aircrafts, geomechanics and various other fields.

PUBLIC INTEREST STATEMENT

The result of the problem is useful in the two-dimensional problem of dynamic response due to various thermal and mechanical sources which has various geophysical and industrial applications. The physical applications are encountered in the context of problems such as ground explosions and oil industries. This problem is also useful in the field of geo-mechanics, where the interest is in various phenomena occurring in earthquakes and measurement of displacements, stresses and temperature field due to the presence of certain sources.

timescales. Catteno (1958) and Vernotte (1958) proposed a thermal wave with a single phase lag in which the temperature gradient after a certain elapsed time was given by $\boldsymbol{q} + \tau_q \frac{\partial \boldsymbol{q}}{\partial t} = -k\nabla T$, where τ_q denotes the relaxation time required for thermal physics to take account of hyperbolic effect within the medium. Here, when $\tau_q > 0$, the thermal wave propagates through the medium with a finite speed of $\sqrt{\frac{\alpha}{\tau_q}}$, where α is thermal diffusivity. When τ_q approaches zero, the thermal wave has an infinite speed and thus the single-phase-lag model reduces to the traditional Fourier model. The dual-phase-lag model of heat conduction was proposed by Tzou (1996) $\boldsymbol{q} + \tau_q \frac{\partial \boldsymbol{q}}{\partial t} = -k(\nabla T + \tau_t \frac{\partial}{\partial t} \nabla T)$, where the temperature gradient ∇T at a point P of the material at time $t + \tau_t$ corresponds to the heat flux vector q at the same time at the time $t + \tau_q$. Here, k is thermal conductivity of the material. The delay time τ_t is interpreted as that caused by the microstructural interactions and is called the phase-lag of temperature gradient. The other delay time τ_q interpreted as the relaxation time due to the fast transient effects of thermal inertia and is called the phase-lag of heat flux. This universal model is claimed to be able to bridge the gap between microscopic and macroscopic approaches, covering a wide range of heat transfer models. If $\tau_t = 0$, Tzou (1996) refers to the model as single-phase model. Numerous efforts have been invested in the development of an explicit mathematical solution to the heat conduction equation under dual-phase-lag model. Quintanilla and Racke (2006) compared two different mathematical hyperbolic models. Kumar and Mukhopadhaya (2010a, 2010b) Investigated the propagation of harmonic waves of assigned frequency by employing the thermoelasticity theory with three-phase-lags. Chou and Yang (2009) discussed two dimensional dual-phase-lag thermal behaviour in single-/multi layer structures. Zhou, Zhang, and Chen (2009) proposed an axisymmetric dual-phase-lag bioheat model for laser heating of living tissues. Kumar, Chawla, and Abbas, (2012) discussed effect of viscosity on wave propagation in anisotropic thermoelastic medium with three-phase-lag model. Ying and Yun (2015) built a fractional dual-phase-lag model and the corresponding bioheat transfer equation. Abdallah (2009) used uncoupled thermoelastic model based on dual-phase-lag to investigate the thermoelastic properties of a semi-infinite medium. Rukolaine (2014) employed dual-phase-lag models to study unphysical problems. Tripathi, Kedar, and Deshmukh (2015) discussed generalized thermoelastic diffusion problem in a thick circular plate with axisymmetric heat supply.

Chen and Gurtin (1968), Chen, Gurtin, and Williams (1968, 1969) have formulated a theory of heat conduction in deformable bodies which depends upon two distinct temperatures, the conductive temperature φ and the thermodynamical temperature T. For time independent situations, the difference between these two temperatures is proportional to the heat supply, and in absence of heat supply, the two temperatures are identical. For time dependent problems, the two temperatures are different, regardless of the presence of heat supply. The two temperatures T, φ and the strain are found to have representations in the form of a travelling wave plus a response, which occurs instantaneously throughout the body (Boley & Tolins, 1962). The wave propagation in the two-temperature theory of thermoelasticity was investigated by Warren and Chen (1973). Youssef (2011), constructed a new theory of generalized thermoelasticity by taking into account two-temperature generalized thermoelasticity theory for a homogeneous and isotropic body without energy dissipation. Several researchers studied various problems involving dual-phase-lags (e.g. Abbas 2015a, 2015b, 2015c; Abbas, Kumar, & Reen, 2014; Abbas & Zenkour, 2013, 2014, 2015; Atwa & Jahangir, 2014; Ezzat & Awad, 2010; Kaushal, Kumar, & Miglani, 2011; Kaushal, Sharma, & Kumar, 2010; Kumar & Mukhopadhaya, 2010a, 2010b; Kumar, Sharma, & Garg, 2014; Sharma & Marin, 2013; Youssef, 2006).

In this investigation, the thermoelastic interactions for the dual-phase-lag heat conduction in a thick circular plate are studied in the light of two-temperature thermoelasticity theory. The components of displacements, stresses, conductive temperature, temperature change and cubic dilatation are computed numerically. Numerically computed results are depicted graphically. The effect of dual-phase-lag and two-temperature are shown on the various components.

2. Basic equations

The basic equations of motion, heat conduction in a homogeneous isotropic thermoelastic solid with dual-phase-lag and two-temperature in the absence of body forces, heat sources are:

$$(\lambda + \mu)\nabla(\nabla \cdot u) + \mu\nabla^2 u - \beta_1 \nabla = \rho \ddot{u} \tag{1}$$

$$\left(1 + \tau_t \frac{\partial}{\partial t}\right) K T_{,ii} = \left(1 + \tau_q \frac{\partial}{\partial t} + \tau_q^2 \frac{\partial^2}{\partial t^2}\right)\left[\rho C_E \dot{T} + \beta_1 T_0 \dot{e}_{kk}\right] \tag{2}$$

$$T = (1 - a\nabla^2)\varphi$$

and the constitutive relations are:

$$\sigma_{ij} = 2\mu e_{ij} + \delta_{ij}(\lambda e_{kk} - \beta_1 T) \tag{3}$$

$$\rho T_0 S = \left(1 + \tau_q \frac{\partial}{\partial t} + \tau_q^2 \frac{\partial^2}{\partial t^2}\right)(\rho C_E T + \beta_1 T_0 e_{kk}) \tag{4}$$

where λ, μ are Lame's constants, ρ is the density assumed to be independent of time, u_i are components of displacement vector u, K is the coefficient of thermal conductivity, C_E is the specific heat at constant strain, T is the absolute temperature of the medium, σ_{ij} and e_{ij} are the components of stress and strain respectively, e_{kk} is dilatation, S is the entropy per unit mass, $\beta_1 = (3\lambda + 2\mu)\alpha_t$, α_t is the coefficient of thermal linear expansion. τ_t, τ_q are, respectively, phase-lag of temperature gradient, the phase-lag of heat flux, a is the two-temperature parameter. In the above equations, a comma followed by suffix denotes spatial derivative and a superposed dot denotes derivative with respect to time.

3. Formulation and solution of the problem

Consider a thick circular plate of thickness $2b$ occupying the space D defined by $0 \leq r \leq \infty, -b \leq z \leq b$. Let the plate be subjected to an axisymmetric heat supply depending on the radial and axial directions of the cylindrical coordinate system. The initial temperature in the thick plate is given by a constant temperature T_0, and the heat flux $g_0 F(r, z)$ is prescribed on the upper and lower boundary surfaces. Under these conditions, the thermoelastic quantities in a thick circular plate are required to be determined. We take a cylindrical polar coordinate system (r, θ, z) with symmetry about z-axis. As the problem considered is plane axisymmetric, the field component $u_\theta = 0$, and u_r, u_z, T and C are independent of θ and restrict our analysis to the two dimensional problem with

$$\mathbf{u} = (u_r, 0, u_z) \tag{5}$$

Equations (1) and (2) with the aid of (5) take the form:

$$(\lambda + \mu)\frac{\partial e}{\partial r} + \mu\left(\nabla^2 - \frac{1}{r^2}\right)u_r - \beta_1 \frac{\partial T}{\partial r} = \rho \frac{\partial^2 u_r}{\partial t^2} \tag{6}$$

$$(\lambda + \mu)\frac{\partial e}{\partial z} + \mu\nabla^2 u_z - \beta_1 \frac{\partial T}{\partial z} = \rho \frac{\partial^2 u_z}{\partial t^2} \tag{7}$$

$$\left(1 + \tau_t \frac{\partial}{\partial t}\right) K \nabla^2 T = \left(1 + \tau_q \frac{\partial}{\partial t} + \frac{\tau_q^2}{2} \frac{\partial^2}{\partial t^2}\right)\left[\rho C_E \frac{\partial}{\partial t}\left(1 - a\nabla^2\right)\varphi + \beta_1 T_0 \frac{\partial}{\partial t} \operatorname{div} u\right] \tag{8}$$

and constitutive relations

$$\sigma_{rr} = 2\mu e_{rr} + \lambda e - \beta_1(1 - a\nabla^2)\varphi \tag{9}$$

$$\sigma_{\theta\theta} = 2\mu e_{\theta\theta} + \lambda e - \beta_1(1 - a\nabla^2)\varphi \tag{10}$$

$$\sigma_{zz} = 2\mu e_{zz} + \lambda e - \beta_1\left(1 - a\nabla^2\right)\varphi \tag{11}$$

$$\sigma_{rz} = \mu e_{rz},\ \sigma_{r\theta} = 0,\ \sigma_{z\theta} = 0 \tag{12}$$

where $e = \frac{\partial u_r}{\partial r} + \frac{u_r}{r} + \frac{\partial u_z}{\partial z}, e_{rr} = \frac{\partial u_r}{\partial r}, e_{\theta\theta} = \frac{u_r}{r}, e_{zz} = \frac{\partial u_z}{\partial z}, e_{rz} = \frac{1}{2}\left(\frac{\partial u_r}{\partial z} + \frac{\partial u_z}{\partial r}\right)$

To facilitate the solution, the following dimensionless quantities are introduced:

$$r' = \frac{\omega_1}{c_1}r,\ z' = \frac{\omega_1}{c_1}z,\ (u_r',\ u_z') = \frac{\omega_1}{c_1}(u_r,\ u_z),\ t' = \omega_1 t,\ \omega_1 = \frac{\rho C_E c_1^2}{K},\ c_1^2 = \frac{\lambda + 2\mu}{\rho}\ (\sigma_{rr}', \sigma_{\theta\theta}', \sigma_{zz}', \sigma_{rz}') = \frac{1}{\beta_1 T_0}(\sigma_{rr}, \sigma_{\theta\theta}, \sigma_{zz}, \sigma_{rz}),\ (T', \varphi') = \frac{\beta_1}{\rho c_1^2}(T,\ \varphi)\ (\tau_q^1,\ \tau_t^1) = \omega_1(\tau_q, \tau_t) \tag{13}$$

in Equations (6)–(8) and after that suppressing the primes and then applying the Laplace transform defined by (14):

$$\bar{f}(r,\ z,\ s) = \int_0^\infty f(r,\ z,\ t)e^{-st}\,dt \tag{14}$$

$$\bar{f}^*(\xi,\ z,\ s) = \int_0^\infty f(r,\ z,\ s)rJ_n(r\xi)\,dr \tag{15}$$

On the resulting quantities and simplifying we obtain

$$\left(\nabla^2 - s^2\right)\bar{e} - \nabla^2\bar{\varphi} + \delta_1\nabla^4\bar{\varphi} = 0 \tag{16}$$

$$\tau_q^1\zeta_2\bar{e} + \tau_q^1\zeta_1\bar{\varphi} - (\tau_q^1\delta_1 - \tau_t^1 K)\nabla^2\bar{\varphi} = 0 \tag{17}$$

where $\tau_q^1 = 1 + s\tau_q + \frac{s^2\tau_q^2}{2}, \tau_t^1 = 1 + s\tau_t, \zeta_1 = \frac{\rho C_E c_1^2}{\omega_1}, \zeta_2 = \frac{\beta_1^2 T_0}{\rho\omega_1}s, \delta_1 = \frac{a{\omega_1}^2}{c_1^2}$

Eliminating $\bar{\varphi}$ and $\bar{e}$ from Equations (16) and (17), we obtain:

$$(\nabla^2 - k_1^2)(\nabla^2 - k_2^2)(\bar{e},\ \bar{\varphi}) = 0 \tag{18}$$

The solutions of Equation (18) can be written in the form:

$\bar{\varphi} = \sum_{i=1}^{3} \bar{\varphi}_i, \bar{e} = \sum_{i=1}^{3} \bar{e}_i$ where, $\bar{e}_i$, and $\bar{\varphi}_i$ are solutions of the following equation:

$$(\nabla^2 - k_i^2)(\bar{e}_i,\ \bar{\varphi}_i) = 0, \quad i = 1,\ 2 \tag{19}$$

On taking Hankel transform of (19) defined by (15), we obtain:

$$\left(D^2 - \xi^2 - k_i^2\right)\left(\bar{\varphi}_i^*,\ \bar{e}_i^*\right) = 0 \tag{20}$$

The solution of (20) has the form

$$\bar{e}^* = \sum_{i=1}^{2} A_i(\xi,\ s)\cosh(q_i z) \tag{21}$$

$$\bar{\varphi}^* = \sum_{i=1}^{2} d_i A_i(\xi, s) \cosh(q_i z) \tag{22}$$

where $q_i = \sqrt{\xi^2 + k_i^2}, d_i = \frac{\tau_q^1 \zeta_2}{\tau_q^1 \zeta_1 - \zeta_3 q_i^2}, \zeta_3 = \tau_q^1 \delta_1 - \tau_t^1 k$

Applying inversion of Hankel transform on (23), and (24), we get:

$$\bar{e} = \int_0^\infty \left\{ \sum_{i=1}^{2} A_i(\xi, s) \cosh(q_i z) \right\} \xi J_0(\xi r)\, d\xi \tag{23}$$

$$\bar{\varphi} = \int_0^\infty \left\{ \sum_{i=1}^{2} d_i A_i(\xi, s) \cosh(q_i z) \right\} \xi J_0(\xi r)\, d\xi \tag{24}$$

Using (6)–(8), (13) and (23), (24), we obtain the displacement components in the transformed domain as:

$$\begin{aligned} \bar{u}_r(r, z, s) = \int_0^\infty & E(\xi, s) \cosh(qz) \xi J_0(\xi r) + \sum_{i=1}^{2} \left[(-\eta_i + \mu_i) q_i^2 \xi^2 \cosh(q_i z) \right] J_1(\xi r) \\ & + \delta_1 \mu_i \cosh(q_i z) \left(\frac{\xi^3}{r} J_2 - J_1 \left(\xi^4 - \frac{\xi^2}{r^2} + \xi^2 q_i^2 \right) + \frac{\xi^3}{r} J_0 \right)]\, d\xi \end{aligned} \tag{25}$$

$$\begin{aligned} \bar{u}_z(r, z, s) = \int_0^\infty & G(\xi, z) \sinh(qz) \xi J_0(\xi r) + \sum_{i=1}^{2} [(-\eta_i + \mu_i) \sinh(q_i z) \xi J_0(\xi r) \\ & - \delta_1 \mu_i \sinh(q_i z) \left(\frac{\xi^2}{r} J_2 - J_1 \left(\xi^3 + \frac{\xi}{r} \right) + \xi q_i^2 J_0 \right) \Big] \end{aligned} \tag{26}$$

where

$$G(\xi, s) = \frac{\xi^2 E(\xi, s)}{q},\ q = \sqrt{\xi^2 + \frac{\rho c_1^2}{\mu} s^2},\ \eta_i = \frac{\frac{\lambda+\mu}{\rho c_1^2} A_i}{\left(\frac{\mu q_i^2}{\rho c_1^2} - s^2 \right)},\ \mu_i = \frac{d_i A_i}{\left(\frac{\mu q_i^2}{\rho c_1^2} - s^2 \right)},\ \lambda^0 = \frac{\lambda}{\beta_1 T_0},\ \zeta = \frac{\rho c_1^2}{\beta_1 T_0}$$

Using (6)–(8), (13), (23)–(26) we obtain the stress components and conductive temperature, temperature change T, cubic dilatation in the Laplace transform domain as:

$$\begin{aligned} \overline{\sigma_{zz}} = \frac{2\mu}{\beta_1 T_0} \int_0^\infty & \xi J_0(\xi r) \left[G(\xi, s) q \cosh(qz) + \left(\sum_{i=1}^{2} (-\eta_i + \mu_i) q_i^2 - \zeta d_i A_i + \lambda^0 A_i \right) \cosh(q_i z) \right] \\ & - [\delta_1 (\mu_i q_i - \zeta B_i) \cosh(q_i z)(\xi^3 J_0(\xi r) - J_1(\xi r) \left(\frac{\xi - 1}{r} \right) + \xi q_i^2 J_0(\xi r)]\, d\xi \end{aligned} \tag{27}$$

$$\begin{aligned} \overline{\sigma_{rz}} = \frac{\mu}{2\beta_1 T_0} \int_0^\infty & \xi^2 J_1(\xi r) \left[\left\{ \left(\frac{q^2 - \xi^2}{q} \right) E(\xi, s) q \sinh(qz) + 2 \sum_{i=1}^{2} (\eta_i - \mu_i) q_i \sinh(q_i z) \right\} \right. \\ & + \delta_1 \mu_i \sinh(q_i z) \{ (q_i (\frac{\xi^3}{r} J_2(\xi r)) - J_1(\xi r) \left(\xi^4 + q_i^2 \xi^2 + \frac{4}{\xi^2 r^2} + q_i^2 \xi \right) \\ & \left. + \left(\frac{2}{\xi r} - \xi^4 - \frac{\xi^2(\xi - 1)}{r} + q_i^2 \xi^2 \right) J_0(\xi r) \} \right] d\xi \end{aligned} \tag{28}$$

$$\overline{\sigma_{rr}} = \frac{2\mu}{\beta_1 T_0}\int_0^\infty \Bigg[-\xi^2 J_1(\xi r)E(\xi,s)\cosh(qz) + (\sum_{i=1}^{2}(-\eta_i+\mu_i)q_i^2(\xi^2 J_1(\xi r)$$

$$-\xi^3 J_0(\xi r)) + \delta_1\mu_i\{J_1(\xi r)(-\xi^2\left(q_i^2+\xi^2-3\right) + J_0(\xi r)\xi^3\left(q_i^2+\frac{1}{r^2}+\xi^2\right)d\xi\Bigg]$$

$$+\int_0^\infty \xi J_0(\xi r)\cosh(q_i z)[\lambda' - \zeta d_i\left(1+\delta_1 q_i^2 - \delta_1\xi J_0(\xi r)\right) + \delta_1 J_1(\xi r)\left(1-\frac{1}{r}\right)d\xi$$

$$\bar{\varphi} = \int_0^\infty (d_1 A_1(\xi,\ s)\cosh(q_1 z) + d_2 A_2(\xi,\ s)\cosh(q_2 z))\xi J_0(\xi r)\,d\xi \tag{29}$$

$$\bar{T} = \int_0^\infty \sum_{i=1}^{2}(d_i A_i(\xi,\ s)\cosh(q_i z)\left[\xi J_0(\xi r)\left(1+\delta_1 q_i^2-\delta_1\xi\right) + J_1(\xi r)\delta_1\xi\left(1-\frac{1}{r}\right)\right]d\xi \tag{30}$$

4. Boundary conditions

We consider a cubical thermal source and normal force of unit magnitude along with vanishing of tangential stress components at the stress-free surface at $z = \pm b$. Mathematically, these can be written as:

$$\frac{\partial\varphi}{\partial z} = \pm g_0 F(r,\ z) \tag{31}$$

$$\sigma_{zz} = \delta(t)H(a-r) \tag{32}$$

$$\sigma_{rz} = 0 \tag{33}$$

where $F(r, z)=z^2\, e^{(-\omega r)}$, $\delta(t)$ is the Dirac delta function and $H(a\text{-}r)$ is Heaviside function

Applying Laplace transform and Hankel transform on both sides of the boundary conditions (31)–(33),

where $\bar{F}^*(\xi, z) = \dfrac{z^2\omega}{\left(\xi^2+\omega^2\right)^{3/2}}$

we o btain the values of unknown parameters as

$$A_1 = \frac{\Delta_1}{\Delta},\ A_2 = \frac{\Delta_2}{\Delta},\ E(\zeta,\ s) = \frac{\Delta_3}{\Delta} \text{ where}$$

$$\Delta = -\frac{2\mu}{\beta_1 T_0}\cosh(qb)\left(\Delta_{11}\Delta_{32}-\Delta_{12}\Delta_{31}\right) + \sinh(qb)(\Delta_{11}\Delta_{22}-\Delta_{12}\Delta_{21})$$

$$\Delta_1 = g_0\bar{F}^*(\xi,b)\left(\Delta_{21}\Delta_{32}-\Delta_{22}\Delta_{31}\right) - \frac{aJ_1(\xi a)}{\xi}\left(\Delta_{11}\Delta_{32}-\Delta_{12}\Delta_{31}\right)$$

$$\Delta_2 = -g_0\bar{F}^*(\xi,\ b)\left(\frac{2\mu}{\beta_1 T_0}\cosh(qb)\Delta_{32} -_D elta22\sinh(qb)\right) + \frac{aJ_1(\xi a)}{\xi}\left(-\Delta_{12}\sinh(qb)\right)$$

$$\Delta_3 = g_0\bar{F}^*(\xi,\ b)\left(\frac{2\mu}{\beta_1 T_0}\cosh(qb)\Delta_{31} - \Delta_{21}\sinh(qb)\right) + \frac{aJ_1(\xi a)}{\xi}\left(\Delta_{11}\sinh(qb)\right)$$

$\Delta_{1i} = d_i q_i \sinh(q_i b),\ \Delta_{2i} = ((\mu_i-\eta_i)q_i^2 - \delta_1\mu_i q_i - \zeta d_i(1+\delta_1) + \lambda^*)\cosh(q_i b),$
$\Delta_{3i} = (2(\eta_i-\mu_i)q_i + \delta_1\mu_i)\sinh(q_i b) \quad i = 1,\ 2$

5. Inversion of double transform

Due to the complexity of the solution in the Laplace transform domain, the inverse of the Laplace transform is obtained by using the Gaver–Stehfest algorithm. Graver (1996) and Stehfest (1970a, 1970b) derived the formula given below. By this method, the inverse $f(t)$ of Laplace transform $\bar{f}(s)$ is approximated by

$$f(t) = \frac{\log 2}{t} \sum_{j=1}^{k} D(j, K) F\left(j \frac{\log 2}{t}\right)$$

with

$$D(j, K) = (-1)^{j+M} \sum_{n=m}^{\min(j, M)} \frac{n^M (2n)!}{(M-n)!n!(n-1)!(j-n)!(2n-j)!}$$

where K is an even integer, whose value depends on the word length of computer used. $M = K/2$, and m is an integer part of $(j + 1)/2$. The optimal value of K was chosen as described in Gaver–Stehfest algorithm, for the fast convergence of results with desired accuracy. The Romberg numerical integration technique (Press, Flannery, Teukolsky, & Vatterling, 1986) with variable step size was used to evaluate the results involved.

6. Particular cases

(1) If $a = 0$, from Equations (25)–(30), we obtain the corresponding expressions for displacements, and stresses, conductive temperature, temperature change and cubic dilatation for thermoelastic solid without two-temperature and due to dual-phase-lag.

(2) If $\tau_q = \tau_t = 0$, we obtain the coupled expression in thermoelasticity with two-temperature model.

(3) $\tau_q = 0$ then dual-phase-lag thermal model (DPLT) model reduce to single-phase-lag thermal model (SPLT).

7. Numerical results and discussion

The graphs have been plotted to study the effect of phase-lags and two-temperatures on the various quantities in the range $0 \le r \le 10$.

The mathematical model is prepared with copper material for the purpose of numerical computation. The material constants for the problem are taken from Dhaliwal and Singh (1980).

$$\lambda = 7.76 \times 10^{10}\,\text{N m}^{-2},\ \mu = 3.86 \times 10^{10}\,\text{N m}^{-2},\ K = 386\,\text{J K}^{-1}\,\text{m}^{-1}\,\text{s}^{-1},$$
$$\beta_1 = 5.518 \times 10^{6}\,\text{N m}^{-2}\,\text{deg}^{-1},\ \rho = 8{,}954\,\text{kg m}^{-3},\ a = 1.2 \times 10^{4}\,\text{m}^2/\text{s}^2\,\text{k},$$
$$b = 0.9 \times 10^{6}\,\text{m}^5/\text{kg s}^2, D = 0.88 \times 10^{-8}\,\text{kg s/m}^3, \beta_2 = 61.38 \times 10^{6}\,\text{N m}^{-2},$$
$$T_0 = 293\,\text{K}, C_E = 383.1\,\text{J kg}^{-1}\,K^{-1}$$

(1) In the figures solid line corresponds to the dual-phase-lag of heat transfer with two-temperature with $\tau_q > \tau_t$, $a = .07$ ($\tau_q = 1.2$, $\tau_t = .06$)

(2) In the figures small dashed line corresponds to the dual-phase-lag of heat transfer with two-temperature with $\tau_q < \tau_t$, $a = .07$ ($\tau_t = 1.2$, $\tau_q = .06$)

(3) Solid line with centre symbol circle corresponds to $\tau_q > \tau_t$, $a = 0$

(4) Small dashed line with centre symbol diamond corresponds to $\tau_q < \tau_t$, $a = 0$

Figure 1 exhibits variations of displacement component u_r with distance r. Near the loading surface, there is a sharp decrease for the range $0 \le r \le 2$ and behaviour is oscillatory in the rest for all the cases with amplitudes of oscillations decreasing as r increases. Figure 2 shows variations of displacement component u_z with distance *r*. Here behaviour is oscillatory for the whole range except for $0 \leq r \leq 1.5$, as for this range, a sharp decrease is noticed. Figure 3 shows variation of stress component σ_{zz} with distance *r*. We find that there is a sharp increase for the range $0 \leq r \leq 3$ corresponding to all the cases and similar oscillatory trend is observed afterwards. Small variations near boundary

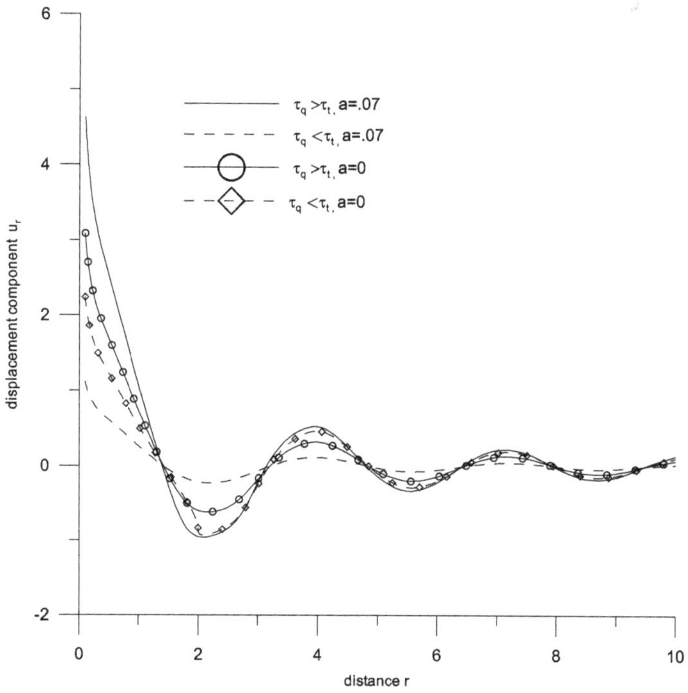

Figure 1. Variation of displacement component u_r with distance *r*.

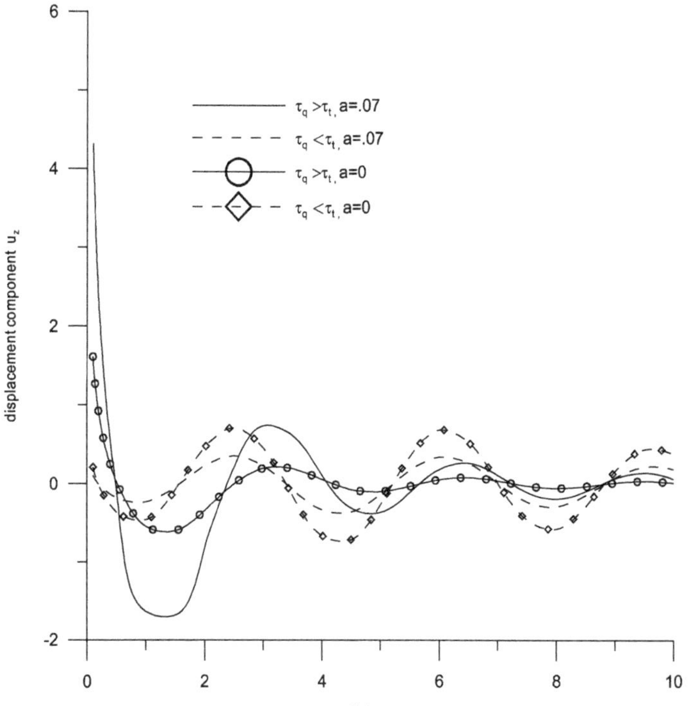

Figure 2. Variation of displacement component u_z with distance *r*.

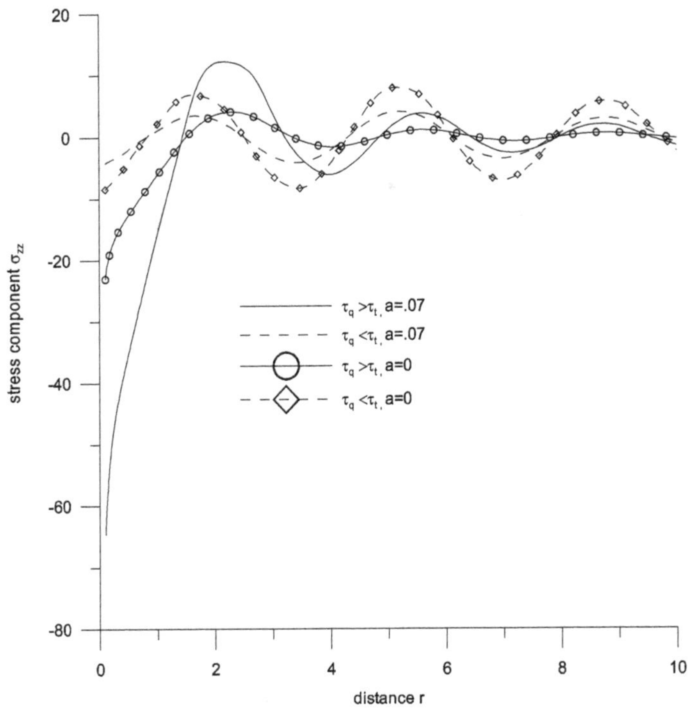

Figure 3. Variation of vertical stress component σ_{zz} with distance *r*.

surface are observed corresponding to the case $\tau_q > \tau_t$, $a = 0$ for the range $3 \le r \le 10$. Figure 4 gives variation of conductive temperature φ with distance *r*. Here, we notice that either there are sudden increases and decreases or there are small variations. Here descents are noticed at the points $r = .5$

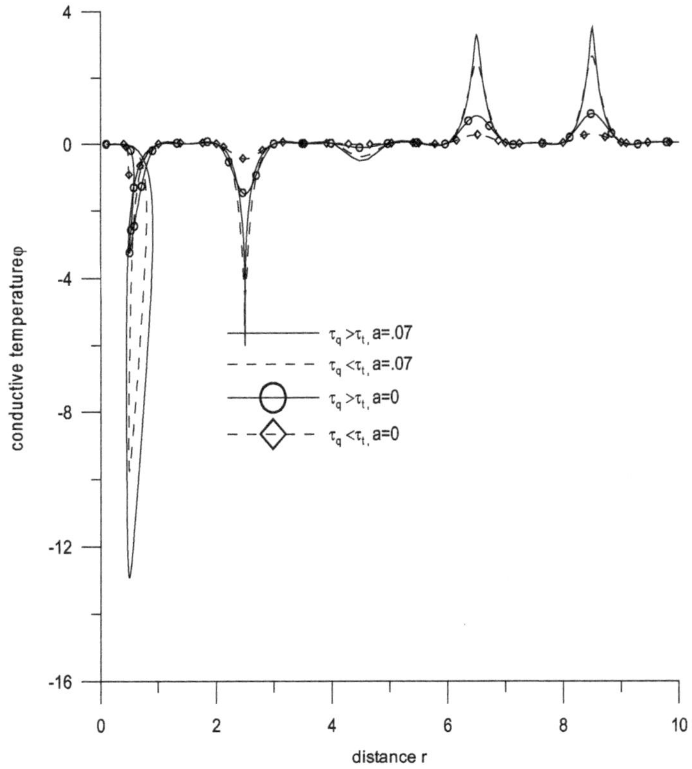

Figure 4. Variation of conductive temperature φ with distance *r*.

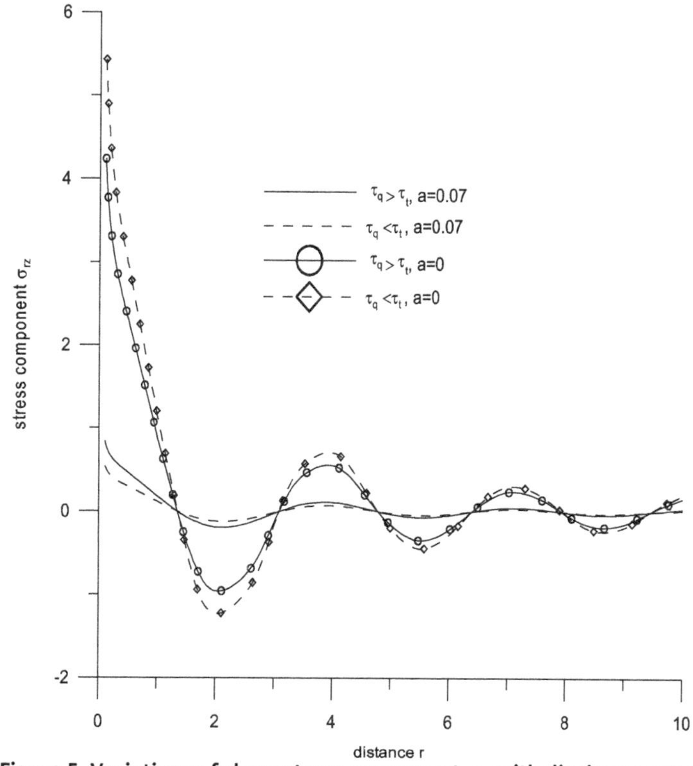

Figure 5. Variations of shear stress component σ_{rz} with displacement *r*.

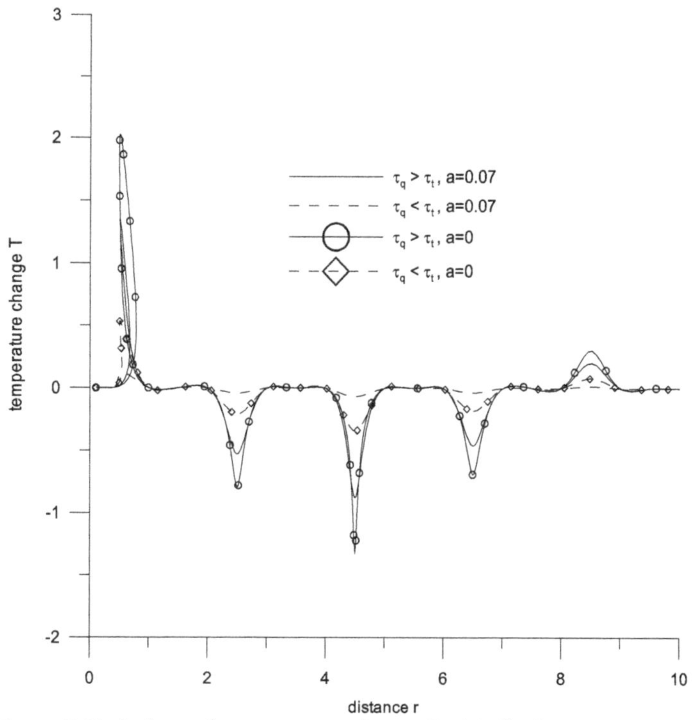

Figure 6. Variations of temperature change *T* with displacement *r*.

and $r = 2.5$, whereas hikes are seen at the points $r = 6.5$ and $r = 9$. With two temperatures, there are hikes and descents while without two temperature there are small variations. Figure 5 gives variations of stress component σ_{rz} with distance *r*. It is evident from this figure that the behaviour is

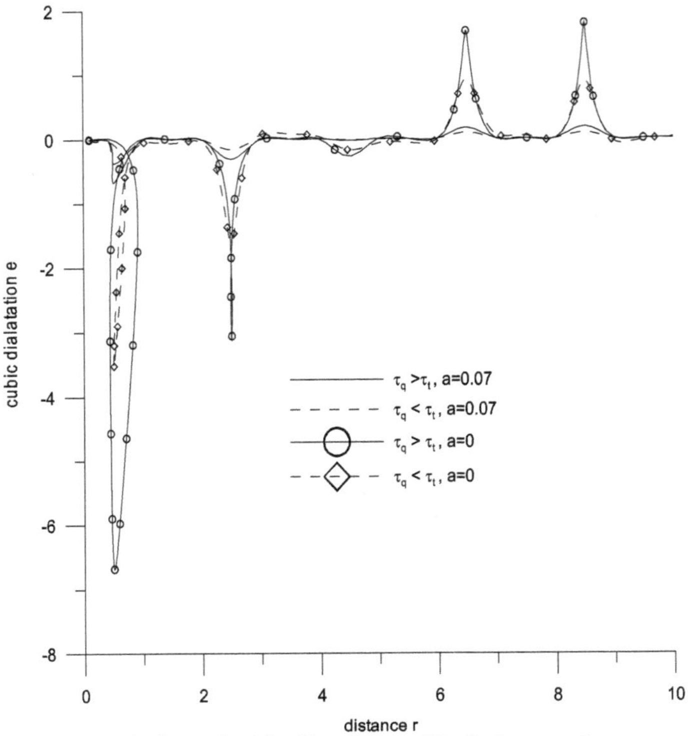

Figure 7. Variations of cubic dilatation e with displacement *r*.

descending oscillatory corresponding to the case $\tau_q > \tau_t$, $a = 0$ and $\tau_q < \tau_t$, $a = 0$, i.e. without two temperatures, whereas small variations are observed corresponding to the rest. Figure 6 exhibits variations of temperature change T with displacement r. Here, there is a hike at the point $r = 1$ and

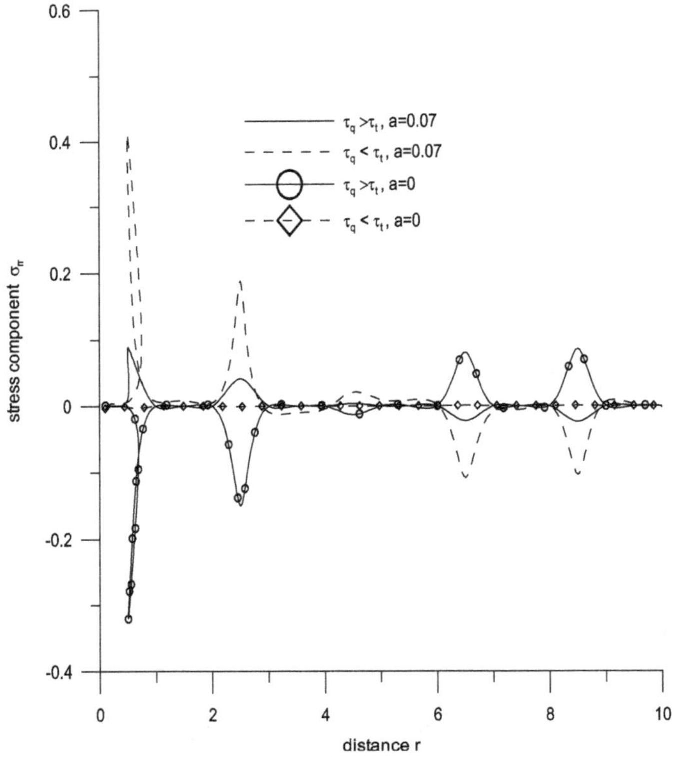

Figure 8. Variations of radial stress component σ_{rr} with displacement *r*.

descents at the points $r = 2.5$, $r = 4.5$, $r = 6.5$ and a small hike is noticed at $r = 9$ and small variations are observed for the remaining range except for the small neighbourhoods of these points. Maximum hikes and descents are noticed corresponding to case 3. Figure 7 shows variations of cubic dilatation e with distance r. Here we notice that corresponding to the cases 3 and 4 there are hikes and descents, whereas corresponding to cases 1 and 2 there are small variations. Figure 8 displays variations in stress component σ_{rr} with displacement r. Here opposite trends are noticed corresponding to the cases of without two temperature and with two temperature. As it is evident that without two temperature there is a descent at the point .5, whereas there is a hike at the same point corresponding to case of two temperature. While comparing the effect of phase lags, the trends are similar corresponding to both the cases.

8. Conclusion

From the graphs, it is evident that there is a significant impact on the deformation of various components of stresses, components of displacement, conductive temperature, temperature change and cubic dilatation in the thick circular plate while comparing the effect of thermal phase-lags and two temperatures. Amplitude of oscillation is slightly greater in case of $\tau_q > \tau_t$, $a = .07$ as compared with the case $\tau_q < \tau_t$, $a = .07$. More variations are observed in the case of without two-temperature than with two-temperature. As disturbance travels through the constituents of the medium, it suffers sudden changes resulting in an inconsistent/non uniform pattern of graphs in case of conductive temperature, temperature change, cubic dilatation. The use of thermal phase-lags in the heat conduction equation gives a more realistic model of thermoelastic media as it allows a delayed response to the relative heat flux vector. The result of the problem is useful in the two-dimensional problem of dynamic response due to various thermal and mechanical sources which has various geophysical and industrial applications.

Funding
The authors received no direct funding for this research.

Author details
Rajneesh Kumar[1]
E-mail: rajneesh_kuk@rediffmail.com
Nidhi Sharma[2]
E-mail: nidhi_kuk26@rediffmail.com
Parveen Lata[3]
E-mail: parveenlata@pbi.ac.in
ORCID ID: http://orcid.org/0000-0003-2592-0885
[1] Department of Mathematics, Kurukshetra University, Kurukshetra, Haryana, India.
[2] Department of Mathematics, MM University, Mullana, Ambala, Haryana, India.
[3] Department of Basic and Applied Sciences, Punjabi University, Patiala, Punjab, India.

References

Abbas, I. A. (2015a). Eigenvalue approach to fractional order generalized magneto-thermoelastic medium subjected to moving heat source. *Journal of Magnetism and Magnetic Materials, 377*, 452–459. http://dx.doi.org/10.1016/j.jmmm.2014.10.159

Abbas, I. A. (2015b). A dual phase lag model on thermoelastic interaction in an infinite fiber-reinforced anisotropic medium with a circular hole. *Mechanics Based Design of Structures and Machines*, 43, 501–513. http://dx.doi.org/10.1080/15397734.2015.1029589

Abbas, I. A. (2015c). Analytical solution for a free vibration of a thermoelastic hollow sphere. *Mechanics Based Design of Structures and Machines*, 43, 265–276. http://dx.doi.org/10.1080/15397734.2014.956244

Abbas, I. A., Kumar, R., & Reen, L. S. (2014). Response of thermal source in transversely isotropic thermoelastic materials without energy dissipation and with two temperatures. *Canadian Journal of Physics, 92*, 1305–1311. http://dx.doi.org/10.1139/cjp-2013-0484

Abbas, I. A., & Zenkour, A. M. (2013). LS model on electro-magneto-thermoelastic response of an infinite functionally graded cylinder. *Composite Structures, 96*, 89–96. http://dx.doi.org/10.1016/j.compstruct.2012.08.046

Abbas, I. A., & Zenkour, A. M. (2014). Two-temperature generalized thermoelastic interaction in an infinite fiber-reinforced anisotropic plate containing a circular cavity with two relaxation times. *Journal of Computational and Theoretical Nanoscience, 11*(1), 1–7. http://dx.doi.org/10.1166/jctn.2014.3309

Abbas, I. A., & Zenkour, A. M. (2015). The effect of magnetic field on thermal shock problem for a fiber-reinforced anisotropic half-space using Green-Naghdi's Theory. *Journal of Computational and Theoretical Nanoscience, 12*, 438–442. http://dx.doi.org/10.1166/jctn.2015.3749

Abdallah, I. A. (2009). Dual phase lag Heat Conduction and thermoelastic properties of a semi infinite medium Induced by Ultrashort Pulsed layer. *Progress in Physics, 3*, 60–63.

Atwa, S. Y., & Jahangir, A. (2014). Two-temperature effects on plane waves in generalized thermo-microstretch elastic solid. International Journal of Thermophysics, *35*, 175–193. http://dx.doi.org/10.1007/s10765-013-1541-9

Boley, B. A., & Tolins, I. S. (1962). Transient coupled thermoelastic boundary value problems in the half-

SPACE. *Journal of Applied Mechanics, 29*, 637–646. http://dx.doi.org/10.1115/1.3640647

Catteno, C. (1958). A form of heat conduction equation which eliminates the paradox of instantaneous propagation. *Compute Rendus, 247*, 431–433.

Chen, P. J., & Gurtin, M. E. (1968). On a theory of heat conduction involving two temperatures. *Zeitschrift für angewandte Mathematik und Physik ZAMP, 19*, 614–627. http://dx.doi.org/10.1007/BF01594969

Chen, P. J., Gurtin, M. E., & Williams, W. O. (1968). A note on non-simple heat conduction. *Zeitschrift für angewandte Mathematik und Physik ZAMP, 19*, 969–970. http://dx.doi.org/10.1007/BF01602278

Chen, P. J., Gurtin, M. E., & Williams, W. O. (1969). On the thermodynamics of non simple elastic materials with two temperatures. *Journal of Applied Mathematics and Physics (ZAMP), 20*, 107–112.

Chou, Y., & Yang, R.-J. (2009). Two-dimensional dual-phase-lag thermal behavior in single-/multi-layer structures using CESE method. *International Journal of Heat and Mass Transfer, 52*, 239–249. http://dx.doi.org/10.1016/j.ijheatmasstransfer.2008.06.025

Dhaliwal, R. S., & Singh, A. (1980). *Dynamic coupled thermoelasticity* (p. 726). New Delhi: Hindustan Publisher.

Ezzat, M. A., & Awad, E. S. (2010). Constitutive relations, uniqueness of solution and thermal shock application in the linear theory of micropolar generalized thermoelasticity involving two temperatures, *Journal of Thermal Stresses, 33*, 225–250.

Gaver, D. P. (1966). Observing stochastic processes, and approximate transform inversion. *Operations Research, 14*, 444–459. http://dx.doi.org/10.1287/opre.14.3.444

Kaushal, S., Sharma, N., & Kumar, R. (2010). Propagation of waves in generalized thermoelastic continua with two temperature. *International Journal of Applied Mechanics and Engineering, 15*, 1111–1127.

Kaushal, S., Kumar, R., & Miglani, A. (2011). Wave propagation in temperature rate dependent thermoelasticity with two temperatures, *Mathematical Sciences, 5*, 125–146.

Kumar, R., & Mukhopadhyay, S. (2010a). Effects of thermal relaxation time on plane wave propagation under two-temperature thermoelasticity. *International Journal of Engineering Science, 48*, 128–139. http://dx.doi.org/10.1016/j.ijengsci.2009.07.001

Kumar, R., & Mukhopadhyay, S. (2010b). Analysis of the effects of phase-lags on propagation of harmonic plane waves in thermoelastic media. *Computational Methods in Science and Technology, 16*, 19–28. http://dx.doi.org/10.12921/cmst

Kumar, R., Chawla, V., & Abbas, I. A. (2012). Effect of viscosity on wave propagation in anisotropic thermoelastic medium with three-phase-lag model. *Theoretical and Applied Mechanics, 39*, 313–341. http://dx.doi.org/10.2298/TAM1204313K

Kumar, R., Sharma, K. D., & Garg, S. K. (2014). Effect of two temperature on reflection coefficient in micropolar thermoelastic media with and without energy dissipation. *Advances in Acoustics and Vibrations*, ID 846721, 11.

Press, W. H., Flannery, B. P., Teukolsky, S. A., & Vatterling, W. A. (1986). *Numerical recipes*. Cambridge: Cambridge University Press. The art of scientific computing.

Quintanilla, R., & Racke, R. (2006). A note on stability in dual-phase-lag heat conduction. *International Journal of Heat and Mass Transfer, 9*, 1209–1213. http://dx.doi.org/10.1016/j.ijheatmasstransfer.2005.10.016

Rukolaine, S. A. (2014). Unphysical effects of the dual-phase-lag model of heat conduction. *International Journal of Heat and Mass Transfer, 78*, 58–63. http://dx.doi.org/10.1016/j.ijheatmasstransfer.2014.06.066

Sharma, K., & Marin, M. (2013). Effect of distinct conductive and thermodynamic temperatures on the reflection of plane waves in micropolar elastic half-space, *University Politehnica of Bucharest Scientific Bulletin-Series A-Applied Mathematics and Physics, 75*, 121–132.

Stehfast, H. (1970a). Algorithm 368, Numerical inversion of Laplace Transforms. *Communications of the ACM, 13*, 47–49.

Stehfast, H. (1970b). Remark on Algorithm 368, numerical inversion of Laplace transforms. *Communications of the ACM, 3*, 624.

Tripathi, J. J., Kedar, G. D., & Deshmukh, K. C. (2015). Generalized thermoelastic diffusion problem in a thick circular plate with axisymmetric heat supply. *Acta Mechanica, 226*, 2121–2134. http://dx.doi.org/10.1007/s00707-015-1305-7

Tzou, D. Y. (1996). *Macro to microscale heat transfer, the lagging behaviour*. Washington, DC: Taylor and Francis.

Vernotte, P. (1958). Les paradox de la theorie continue de l'equation de la chaleur. *Compute Rendus, 246*, 3145–D3155.

Warren, W. E., & Chen, P. J. (1973). Wave propagation in the two temperature theory of thermoelasticity. *Acta Mechanica, 16*, 21–33. http://dx.doi.org/10.1007/BF01177123

Ying, X. H, & Yun, J. X. (2015). Time fractional dual-phase-lag heat conduction equation. *Chinese Physics B, 24*, 034401.

Youssef, H. M. (2006). Theory of two temperature generalized thermoelasticity. *IMA Journal of Applied Mathematics, 71*, 383–390. http://dx.doi.org/10.1093/imamat/hxh101

Youssef, H. M. (2011). Theory of two-temperature thermoelasticity without energy dissipation. *Journal of Thermal Stresses, 34*, 138–146. http://dx.doi.org/10.1080/01495739.2010.511941

Zhou, J., Zhang, Y., & Chen, J. -K. (2009). An axisymmetric dual-phase-lag bioheat model for laser heating of living tissues. *International Journal of Thermal Sciences, 48*, 1477–1485. http://dx.doi.org/10.1016/j.ijthermalsci.2008.12.012Receive

11

Approximate controllability of abstract semilinear stochastic control systems with nonlocal conditions

Divya Ahluwalia[1], N. Sukavanam[2] and Urvashi Arora[2*]
*Corresponding author: Urvashi Arora, Department of Mathematics, Indian Institute of Technology Roorkee, Roorkee 247667, India
E-mail: urvashiaroraiitr@gmail.com
Reviewing editor: Amar Debbouche, Guelma University, Algeria

Abstract: This paper studies the approximate controllability issue of an abstract semilinear stochastic control system with nonlocal conditions. Sufficient conditions are formulated and proved for the approximate controllability of such systems by splitting the given semilinear system into two systems, namely a semilinear deterministic system and a linear stochastic system. To prove the approximate controllability of semilinear deterministic system, Schauder fixed point theorem has been used. At the end, an example has been presented to illustrate the feasibility of the proposed result.

Subjects: Engineering & Technology; Mathematics & Statistics; Science; Technology

Keywords: semilinear control systems; approximate controllability; Schauder fixed point theorem; nonlocal conditions

AMS subject classifications: 34K30; 34K35; 93C25

1. Introduction

Control theory is an area of application-oriented mathematics which deals with basic principles underlying the analysis and design of control systems. The study of controllability plays an important role in the control theory and engineering. Originated by Kalman (1960) for finite-dimensional linear control systems, controllability study was started systematically at the beginning of sixties. Since

ABOUT THE AUTHORS

Divya Ahluwalia received her PHD degree in Mathematics from Indian Institute of Technology, Roorkee (India) in 2004. Currently, she is an associate professor in the University of Petroleum and Energy Studies, Dehradun, India. Her research area includes controllability of deterministic systems.

N. Sukavanam received his BSc (Maths) degree from University of Madras, India in 1977 and MSc (Maths) from the same university in 1979 and PhD(Maths) from IISC Banglore, India in 1985. At present, he is working as a professor in the Department of Mathematics, Indian Institute of Technology, Roorkee (India). His research interests include nonlinear analysis, control theory and robotics and control.

Urvashi Arora received her MSc (Maths) degree from University of Kurukshetra, India in 2011. Presently she is working as a research scholar in the Department of Mathematics, IIT Roorkee, India. Her research area includes controllability of deterministic and stochastic systems.

PUBLIC INTEREST STATEMENT

Controllability is an important concept pertaining to any control system. It determines whether the state of the system can be steered to a given target state in a prescribed time interval or not. Therefore, it plays a very important role in the analysis and design of control systems. Also, the noise or stochastic perturbation is omnipresent and unavoidable in nature as well as in man made systems. So, we have to move from deterministic systems to stochastic systems. On the other hand, nonlocal conditions have a better effect on the solution and is more precise for the physical measurements than the claasical initial conditions. Therefore, in this paper, we discuss the approximate controllability of semilinear stochastic systems with nonlocal conditions using splitting technique.

then various researches have been carried out in the context of finite-dimensional deterministic systems and infinite-dimensional systems (Curtain & Zwart, 1995; Zabczyk, 1992). Approximate controllability of abstract semilinear systems has been studied by Zhou (1983) under certain inequality conditions involving the system operators. Lions (1988) has proved some exact and approximate controllability results for semilinear control systems. In Joshi and Sukavanam (1990), Naito (1987), Sukavanam (1993, 2000), sufficient conditions for the controllability of semilinear control systems have been studied.

On the other hand, stochastic differential equation is an emerging field drawing attention from both theoretical and applied disciplines. The deterministic models often fluctuate due to noise, which is random or at least appears to be so. Therefore, the study of stochastic problems is more applicable in dynamical system theory. Therefore, these differential equations are important from point of view of application also. The theory of stochastic differential equation can also be applied to various problems outside mathematics, for example in economics, mechanics epidemiology and several fields in engineering. Differential equations play an important role in formulation and analysis of mechanical, electrical, control engineering and physical sciences. Motivated by these facts many researchers are showing great interest to establish an appropriate system to investigate qualitative properties such as existence, uniqueness and controllability of these systems. Therefore, it becomes important to study the controllability of stochastic systems. Only few authors have studied the extensions of deterministic controllability concepts to stochastic control systems (Bashirov, 1996, 1997; Bashirov & Mahmudov, 1999; Mahmudov, 2001; Klamka, 2007; Klamka & Socha, 1977, Ren, Dai, & Sakthivel, 2013). In Bashirov (1996, 1997), the Controllability concepts were weakened and the controllability notions for partially observed linear Gaussian stochastic systems were introduced and their relation with complete and approximate controllabilities was studied.

On the other hand, the first results concerning the existence and uniqueness of mild solutions to abstract Cauchy problems with nonlocal conditions were formulated and proved by Byszewski and Lakshmikantham (1991). He argued that the corresponding models more accurately describe the phenomena since more information was taken into account at the onset of the experiment, thereby reducing the ill effects incurred by a single initial measurement. Also, it has a better effect on the solution and is more precise for physical measurements than classical condition $x(0) = x_0$ alone. Chang, Nieto, and Li (2009) studied the controllability of semilinear differential systems with nonlocal initial conditions using Sadovskii's Fixed Point theorem. Kumar and Sukavanam (2014) proved the controllability of second-order systems with nonlocal conditions using Sadovskii's Fixed point theorem. Shukla, Arora, and Sukavanam (2015) established sufficient conditions for the approximate controllability of first-order semilinear retarded stochastic system with nonlocal conditions using Banach Fixed Point Theorem. Arora and Sukavanam (2015) established sufficient conditions for the approximate controllability of second-order semilinear stochastic system with nonlocal conditions using Sadovskii's Fixed point theorem.

Splitting Technique introduced by Sukavanam and Kumar (2010) provides a better tool for discussing the controllability of the stochastic systems. This method may also be appropriate in the variational formulation to find the solutions of the equation under weaker assumptions on the data. Using this technique, Sukavanam and Kumar (2010) obtained sufficient conditions for the S- controllability of semilinear first-order system.

Up to now, most of the existing articles in the literature concentrate on finding the approximate controllability of stochastic systems using Banach Fixed Point Theorem, Sadovskii's Fixed Point Theorem and many other techniques. But there is no work reported on the approximate controllability of abstract semilinear stochastic system with nonlocal conditions using Splitting technique. Motivated by the above analysis, in this paper we establish sufficient conditions for the approximate controllability of an abstract semilinear stochastic system with nonlocal conditions with the help of new strategy which depends on splitting the given system into two systems, one is semilinear deterministic system and the other one is linear stochastic system.

2. Preliminaries

First, we introduce some notations. For a given operator A, $D(A)$, $R(A)$ and $N_0(A)$ denote the domain, range and null space of A, respectively. $\bar{E}$ and $E^{\perp}$ are the closure and orthogonal complement of a set E, respectively. $cov\ (x, y)$ is the covariance operator of the random variables x, y and $cov\ (x) = cov\ (x, x)$.

Let V, $\hat{V}$ and E be separable Hilbert Spaces and $Z = L_2[0,T;V]$ and $Y = L_2[0,T;\hat{V}]$ be the corresponding function spaces defined on $J = [0,T]$, $0 \le T < \infty$. Let (Ω, ζ, P) be a complete probability space equipped with a normal filtration $\{\zeta_t | t \in [0,b]\}$ generated by a Wiener Process $\{m(s) | 0 \le s \le t\}$. Suppose m is a Q-valued Wiener process on (Ω, ζ, P) with the covariance operator Q such that $trQ < \infty$. Let $L_2^0 = L_2(Q^{1/2}E, V)$ be the space of all Hilbert-Schmidt operators from $Q^{1/2}E$ to V. Then the space L_2^0 is a separable Hilbert space equipped with the norm $<\psi, \pi>= tr[\psi Q \pi^*]$. Let $L_2(\Omega, \zeta_T, V)$ be the Hilbert space of all ζ_T measurable square integrable random variables. $L_2^{\zeta}(J, V)$ is the space of all ζ_T-adapted, V-valued measurable square integrable processes on $J \times \Omega$.

Now, we consider the semilinear stochastic control system with nonlocal conditions

$$\left.\begin{aligned} dx_u(t) &= [Ax_u(t) + Bu(t) + f(x_u(t))]dt + dm(t), \quad 0 < t \le T, \\ x_u(0) &= x_0 + g(x) \end{aligned}\right\} \tag{2.1}$$

where $x_u(t)$ is the state value at time $t \in [0,T]$ corresponding to the control u taken from the set of admissible controls Y. x_0 is a Gaussian random variable with $covx_0 = P_0$ and x_0, m are mutually independent. $A{:}D(A) \subseteq V \to V$ is a closed linear operator with dense domain $D(A)$ generating a C_0-semigroup $S(t)$, $f{:}[0,T] \times V \to V$ is a nonlinear operator which satisfies Caratheodory conditions (Krasnoselskii, 1963) and $B{:}\hat{V} \to V$ is a bounded linear operator. Here g is a continuous function from $C(J, V)$ to V.

Splitting the system (Equation 2.1), we get the following pair of coupled systems

$$\left.\begin{aligned} dy_v(t) &= [Ay_v(t) + Bv(t) + f(y_v(t) + z_w(t))]dt, \quad 0 < t \le T, \\ y_v(0) &= y_0 + g(x) = \mathbb{E}x_0 + g(x) \end{aligned}\right\} \tag{2.2}$$

and

$$\left.\begin{aligned} dz_w(t) &= [Az_w(t) + Bw(t)]dt + dm(t), \quad 0 < t \le T, \\ z_w(0) &= z_0 = x_0 - \mathbb{E}x_0 \end{aligned}\right\} \tag{2.3}$$

where v and w are Y-valued control functions and $u = v + w$.

The solution $y_v(t)$ of the semilinear system (Equation 2.2) depends on the solution $z_w(t)$ of linear stochastic system (Equation 2.3). The mild solution of (Equation 2.1), (Equation 2.2) and (Equation 2.3) can be written as for $0 < t \le T$

$$\begin{aligned} x_u(t) = S(t)(x_0 + g(x)) &+ \int_0^t S(t-s)Bu(s)ds + \int_0^t S(t-s)f(x_u(s))ds \\ &+ \int_0^t S(t-s)dm(s) \end{aligned} \tag{2.4}$$

$$y_v(t) = S(t)(y_0 + g(x)) + \int_0^t S(t-s)Bv(s)ds + \int_0^t S(t-s)f(y_v(s) + z_w(s))ds \tag{2.5}$$

$$z_w(t) = S(t)z_0 + \int_0^t S(t-s)Bw(s)ds + \int_0^t S(t-s)dm(s) \tag{2.6}$$

It is clear that the systems (Equation 2.2) and (Equation 2.3) together give the solution of the given system (Equation 2.1). For each realization $z_w(t)$ of (Equation 2.3), the system (Equation 2.2) is a deterministic system.

Definition 2.1 The set $K_T(f) = \{x_u(T) \in V : x_u(.) \in Z$ is a mild solution of (Equation 2.1) for $u \in Y\}$ is called the reachable set of the system (Equation 2.1). $K_T(0)$ denotes the reachable set of the corresponding linear system of (Equation 2.1).

Definition 2.2 The system (Equation 2.1) is said to be approximate controllable if $K_T(f)$ is dense in V and the corresponding linear system is approximate controllable if $K_T(0)$ is dense in V.

3. Controllability results

In this section, sufficient conditions have been established for the approximate controllability of the system (Equation 2.1).

We define an operator $F : Z \to Z$ by

$$(Fx)(t) = f(t, x(t)), \quad 0 \le t \le T$$

where F is called the Nemytskii operator of f.

Now, in order to obtain the desired result, we assume the following conditions:

(a) For every $p \in Z$ there exists a $q \in \overline{R(B)}$ such that $Lp = Lq$ where L is the operator defined as in (3.3).

(b) A generates a compact semigroup $S(t)$.

(c) The operator $f(t, x)$ satisfies Lipschitz continuity in x, i.e.

$$||f(t,x) - f(t,y)|| \le l||x - y||_V$$

for some constant $l > 0$.

(d) f is uniformly bounded on V, i.e., $||f(t,x)|| \le k$, a constant.

The corresponding linear system of (Equation 2.2) which is a deterministic system is given by

$$\left.\begin{aligned} \frac{dp_r}{dt} &= Ap_r(t) + Br(t) \\ p_r(0) &= p_0 = y_0 + g(x) \end{aligned}\right\} \tag{3.1}$$

The above system (Equation 3.1) is approximate controllable under the condition (a) (see lemma 2 Naito, 1987).

Let K be an operator from Z into itself defined as

$$[Kz](t) = \int_0^t S(t-s)z(s)ds \tag{3.2}$$

and L and N be operators from Z into V defined as

$$Lz = \int_0^T S(T-s)z(s)ds \tag{3.3}$$

$$Nz = \int_0^T S(T-s)[Fz](s)ds \tag{3.4}$$

It is evident that hypothesis (a) is equivalent to the condition $Z = N_0(L) + \overline{R(B)}$. Moreover, Z can be decomposed as $Z = N_0(L) + N_0^{\perp}(L)$. Also, under hypothesis (a) we can define a map $P{:}N_0^{\perp}(L) \to \overline{R(B)}$ as follows :

Let $u \in N_0^{\perp}(L)$, $P(u) = u_0$ where u_0 is the unique minimum norm element in the set satisfying

$$||Pu|| = ||u_o|| = min\{||v||{:}v \in \{u + N_0(L)\} \cap \overline{R(B)}\}.$$

The operator P is well-defined as it follows from hypothesis (a) that for each $u \in N_0^{\perp}(L)$, the set $\{u + N_0(L)\} \cap \overline{R(B)}$ is nonempty.

Lemma 1 *The operator P from $N_0^{\perp}(L)$ to $\overline{R(B)}$ is linear and continuous.*

Proof See Lemma 1 (Naito, 1987). □

From continuity of P it follows that $||Pu||_Z \leq c||u||_Z$ for some constant $c \geq 0$.

Lemma 2 *If an element $z \in Z$ can be uniquely decomposed as $z = n + q{:}n \in N_0(L), q \in \overline{R(B)}$, then $||n|| \leq (1 + c)||z||$.*

Proof Let $u \in N_0^{\perp}(L)$, then $Pu = (n_0 + u) \in \overline{R(B)}$ for some $n_0 \in N_0(L)$.

Now if $z \in Z$ has unique decomposition, namely, $z = n_1 + u{:}n_1 \in N_0(L)$ and $u \in N_0^{\perp}(L)$, then z can be uniquely decomposed as: $z = n + q{:}n \in N_0(L)\,;q \in \overline{R(B)}$ where $q = Pu$ and $n = n_1 - n_0$.

Now,

$$z = n_1 + u, \Rightarrow ||z||^2 = ||n_1||^2 + ||u||^2 \text{ or } ||u|| \leq ||z||. \tag{3.5}$$

Also $n = z - q, \Rightarrow ||n|| = ||z - Pu||$

Hence using (Equation 3.5), we get □

$$||n|| \leq ||z|| + c||u|| \leq (1 + c)||z|| \tag{3.6}$$

Let M_0 be the subspace of Z such that $M_0 = \{m \in Z{:}m = K(n), n \in N_0(L)\}$.

For each solution $p_r(t)$ of (Equation 3.1) with control r, define the random operator $f_{p_r}{:}\overline{M_0} \to M_0$ as

$$f_{p_r}(m) = K(n) \tag{3.7}$$

where n is given by the unique decomposition

$$F(p_r(t) + z_w(t) + m) = n + q{:}n \in N_0(L), q \in \overline{R(B)} \tag{3.8}$$

For approximate controllability of (Equation 2.3), let us introduce some operators and lemmas.

Define the linear operator $L_0^T{:}L_2^{\zeta}[0,T;\hat{V}] \to L_2[\Omega, \zeta_t, V]$, the controllability operator $\Pi_s^T{:}L_2[\Omega, \zeta_t, V] \to L_2[\Omega, \zeta_t, V]$ associated with system (Equation 2.6), and the controllability operator $\Gamma_s^T{:}V \to V$ associated with the corresponding deterministic system of (Equation 2.6) as

$$L_0^T u = \int_0^T S(T-s)Bu(s)ds$$

$$\Pi_s^T\{.\} = \int_s^T S(T-t)BB^*S^*(T-t)\mathbb{E}\{.|\zeta_t\}dt$$

$$\Gamma_s^T = \int_s^T S(T-t)BB^*S^*(T-t)dt$$

It is easy to see that the operators L_0^T, Π_s^T, Γ_s^T are linear bounded operators, and the adjoint $(L_0^T)^*:L_2[\Omega,\zeta_t,V] \to L_2^{\zeta}[0,T;\widehat{V}]$ of L_0^T is defined by

$$(L_0^T)^*z = B^*S^*(T-t)\mathbb{E}\{z|\zeta_t\}$$
$$\Pi_0^T = L_0^T(L_0^T)^*.$$

Before studying the approximate controllability of system (Equation 2.3), let us first investigate the relation between Π_s^T and $\Gamma_s^T, s \leq r < T$ and resolvent operator $R(\lambda,\Pi_s^T) = (\lambda I + \Pi_s^T)^{-1}$ and $R(\lambda,\Gamma_r^T) = (\lambda I + \Gamma_r^T)^{-1}, s \leq r < T$ for $\lambda > 0$, respectively.

LEMMA 3 *(Mahmudov, 2001) For every $z \in L_2[\Omega,\zeta_t,V]$ there exists $\varphi(.) \in L_2^{\zeta}(J,L_2^0)$ such that*

(i) $\mathbb{E}\{z|\zeta_t\} = \mathbb{E}\{z\} + \int_0^t \varphi(s)dm(s)$,

(ii) $\Pi_s^T z = \Gamma_s^T \mathbb{E}z + \int_s^T \Gamma_r^T \varphi(r)dm(r)$,

(iii) $R(\lambda,\Pi_s^T)z = R(\lambda,\Gamma_s^T)\mathbb{E}\{z|\zeta_t\} + \int_s^T \Gamma_r^T \varphi(r)dm(r)$.

THEOREM 3.1 *(Mahmudov, 2001) The control system (Equation 2.3) is approximate controllable on [0, T] if and only if one of the following conditions hold.*

(i) $\Pi_0^T > 0$.

(ii) $\lambda R(\lambda,\Pi_0^T)$ converges to the zero operator as $\lambda \to 0^+$ in the strong operator topology.

(iii) $\lambda R(\lambda,\Pi_0^T)$ converges to the zero operator as $\lambda \to 0^+$ in the weak operator topology.

Proof The proof is a straightforward adaptation of the proof of Theorem 4.1 (Mahmudov, 2001). □

THEOREM 3.2 *The following four conditions are equivalent.*

(1) The stochastic system (Equation 2.3) is approximate controllable on [0, *T*].

(2) The corresponding deterministic system of (Equation 2.3) is approximate controllable on every [*s*, *T*], $0 \leq s < T$.

(3) The corresponding deterministic system of (Equation 2.3) is small time approximate controllable.

(4) The stochastic system (Equation 2.3) is small time approximate controllable.

Proof For the proof, refer to the proof of Theorem 4.2 (Mahmudov, 2001). □

From the above results, we conclude that the semilinear stochastic system (Equation 2.3) is approximate controllable since the corresponding deterministic linear system is approximate controllable .

LEMMA 4 *Under the assumptions (b) and (d) ($||S(t)|| \leq M$, a constant) the operator f_{p_r} has a fixed point m_0 for each realization $z_w(t)$ of (Equation 2.3).*

Proof Since $S(t)$ is compact semigroup and $||S(t)|| \leq M$, a constant, for $0 < t < T$, it follows from Pazy (1983) that the operator K and hence f_{p_r} is compact for each p_r. Now let $||m|| < r$, then we have,

$$\begin{aligned} ||f_{p_r}(m)||^2 &= \left|\left| \int_0^t S(t-s)n(s)ds \right|\right|_Z^2 \\ &\leq \int_0^T \left|\left| \int_0^t S(t-s)n(s) \right|\right|_V^2 ds \\ &\leq M^2T^2(1+c)^2||F(x+m)||^2 \\ &\leq M^2T^2k^2(1+c)^2 = R(say) \end{aligned} \tag{3.9}$$

From the compactness of f_{p_r} and (Equation 3.9), it follows from Schauder fixed point theorem that f_{p_r} has a fixed point in M_0 in a ball of radius $r > R$ such that $f_{p_r}(m_0) = m_0$. □

Remark 3.1 If we consider $z_w(t)$ as a random function then the equation (Equation 2.2) becomes a stochastic control system, the functions n and q become random functions and the fixed point m_0 becomes a random fixed point.

THEOREM 3.3 *Under the conditions (a), (b), (c) and (d), the semilinear stochastic control system (Equation 2.2) is approximate controllable for arbitrary $\epsilon > 0$.*

Proof Since f_{p_r} has a fixed point m_0, then operating K on both sides of (Equation 3.8) at $m = m_0$, we get

$$KF(p_r(t) + z_w(t) + m_0) = Kn + Kq = m_0 + Kq$$

Adding $p_r(t)$ on both sides, we get

$$\begin{aligned} &p_r(t) + KF(p_r(t) + z_w(t) + m_0) = p_r(t) + m_0 + Kq \\ &p_r(t) + KF(y(t) + z_w(t)) = y(t) + Kq, \text{ where } p_r(t) + m_0 = y(t) \\ &S(t)(y_0 + g(x)) + KBr + KF(y(t) + z_w(t)) = y(t) + Kq \\ &S(t)(y_0 + g(x)) + K(Br - q) + KF(y(t) + z_w(t)) = y(t) \end{aligned}$$

Thus, it follows that $p_r(t) + m_0 = y(t)$ is a solution of the following system

$$\left.\begin{aligned} &\frac{dy}{dt} = Ay(t) + f(y(t) + z_w(t)) + Br(t) - q(t), \\ &y(0) = y_0 + g(x) \end{aligned}\right\} \tag{3.10}$$

with control $Br(t) - q(t)$. Moreover, $y(T) = p_r(T) + m(T) = p_r(T)$, since $m(T) = 0$. It shows that the reachable set of (Equation 3.10) is a superset of the reachable set $K_T(0)$ of (Equation 3.1) which is dense in V.

Since $q \in \overline{R(B)}$, for any given $\epsilon > 0$ there exists $v_1 \in Y$ such that $||q - Bv_1|| \leq \epsilon$. Now consider the equation

$$\left.\begin{aligned} &\frac{dy_v}{dt} = Ay_v(t) + f(y_v(t) + z_w(t)) + B(r(t) - v_1(t)); \\ &y_v(0) = y_0 + g(x) \end{aligned}\right\} \tag{3.11}$$

Let $y_v(t)$ be the solution of (Equation 3.11) corresponding to control $v = r - v_1$. Then $||y(T) - y_v(T)||$ can be made arbitrarily small by choosing a suitable v_1, which implies $K_T(0) \subset \overline{K_T(f)}$ where $K_T(f)$ denotes the reachable set of (Equation 3.11). Hence the approximate controllability of (Equation 2.2) follows. □

From the above discussion, we have that the linear stochastic system (2.3) is approximate controllable and for each realization $z_w(t)$ of (Equation 2.3), the semilinear system (Equation 2.2) is approximate controllable. So, we conclude that the semilinear stochastic system (Equation 2.1) is approximate controllable.

4. Example

Let $V = L_2(0,\pi)$ and $A = -\frac{d^2}{d\theta^2}$ with $D(A)$ consisting of all $y \in V$ with $\frac{d^2y}{dx^2} \in V$ and $y(0) = 0 = y(\pi)$. Put $\phi_n(\theta) = (2/\pi)^{1/2}\sin(n\theta), 0 \le \theta \le \pi, n = 1, 2, \cdots$, then $\{\phi_n : n = 1, 2 \cdots\}$ is an orthonormal basis for V and ϕ_n is an eigenfunction corresponding to the eigenvalue $\lambda_n = -n^2$ of the operator $-A, n = 1, 2 \cdots$. Then the C_0-semigroup $S(t)$ generated by $-A$ has $e^{\lambda_n t}$ as the eigenvalues and ϕ_n as their corresponding eigenfunctions (Joshi & Sukavanam, 1990). Now define an infinite-dimensional space $\widehat{V}$ by

$$\widehat{V} = \{u : u = \sum_{n=2}^{\infty} u_n\phi_n \text{ with } \sum_{n=2}^{\infty} u_n^2 < \infty\}$$

The norm in $\widehat{V}$ is defined by $||u|| = \left(\sum_{n=2}^{\infty} u_n^2\right)^{1/2}$.

Define a continuous linear mapping B from $\widehat{V}$ to V as follows

$$Bu = 2u_2\phi_1 + \sum_{n=2}^{\infty} u_n\phi_n \text{ for } u = \sum_{n=2}^{\infty} u_n\phi_n \in \widehat{V}$$

Consider the control system governed by the semilinear heat equation

$$\left.\begin{aligned} &\frac{\partial x_u(t,\theta)}{\partial t} = \frac{\partial^2 x_u(t,\theta)}{\partial\theta^2} + Bu(t,\theta) + f(t, x_u(t,\theta)) + dm(t); 0 < t < T, 0 < \theta < \pi \\ &x_u(t,0) = x_u(t,\pi) = 0; 0 \le t \le T \\ &x_u(0,\theta) + \sum_{i=1}^{n} \alpha_i x_u(t_i,\theta) = x_u(\theta), 0 \le \theta \le \pi, 0 < t_i \le T \end{aligned}\right\} \tag{4.1}$$

The approximate controllability of the corresponding deterministic semilinear heat equation of (Equation 4.1) was considered by Naito (1987). Here the approximate controllability of stochastic semilinear heat control system with nonlocal conditions is considered.

Now we define the bounded linear function $\widetilde{B}$ from $L_2(0, T:\widehat{V})$ to $L_2(0, T:V)$ by

$$(\widetilde{B}u)(t) = Bu(t) \text{ for } u \in L_2(0, T; \widehat{V}).$$

The nonlinear operator f is assumed to satisfy the conditions (c) and (d).

To the system (Equation 4.1) we can associate two control systems. The one is the deterministic control system with nonlocal conditions

$$\left.\begin{aligned} &\frac{\partial y_v(t,\theta)}{\partial t} = \frac{\partial^2 y_v(t,\theta)}{\partial\theta^2} + Bv(t,\theta) + f(y_v(t,\theta) + z_w(t,\theta)); 0 < t < T, 0 < \theta < \pi \\ &y_v(t,0) = y_v(t,\pi) = 0; 0 \le t \le T \\ &y_v(0,\theta) + \sum_{i=1}^{n} \alpha_i x_u(t_i,\theta) = y_v(\theta); 0 \le \theta \le \pi, 0 < t_i \le T \end{aligned}\right\} \tag{4.2}$$

and the second is the stochastic linear control system

$$\left.\begin{aligned}&\frac{\partial z_w(t,\theta)}{\partial t} = \frac{\partial^2 z_w(t,\theta)}{\partial \theta^2} + Bw(t,\theta) + dm(t);\\ &z_w(t,0) = z_w(t,\pi) = 0; 0 \le t \le T\\ &z_w(0,\theta) = z_w(\theta); 0 \le \theta \le \pi\end{aligned}\right\} \tag{4.3}$$

Now for each realization $z_w(t)$ of the system (Equation 4.3), the system (Equation 4.2) is a deterministic system. From Theorem 3 and using the conditions (a)–(d), it is clear that for each realization $z_w(t)$ of the system (Equation 4.3), the system (Equation 4.2) is approximate controllable. It follows that the system (Equation 4.1) is approximate controllable.

5. Conclusion

In this paper, sufficient conditions have been established for the approximate controllability of an abstract semilinear stochastic system with nonlocal conditions using splitting technique. Then by using this approach, approximate controllability of heat equation has been discussed in the example. Further, we can find the approximate controllability of stochastic integrodifferential systems with nonlocal conditions under weaker conditions by using this technique.

Funding
Urvashi Arora, the corresponding author would like to thank Ministry of Human Resource Development [grant number MHR-02-23-200-304] for their support to carry out her research work.

Author details
Divya Ahluwalia[1]
E-mail: divya.ahluwalia@yahoo.co.in
N. Sukavanam[2]
E-mail: nsukavanam@gmail.com
Urvashi Arora[2]
E-mail: urvashiaroraiitr@gmail.com
[1] Department of Mathematics, University of Petroleum and Energy Studies, Dehradun, India.
[2] Department of Mathematics, Indian Institute of Technology Roorkee, Roorkee 247667, India.

References

Arora, U., & Sukavanam, N. (2015). Approximate controllability of second order semilinear stochastic system with nonlocal conditions. *Applied Mathematics and Computation, 258*, 111–119.

Bashirov, A. E. (1996). On weakening of the controllability concepts. In *Proceedings of the 35th IEEE Conference on Decision and Control* (pp. 640–645). Kobe.

Bashirov, A. E., & Mahmudov, N. I. (1999). On concepts of controllability for deterministic and stochastic systems. *SIAM Journal on Control and Optimization, 37*, 1808–1821.

Bashirov, A. E., & Kerimov, K. R. (1997). On controllability conception for stochastic systems. *SIAM Journal on Control and Optimization, 35*, 384–398.

Byszewski, L., & Lakshmikantham, V. (1991). Theorem about the existence of a solution of a nonlocal abstract Cauchy problem in a Banach space. *Journal of Applied Analysis, 40*, 11–19.

Chang, Y. K., Nieto, J. J., & Li, W. S. (2009). Controllability of semilinear differential systems with nonlocal initial conditions in Banach spaces. *Journal of Optimization Theory and Applications, 142*, 267–273.

Curtain, R. F., & Zwart, H. J. (1995). *An introduction to infinite dimensional linear systems theory*. New York, NY: Springer-Verlag.

Joshi, M. C., & Sukavanam, N. (1990). Approximate solvability of semilinear operator equations. *Nonlinearity, 3*, 519–525.

Kalman, R. E. (1960). A new approach to linear filtering and prediction problems. *Transactions of the ASME Journal of Basic Engineering (Series D), 82*, 35–45.

Klamka, J. (2007). Stochastic controllability of linear systems with delay in contol. *Bulletin of the Polish Academy of Sciences, 55*, 23–29.

Klamka, J., & Socha, L. (1977). Some remarks about stochastic controllability. *IEEE Transactions on Automatic Control, 5*, 880–881.

Krasnoselskii, M. A. (1963). *Topological methods in the theory of nonlinear integral equations*. Oxford: Pergamon.

Kumar, S., & Sukavanam, N. (2014). Controllability of second-order systems with nonlocal conditions in Banach spaces. *Numerical Functional Analysis and Optimization, 35*, 423–431.

Lions, J. L. (1988). Exact controllability, stabilizability and perturbation for distributed systems. *SIAM Review, 30*, 1–68.

Mahmudov, N. I. (2001). Controllability of linear stochastic systems in Hilbert spaces. *Journal of Mathematical Analysis and Applications, 259*, 64–82.

Naito, K. (1987). Controllability of semilinear control systems dominated by the linear part. *SIAM Journal on Control and Optimization, 25*, 715–722.

Pazy, A. (1983). *Semigroup of linear operators and applications to partial differential equations*. New York, NY: Springer-Verlag.

Ren, Y., Dai, H., & Sakthivel, R. (2013). Approximate controllability of stochastic differential systems driven by a Levy process. *International Journal of Control, 86*, 1158–1164.

Shukla, A., Arora, U., & Sukavanam, N. (2015). Approximate controllability of retarded semilinear stochastic system with non local conditions. *Journal of Applied Mathematics and Computing*. doi:10.1007/s12190-014-0851-9

Sukavanam, N. (1993). Approximate controllability of semilinear control systems with growing nonlinearity. In *Mathematical Theory of Control Proceedings of International Conference* (pp. 353–357). New York: Marcel Dekker.

Sukavanam, N. (2000). Solvability of semilinear operator equations with growing nonlinearity. *Journal of Mathematical Analysis and Applications, 241*, 39–45.

Sukavanam, N., & Kumar, M. (2010). S-controllability of an abstract first order semilinear control system. *Numerical Functional Analysis and Optimization, 31*, 1023–1034.

Zabczyk, J. (1992). *Mathematical control theory: An introduction*. Berlin: Birkhauser.

Zhou, H. X. (1983). Approximate controllability for a class of semilinear abstract equations. *SIAM Journal on Control and Optimization, 21*, 551–565.

12

Two-sided bounds on the mean vector and covariance matrix in linear stochastically excited vibration systems with application of the differential calculus of norms

Ludwig Kohaupt[1*]

*Corresponding author: Ludwig Kohaupt, Department of Mathematics, Beuth University of Technology Berlin, Luxemburger Str. 10, D-13353 Berlin, Germany

Email: kohaupt@beuth-hochschule.de

Reviewing editor: Cedric K.F. Yiu, Hong Kong Polytechnic University, Hong Kong

Abstract: For a linear stochastic vibration model in state-space form, $\dot{x}(t) = Ax(t) + b(t),\ x(0) = x_0$, with system matrix A and white noise excitation $b(t)$, under certain conditions, the solution $x(t)$ is a random vector that can be completely described by its mean vector, $m_x(t) := m_{x(t)}$, and its covariance matrix, $P_x(t) := P_{x(t)}$. If matrix A is asymptotically stable, then $m_x(t) \to 0\ (t \to \infty)$ and $P_x(t) \to P\ (t \to \infty)$, where P is a positive (semi-)definite matrix. As the main new points, in this paper, we derive two-sided bounds on $\|m_x(t)\|_2$ and $\|P_x(t) - P\|_2$ as well as formulas for the right norm derivatives $D_+^k \|P_x(t) - P\|_2,\ k = 0, 1, 2$, and apply these results to the computation of the best constants in the two-sided bounds. The obtained results are of special interest to applied mathematicians and engineers.

Subjects: Applied Mathematics; Computer Mathematics; Engineering & Technology; Engineering Mathematics; Mathematics & Statistics; Mathematics & Statistics for Engineers; Mathematics Education; Science; Technology

Keywords: linear stochastic vibration system excited by white noise; mean vector; covariance matrix; two-sided bounds; differential calculus of norms

AMS subject classifications: 34D05; 34F05; 65L05

ABOUT THE AUTHOR

Ludwig Kohaupt received the equivalent to the Master Degree (Diplom-Mathematiker) in Mathe\-matics in 1971 and the equivalent to the PhD (Dr phil nat) in 1973 from the University of Frankfurt/Main.

From 1974 until 1979, he was a teacher in Mathematics and Physics at a Secondary School. During that time (from 1977 until 1979), he was also an auditor at the Technical University of Darmstadt in Engineering Subjects, such as Mechanics, and especially Dynamics.

From 1979 until 1990, he joined the Mercedes-Benz car company in Stuttgart as a Computational Engineer, where he worked in areas such as Dynamics (vibration of car models), Cam Design, Gearing, and Engine Design. Some of the results were published in scientific journals (on the whole, 12 papers and 1 monograph).

Then, in 1990, he combined his preceding experiences by taking over a professorship at the Beuth University of Technology Berlin. He retired on 01 April 2014.

PUBLIC INTEREST STATEMENT

In recent years, the author has developed a differential calculus for norms of vector and matrix functions. More precisely, differentiability properties of these quantities were derived for various vector and matrix norms, and formulas for the pertinent (right-hand, resp. left-hand) derivatives were obtained. These results have been applied to a number of linear and non-linear problems by computing the best constants in two-sided bounds on the solution of the pertinent initial value problems. In the present paper, the application area is extended to stochastically excited vibration systems. Specifically, new two-sided estimates on the mean vector and the co-variance matrix are derived, and the optimal constants in these bounds are computed in a numerical example employing the differential calculus of norms.

1. Introduction

In this paper, linear stochastic vibration models of the form $\dot{x}(t) = Ax(t) + b(t)$, $x(0) = x_0$, with real *system matrix* A and *white noise excitation* $b(t)$ are investigated, in which the initial vector x_0 can be completely characterized by its mean vector m_0 and its covariance matrix P_0. Likewise, the solution $x(t)$, also called *response*, is a random vector that can be described by its mean vector $m_x(t) := m_{x(t)}$, and its covariance matrix, $P_x(t) := P_{x(t)}$. For asymptotically stable matrices A, it is known that $m_x(t) \to 0\,(t \to \infty)$ and $P_x(t) \to P\,(t \to \infty)$, where P is a positive (semi-)definite matrix. This leads to the question of the asymptotic behavior of $m_x(t)$ and $P_x(t) - P$. As appropriate norms for the investigation of this problem, the Euclidean norm for $m_x(t)$ and the spectral norm for $P_x(t) - P$ is the respective natural choice; both norms are denoted by $\|\cdot\|_2$.

The *main new points* of the paper are

- the determination of two-sided bounds on $\|m_x(t)\|_2$ and $\|P_x(t) - P\|_2$,
- the derivation of formulas for the right norm derivatives $D_+^k \|P_x(t) - P\|_2$, $k = 0, 1, 2$, and
- the application of these results to the computation of the best constants in the two-sided bounds.

The paper is structured as follows.

In Section 2, the linear stochastically excited vibration model in state-space form is presented. Then, in Section 3, new two-sided bounds on $\|m_x(t)\|_2$ are determined. In Section 4, preliminary work for two-sided bounds on $\|P_x(t) - P\|_2$ is made that is employed in Section 5 to derive new two-sided bounds on $\|P_x(t) - P\|_2$ itself. In Section 6, the local regularity of $\|P_x(t) - P\|_2$ is studied. In Section 7, as the new result, formulas for the right norm derivatives $D_+^k \|P_x(t) - P\|_2$, $k = 0, 1, 2$ are obtained. In Section 8, for the specified data in the stochastically exited model, the differential calculus of norms is applied to compute the best constants in the new two-sided bounds on $\|m_x(t)\|_2$ and $\|P_x(t) - P\|_2$. In Section 9, conclusions are drawn. Finally, in Appendix A, more details on some items are given.

2. The linear stochastically excited vibration system

In order to make the paper as far as possible self-contained, we summarize the known facts on linear stochastically excited systems. In the presentation, we follow closely the line of Müller and Schiehlen (1976, Sections 9.1 and 9.2).

So, let us depart from the *deterministic model in state-space form*

$$\dot{x}(t) = Ax(t) + b(t),\; t \geq 0,\; x(0) = x_0 \tag{1}$$

with *system matrix* $A \in \mathbb{R}^{n\times n}$, the *state vector* $x(t) \in \mathbb{R}^n$ and the *excitation vector* $b(t) \in \mathbb{R}^n$, $t \geq 0$.

Now, we replace the deterministic excitation $b(t)$ by a *stochastic excitation* in the form of *white noise*. Thus, $b(t)$ can be completely described by the *mean vector* $m_b(t)$ and the *central correlation matrix* $N_b(t,\tau)$ with

$$\begin{aligned} m_b(t) &= 0 \\ N_b(t,\tau) &= Q\,\delta(t-\tau) \end{aligned} \tag{2}$$

where $Q = Q_b$ is the $n \times n$ *intensity matrix* of the excitation and $\delta(t-\tau)$ the δ-function (more precisely, the δ-functional).

From the central correlation matrix, one obtains for $\tau = t$ the positive semi-definite *covariance matrix*

$$P_b(t) := N_b(t,t) \tag{3}$$

At this point, we mention that the definition of a real positive semi-definite matrix includes its symmetry.

When the excitation is white noise, the deterministic initial value problem (1) can be formally maintained as the theory of linear stochastic differential equations shows. However, the initial state x_0 must be introduced as Gaussian random vector,

$$x_0 \sim (m_0, P_0) \tag{4}$$

which is to be independent of the excitation (2); here, the sign $\sim$ means that the initial state x_0 is completely described by its mean vector m_0 and its covariance matrix P_0. More precisely: x_0 is a Gaussian random vector whose density function is completely determined by m_0 and P_0 alone.

The *stochastic response* of the system (1) is formally given by

$$x(t) = \Phi(t)x_0 + \int_0^t \Phi(t-\tau)b(\tau)d\tau \tag{5}$$

where besides the fundamental matrix $\Phi(t) = e^{At}$ and the initial vector x_0- a stochastic integral occurs.

It can be shown that the stochastic response $x(t)$ is a non-stationary Gauss-Markov process that can be described by the mean vector $m_x(t) := m_{x(t)}$ and the correlation matrix $N_x(t, \tau) := N_{(x(t), x(\tau))}$. For $\tau = t$, we get the covariance matrix $P_x(t) := P_{x(t)}$.

If the system is asymptotically stable, the properties of first and second order for the stochastic response $x(t)$ we need are given by

$$\begin{aligned} m_x(t) &= \Phi(t)m_0, \\ P_x(t) &= \Phi(t)(P_0 - P)\Phi^T(t) + P \end{aligned} \tag{6}$$

where the positive semi-definite $n \times n$ matrix P satisfies the *Lyapunov matrix equation*

$$AP + PA^T + Q = 0$$

This is a special case of the matrix equation $AX + XB = C$, whose solution can be obtained by a method of Ma, cf. (1966). For the special case of diagonalizable matrices A and B, this is shortly described in Appendix A1.

For asymptotically stable matrix A, one has $\lim_{t\to\infty} \Phi(t) = 0$ and thus from (6),

$$\lim_{t\to\infty} m_x(t) = 0 \tag{7}$$

and

$$\lim_{t\to\infty} P_x(t) = P \tag{8}$$

Therefore, it is of interest to investigate the asymptotic behavior of $m_x(t)$ and $P_x(t) - P$. This investigation will be done in the next sections by giving two-sided bounds on both quantities in appropriate norms.

Even though the two-sided bounds on $m_x(t)$ can be obtained by just applying known estimates, they will be stated for the sake of completeness in Section 3.

As opposed to this, the determination of two-sided bounds on $P_x(t) - P$ leads to a new interesting problem and will be solved in two steps described in Sections 4 and 5.

3. Two-sided bounds on $m_x(t)$

According to Equation (6_1), we have

$$m_x(t) = \Phi(t)m_0,\ t \geq 0$$

From Kohaupt (2006, Theorem 8), one obtains two-sided bounds on $m_x(t)$.

To see this, let $m_0 \neq 0$ and $\|\cdot\|_2$ the Euclidean norm in $\mathbb{R}^n$. Then, there exists a constant $X_0 > 0$ and for every $\varepsilon > 0$ an constant $X_{1,\varepsilon} > 0$ such that

$$X_0\, e^{\nu_{m_0}[A]t} \leq \|m_x(t)\|_2 \leq X_{1,\varepsilon}\, e^{(\nu_{m_0}[A]+\varepsilon)t},\ t \geq 0 \tag{9}$$

where $\nu_{m_0}[A]$ is the spectral abscissa of matrix A with respect to the vector m_0 (see Kohaupt 2006, Section 7, p. 146). We mention that often $\nu_{m_0}[A] = \nu[A]$, cf. (Kohaupt, 2006, p. 154).

4. Preliminary work for two-sided bounds on $P_x(t) - P$

In this section, we derive two-sided bounds that are of general interest beyond their application in Section 5. Therefore, more general assumptions than needed there will be made. We obtain the following lemma.

Lemma 1 *(Two-sided bounds on $\|\Psi^* C\Psi\|_2$)*

Let $C \in \mathbb{C}^{n\times n}$ with $C^* = C$, where C^* is the adjoint of C. Further, let $\|\cdot\|_2$ be the spectral norm of a matrix.

Then, the two-sided bound

$$c_0\, \|\Psi\|_2^2 \leq \|\Psi^*\, C\, \Psi\|_2 \leq c_1\, \|\Psi\|_2^2,\ \ \Psi \in \mathbb{C}^{n\times m} \tag{10}$$

is valid where

$$c_0 = \inf_{\|v\|_2=1} |(C\,v, v)| \tag{11}$$

$$c_1 = \sup_{\|v\|_2=1} |(C\,v, v)| \tag{12}$$

Proof Decisive tool is the fact that for $A \in \mathbb{C}^{n\times n}$ with $A^* = A$ one has the two representations

$$\|A\|_2 = \sup_{\|v\|_2=1} \|Av\|_2 = \sup_{\|v\|_2=1} |(A\,v, v)|$$

In the following, this will be applied to $\Psi^*\, C\, \Psi$.

(ii) Lower bound:

One has

$$\begin{aligned}
\|\Psi^*\, C\, \Psi\|_2 &= \sup_{\|u\|_2=1} |(\Psi^*\, C\, \Psi\, u, u)| \\
&= \sup_{\|u\|_2=1} |(C\, \Psi\, u, \Psi\, u)| \\
&= \sup_{\substack{\|u\|_2=1 \\ \Psi u\neq 0}} |(C\, \frac{\Psi\, u}{\|\Psi\, u\|_2}, \frac{\Psi\, u}{\|\Psi\, u\|_2})|\ \|\Psi\, u\|_2^2 \\
&\geq \sup_{\substack{\|u\|_2=1 \\ \Psi u\neq 0}} \inf_{\|v\|_2=1} |(C\, v, v)|\ \|\Psi\, u\|_2^2 \\
&= \inf_{\|v\|_2=1} |(C\, v, v)|\ \sup_{\|u\|_2=1} \|\Psi\, u\|_2^2 \\
&= c_0\, \|\Psi\|_2^2,\ \Psi \neq 0
\end{aligned}$$

For $\Psi = 0$, this lower bound remains valid.

(iii) Upper bound:

Similarly, one obtains

$$\|\Psi^* C \Psi\|_2 \leq c_1 \|\Psi\|_2^2, \ \Psi \neq 0$$

For $\Psi = 0$, this upper bound remains valid.

Remark In Lemma 1, it is known that

$$c_1 = \sup_{\|v\|_2=1} |(C v, v)| = \sup_{v \neq 0} \frac{\|Cv\|_2}{\|v\|_2} = \sup_{\|v\|_2=1} \|Cv\|_2 = \|C\|_2 \tag{13}$$

where $\|C\|_2 > 0$ for $C \neq 0$.

Similarly, one can derive a chain of relations for $c_0 = \inf_{\|v\|_2=1} |(C v, v)|$, as the next lemma shows.

LEMMA 2 *(Chain of relations for $c_0 = \inf_{\|v\|_2=1} |(C v, v)|$)*

Let $C \in \mathbb{C}^{n \times n}$ with $C^* = C$.

Then,

$$c_0 = \inf_{\|v\|_2=1} |(C v, v)| = \inf_{v \neq 0} \frac{\|Cv\|_2}{\|v\|_2} = \inf_{\|v\|_2=1} \|Cv\|_2 \tag{14}$$

Now, let C be additionally regular, let $\|\cdot\|$ denote any vector norm as well as the associated sub norm.

Then,

$$\inf_{\|v\|=1} \|Cv\| = \frac{1}{\|C^{-1}\|} > 0 \tag{15}$$

this is especially valid also for the spectral norm $\|\cdot\|_2$.

The proof will be given in Appendix A2.

5. Two-sided bounds on $P_x(t) - P = \Phi(t)(P_0 - P)\Phi^T(t)$

In this section, the results of Section 4 are employed to estimate $\|P_x(t) - P\|_2$ above and below by $\|\Phi(t)\|_2^2$ as well as by $e^{2(\nu[A]+\varepsilon)t}$ resp. $e^{2\nu[A]t}$ where $\nu[A]$ is the spectral abscissa of matrix A. New will be the quadratic asymptotic behavior of $\|P_x(t) - P\|_2$.

We show the following lemma.

THEOREM 3 *(Two-sided bounds on $P_x(t) - P = \Phi(t)(P_0 - P)\Phi^T(t)$ based on $\|\Phi(t)\|_2^2$)*

Let $A \in \mathbb{R}^{n \times n}$, let $\Phi(t) = e^{At}$ be the associated fundamental matrix with $\Phi(0) = E$ where E is the identity matrix. Further, let $P_0, P \in \mathbb{R}^{n \times n}$ be the covariance matrices from Section 2.

Then,

$$q_0 \|\Phi(t)\|_2^2 \leq \|P_x(t) - P\|_2 \leq q_1 \|\Phi(t)\|_2^2, \ t \geq 0 \tag{16}$$

where

$$q_0 = \inf_{\|v\|_2=1} |((P_0 - P)v, v)| \tag{17}$$

and

$$q_1 = \sup_{\|v\|_2=1} |((P_0 - P)v, v)| = \|P_0 - P\|_2 \tag{18}$$

If $P_0 \neq P$, then $q_1 > 0$. If $P_0 - P$ *is regular, then*

$$q_0 = \|(P_0 - P))^{-1}\|_2^{-1} > 0 \tag{19}$$

Proof We obtain Theorem 3 by applying Lemmas 1 and 2 with $\Psi = \Phi^*(t) = \Phi^T(t) = e^{A^T t}$, $\Psi^*(t) = \Phi(t) = e^{At}$ and $C = P_0 - P$.

Further, two-sided bounds can be derived by using Kohaupt (2006, Theorem 8). Thus, there is a constant $\varphi_0 > 0$ and for every $\varepsilon > 0$ a constant $\varphi_{1,\varepsilon} < 0$ such that

$$\varphi_0 \, e^{\nu[A]t} \leq \|\Phi(t)\|_2 \leq \varphi_{1,\varepsilon} \, e^{(\nu[A]+\varepsilon)t}, \; t \geq 0 \tag{20}$$

This leads to the following corollary.

Corollary 4 (Two-sided bounds on $P_x(t) - P = \Phi(t)(P_0 - P)\Phi^T(t)$ based on $\nu[A]$)

Let $A \in \mathbb{R}^{n\times n}$, *let* $\Phi(t) = e^{At}$ *be the associated fundamental matrix with* $\Phi(0) = E$ *where E is the identity matrix. Further, let* $P_0, P \in \mathbb{R}^{n\times n}$ *be the covariance matrices from Section 2.*

Then, there exists a constant $p_0 \geq 0$ *and for every* $\varepsilon > 0$ *a constant* $p_1(\varepsilon) \geq 0$ *such that*

$$p_0 \, e^{2\nu[A]t} \leq \|P_x(t) - P\|_2 \leq p_1(\varepsilon) \, e^{2(\nu[A]+\varepsilon)t}, \; t \geq 0. \tag{21}$$

If $P_0 \neq P$, then $p_1(\varepsilon) > 0$. *If* $P_0 - P$ *is regular, then also* $p_0 > 0$.

Remark Due to the equivalence of norms in finite-dimensional spaces, corresponding bounds as in Theorem 3 and Corollary 4 are valid also in all other (not necessarily multiplicative) matrix norms. Of course, besides the spectral norm $\|\cdot\|_2$, also the Frobenius norm $\|\cdot\|_F = |\cdot|_2$ (cf. Kohaupt, 2003) is of special interest in the context of stochastically excited systems.

6. Local regularity of the function $\|P_x(t) - P\|_2$

We have the following lemma which states - loosely speaking - that for every $t_0 \geq 0$, the function $t \mapsto \|\Delta P_x(t)\|_2 := \|P_x(t) - P\|_2 := \|\Phi(t)(P_0 - P)\Phi^T(t)\|_2$ is real analytic in some right neighborhood $[t_0, t_0 + \Delta t_0]$.

Lemma 5 *(Real analyticity of* $t \mapsto \|P_x(t) - P\|_2$ *on* $[t_0, t_0 + \Delta t_0]$*)*

Let $t_0 \in \mathbb{R}_0^+$. *Then, there exists a number* $\Delta t_0 > 0$ *and a function* $t \mapsto \widehat{\Delta P_x}(t)$, *which is real analytic on* $[t_0, t_0 + \Delta t_0]$ *such that*

$$\widehat{\Delta P_x}(t) = \|\Delta P_x(t)\|_2 = \|P_x(t) - P\|_2 = \|\Phi(t)(P_0 - P)\Phi^T(t)\|_2, \; t \in [t_0, t_0 + \Delta t_0]^{\cdot}$$

Proof Based on $\|\Delta P_x(t)\|_2 = \max\{|\lambda_{max}(\Delta P_x(t))|, |\lambda_{min}(\Delta P_x(t))\}|$, the proof is similar to that of Kohaupt (1999, Lemma 1). The details are left to the reader.

7. Formulas for the norm derivatives $D_+^k \|P_x(t) - P\|_2$, $k = 0, 1, 2$

In this section, in a first step, for complex matrices A and C with $C^* = C$, we define a matrix function $\Psi(t) := \Phi(t)C\Phi^*(t)$ and derive formulas for the right norm derivatives $D_+^k \|\Psi(t)\|_2$ based on the representation $\|\Psi(t)\|_2 = \sup_{\|u\|_2=1} |(\Psi u, u)|$ instead of $\|\Psi(t)\|_2 = \sup_{\|u\|_2=1} \|\Psi u\|_2$. Even though the last one is also valid for $C^* \neq C$, the first one leads to simpler formulas. In a second step, the obtained formulas are employed for $C = P_0 - P \in \mathbb{R}^{n\times n}$ and $A \in \mathbb{R}^{n\times n}$ to deliver the formulas for $D_+^k \|P_x(t) - P\|_2$, $k = 0, 1, 2$.

Let $C \in \mathbb{C}^{n\times n}$ with $C^* = C$. Then, the eigenvalues $\lambda_j(C)$, $j = 1, \cdots, n$ of C are real, and for the spectral norm of C, one has the formula

$$\|C\|_2 = \max_{\|u\|_2=1} |(Cu, u)| = \max_{j=1,\cdots,n} |\lambda_j(C)|$$

and thus

$$\|C\|_2 = \max\{|\lambda_{max}(C)|, |\lambda_{min}(C)|\}$$

Now, let $A \in \mathbb{C}^{n\times n}$, $\Phi(t) = e^{At}$ its fundamental matrix, and define

$$\Psi(t) := \Phi(t)C\Phi^*(t),\ t \geq 0$$

Then, $\Psi(t) \in \mathbb{C}^{n\times n}$ with $\Psi^*(t) = \Psi(t)$ and thus

$$\|\Psi(t)\|_2 = \max\{|\lambda_{max}(\Psi(t))|, |\lambda_{min}(\Psi(t))|\},\ t \geq 0$$

cf. (Achieser & Glasman, 1968, Chapter II.2, p. 62) or (Kantorowitsch & Akilow, 1964, p. 255).

We mention that without $C^* = C$, one would have the formula

$$\|\Psi(t)\|_2 = \sqrt{\lambda_{max}(\Psi^*(t)\Psi(t))},\ t \geq 0$$

Of course, this formula remains valid for $C^* = C$, but is more complicated and probably numerically less good than the first representation of $\|\Psi(t)\|_2$. The computation of $D_+^k \|\Psi(t)\|_2$ by the last formula would be similar as in Kohaupt (2001) for $D_+^k \|\Phi(t)\|_2$.

In order to get a formula for $D_+^k \|\Psi(t)\|_2$ in terms of the given matrices A and C, at the beginning, we follow a similar line as in Kohaupt (2001, Section 3, pp. 6–7), however.

Starting point is the series expansion

$$\Psi(t) := \Phi(t)\, C\, \Phi^*(t) = \sum_{j=0}^{\infty} \Phi(t_0)\, B_j\, \Phi^*(t_0)\, \frac{(t-t_0)^j}{j!}$$

with

$$B_j = \sum_{k=0}^{j} \binom{j}{k} A^{*j-k}\, C\, A^k,$$

$j = 0, 1, 2, \cdots$. Thus, e.g.

$$B_0 = C$$
$$B_1 = A\,C + C\,A^*$$
$$B_2 = A^2\, C + 2\,A\,C\,A^* + C\,A^{*2}$$

Consequently,

$$\Psi(t) = T^{(0)} + T^{(1)}(t - t_0) + T^{(2)}(t - t_0)^2 + \cdots$$

with

$$T^{(0)} = \Phi(t_0) B_0 \Phi^*(t_0),$$
$$T^{(1)} = \Phi(t_0) B_1 \Phi^*(t_0),$$
$$T^{(2)} = \Phi(t_0) (\tfrac{1}{2} B_2) \Phi^*(t_0)$$

Then, due to Kato (1966, Theorem 5.11, Chapter II, pp. 115–116) and Kohaupt (1999, Lemma 2.1),

$$\lambda_{max}(\Psi(t)) = v_{0,max} + v_{1,max}(t - t_0) + v_{2,max}(t - t_0)^2 + \cdots, \quad t_0 \leq t \leq t_0 + \Delta t_0$$

where the quantities $v_{j,max}, j = 0, 1, 2$ are given by the formulas for $v_j, j = 0, 1, 2$ in Kohaupt (2001, pp. 6–7) with the operators $T^{(0)}$, $T^{(1)}$, $T^{(2)}$ defined above. This is shortly recapitulated in Appendix A3.

The series expansion

$$\lambda_{\min}(\Psi(t)) = v_{0,\min} + v_{1,\min}(t - t_0) + v_{2,\min}(t - t_0)^2 + \cdots, \quad t_0 \leq t \leq t_0 + \Delta t_0$$

is obtained via the formula

$$\lambda_{\min}(\Psi(t)) = -\lambda_{max}(-\Psi(t))$$

Now, define

$$s(t) := \begin{bmatrix} s_1(t) \\ s_2(t) \end{bmatrix} := \begin{bmatrix} \lambda_{max}(\Psi(t)) \\ \lambda_{\min}(\Psi(t)) \end{bmatrix}$$

Then,

$$s(t) = s(t_0) + Ds(t_0)(t - t_0) + D^2 s(t_0)\frac{(t - t_0)^2}{2} + \cdots, \quad t_0 \leq t \leq t_0 + \Delta t_0$$

with

$$s(t_0) := \begin{bmatrix} v_{0,max} \\ v_{0,\min} \end{bmatrix}$$

$$Ds(t_0) := \begin{bmatrix} v_{1,max} \\ v_{1,\min} \end{bmatrix}$$

$$D^2 s(t_0) := \begin{bmatrix} 2\, v_{2,max} \\ 2\, v_{2,\min} \end{bmatrix}$$

and

$$\|\Psi(t_0)\|_2 = \|s(t_0)\|_\infty$$

Hence, for fixed $t_0 \in \mathbb{R}_0^+$,

$$D_+^k \|\Psi(t_0)\|_2 = D_+^k \|s(t_0)\|_\infty, \; k = 0, 1, 2$$

where the norm derivatives $D_+^k\|s(t_0)\|_\infty$, $k = 0,1,2$ are obtained by the formulas of Kohaupt (2002, Theorem 6). This is shortly recapitulated in Appendix A4.

If one replaces t_0 by t, then one gets the functions $t \mapsto D_+^k\|\Psi(t)\|_2 = D_+^k\|s(t)\|_\infty$, $k = 0,1,2$.

The norm derivatives $D_+^k\|P_x(t) - P\|_2$, $k = 0,1,2$ are obtained as the special case $C \in \mathbb{R}^{n\times n}$ with $C = P_0 - P$ and $A \in \mathbb{R}^{n\times n}$.

These formulas are needed in Section 8.

8. Applications

In this section, we apply the new two-sided bounds on $\|P_x(t) - P\|_2$ obtained in Section 5 as well as the differential calculus of norms developed in Sections 6 and 7 to a linear stochastic vibration model with asymptotically stable system matrix and white noise excitation vector.

In Section 8.1, the stochastic vibration model as well as its state-space form is given, and in Section 8.2 the data are chosen. In Section 8.3, computations with the specified data are carried out, such as the computation of P and $P_0 - P$ as well as the computation of the curves $y = D_+^k\|P_x(t) - P\|_2$, $k = 0,1,2$ and of the curve $y = \|P_x(t) - P\|_2$ along with its best upper and lower bounds. In Section 8.4, computational aspects are shortly discussed.

8.1. The stochastic vibration model and its state-space form

Consider the multi-mass vibration model in Figure 1.

The associated initial-value problem is given by

$$M\ddot{y} + B\dot{y} + Ky = f(t), \quad y(0) = y_0, \ \dot{y}(0) = \dot{y}_0$$

where $y = [y_1, \cdots, y_n]^T$ and $f(t) = [f_1(t), \cdots, f_n(t)]^T$ as well as

$$M = \begin{bmatrix} m_1 & & & & \\ & m_2 & & & \\ & & m_3 & & \\ & & & \ddots & \\ & & & & m_n \end{bmatrix}$$

$$B = \begin{bmatrix} b_1 + b_2 & -b_2 & & & & \\ -b_2 & b_2 + b_3 & -b_3 & & & \\ & -b_3 & b_3 + b_4 & -b_4 & & \\ & \ddots & \ddots & \ddots & & \\ & & -b_{n-1} & b_{n-1} + b_n & -b_n \\ & & & -b_n & b_n + b_{n+1} \end{bmatrix}$$

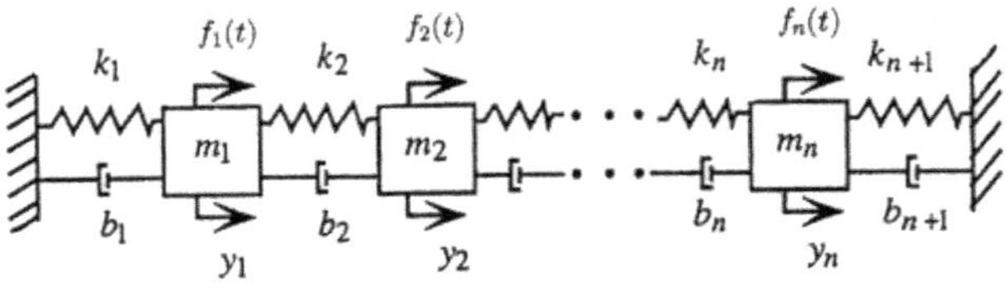

Figure 1. Multi-mass vibration model.

$$K = \begin{bmatrix} k_1+k_2 & -k_2 & & & & \\ -k_2 & k_2+k_3 & -k_3 & & & \\ & -k_3 & k_3+k_4 & -k_4 & & \\ & \ddots & \ddots & \ddots & & \\ & & -k_{n-1} & k_{n-1}+k_n & -k_n \\ & & & -k_n & k_n+k_{n+1} \end{bmatrix}$$

Here, y is the displacement vector, $f(t)$ the applied force, and M, B, and K are the mass, damping, and stiffness matrices, as the case may be. Matrix M is regular.

In the state-space description, one obtains

$$\dot{x}(t) = A\,x(t) + b(t), \quad x(0) = x_0$$

with $x = [y^T, z^T]^T$, $z = \dot{y}$, and $x_0 = [y_0^T, z_0^T]^T$, $z_0 = \dot{y}_0$,

where the initial vector $x_0 = [y_0^T, z_0^T]^T$ is characterized by the *mean vector* m_0 and the *covariance matrix* P_0.

The *system matrix* A and the *excitation vector* $b(t)$ are given by

$$A = \left[\begin{array}{c|c} 0 & E \\ \hline -M^{-1}K & -M^{-1}B \end{array}\right], \qquad b(t) = \left[\begin{array}{c} 0 \\ \hline M^{-1}f(t) \end{array}\right]$$

respectively. The vector $x(t)$ is called *state vector*.

The (symmetric positive semi-definite) intensity matrix $Q = Q_b$ is obtained from the (symmetric positive semi-definite) intensity matrix Q_f by

$$Q = Q_b = \left[\begin{array}{c|c} 0 & 0 \\ \hline 0 & M^{-1}Q_f M^{-1} \end{array}\right]$$

(see Müller, 1976, Formulas (9.65)) and the derivation of this relation in Appendix A5.

8.2. Data for the model

As of now, we specify the values as

$$m_j = 1, j = 1, \cdots, n$$
$$k_j = 1, j = 1, \cdots, n+1$$

and

$$b_j = \begin{cases} 1/2, j\, even \\ 1/4, j\, odd \end{cases}$$

Then,

$$M = E,$$

$$B = \begin{bmatrix} \frac{3}{4} & -\frac{1}{2} & & & & \\ -\frac{1}{2} & \frac{3}{4} & -\frac{1}{4} & & & \\ & -\frac{1}{4} & \frac{3}{4} & -\frac{1}{2} & & \\ & & \ddots & \ddots & \ddots & \\ & & -\frac{1}{4} & \frac{3}{4} & -\frac{1}{2} \\ & & & -\frac{1}{2} & \frac{3}{4} \end{bmatrix}$$

(if n is even), and

$$K = \begin{bmatrix} 2 & -1 & & & & \\ -1 & 2 & -1 & & & \\ & -1 & 2 & -1 & & \\ & & \ddots & \ddots & \ddots & \\ & & & -1 & 2 & -1 \\ & & & & -1 & 2 \end{bmatrix}$$

We choose n=5 in this paper so that the state-space vector $x(t)$ has the dimension $m = 2n = 10$.

Remark In Sections 2–7, we have denoted the dimension of $x(t)$ by n. From the context, the actual dimension should be clear.

For m_0, we take

$$m_0 = [m_{y_0}^T, m_{z_0}^T]^T$$

with

$$m_{y_0} = [-1, 1, -1, 1, -1]^T$$

and

$$m_{z_0} = \begin{cases} [0,0,0,0,0,0]^T & \text{(Case I)} \\ [-1,-1,-1,-1,-1]^T & \text{(Case II)} \end{cases}$$

similarly as in Kohaupt (2002) for y_0 and $\dot{y}_0$. For the 10×10 matrix P_0, we choose

$$P_0 = 0.01\, E$$

The *white-noise force vector* $f(t)$ is specified as

$$f(t) = [0, \cdots, 0; f_n(t)]^T$$

so that its intensity matrix $Q_f \in \mathbb{R}^{n \times n}$ with $q_{f,nn} = :q$ has the form

$$Q_f = \left[\begin{array}{c|c} 0 & 0 \\ \hline 0 & q_{f,nn} \end{array}\right] = \left[\begin{array}{c|c} 0 & 0 \\ \hline 0 & q \end{array}\right] = q \left[\begin{array}{c|c} 0 & 0 \\ \hline 0 & 1 \end{array}\right] = : q\, E^{(n)}$$

We choose

$$q = 0.01$$

With $M = E$, this leads to (see Appendix A5)

$$Q = Q_b = \left[\begin{array}{c|c} 0 & 0 \\ \hline 0 & Q_f \end{array}\right] = \left[\begin{array}{c|c} 0 & 0 \\ \hline 0 & q\, E^{(n)} \end{array}\right] = \left[\begin{array}{c|c} 0 & 0 \\ \hline 0 & q \end{array}\right] \in R^{m \times m}$$

In the Lyapunov equation $BX + XA = C$ of Section 2, we employ the replacements

$$B \to A, \qquad A \to A^T, \qquad C \to -Q$$

to obtain the limiting covariance matrix $X = P = \lim_{t \to \infty} P_x(t)$.

8.3. Computations with the specified data

(i) *Bounds on $y = \Phi(t)m_0$ in the vector norm $\|\cdot\|_2$*

Upper bounds on $y = \Phi(t)m_0$ in the vector norm $\|\cdot\|_2$ for the two cases (I) and (II) of m_0 are already given in Kohaupt (2002, Figures 2 and 3). There, we had a deterministic problem with $f(t) = 0$ and the solution vector $x(t) = \Phi(t)\,x_0$, where x_0 there had the same data as m_0 here. We mention that for the specified data, $\nu_{m_0}[A] = \nu[A]$ in both cases (cf. Kohaupt, 2006, p. 154) for a method to prove this. For the sake of brevity, we do not compute or plot the lower bounds and thus the two-sided bounds, but leave this to the reader.

(ii) *Computation of P and $P_0 - P$ as well as of their associated eigenproblems*

With the data of Section 2, we obtain

$$P = \begin{bmatrix} 0.1624 & 0.2497 & 0.2347 & 0.1645 & 0.0809 & -0.0000 & -0.0084 & -0.0249 & -0.0312 & -0.0058 \\ 0.2497 & 0.4034 & 0.4129 & 0.3134 & 0.1525 & 0.0084 & -0.0000 & -0.0421 & -0.0701 & -0.0260 \\ 0.2347 & 0.4129 & 0.4977 & 0.4453 & 0.2305 & 0.0249 & 0.0421 & 0.0000 & -0.0706 & -0.0821 \\ 0.1645 & 0.3134 & 0.4453 & 0.4826 & 0.3205 & 0.0312 & 0.0701 & 0.0706 & -0.0000 & -0.1426 \\ 0.0809 & 0.1525 & 0.2305 & 0.3205 & 0.4249 & 0.0058 & 0.0260 & 0.0821 & 0.1426 & -0.0000 \\ 0.0000 & 0.0084 & 0.0249 & 0.0312 & 0.0058 & 0.0793 & 0.1023 & 0.0541 & 0.0040 & 0.0007 \\ -0.0084 & 0.0000 & 0.0421 & 0.0701 & 0.0260 & 0.1023 & 0.1506 & 0.1124 & 0.0363 & -0.0103 \\ -0.0249 & -0.0421 & -0.0000 & 0.0706 & 0.0821 & 0.0541 & 0.1124 & 0.1621 & 0.1300 & -0.0282 \\ -0.0312 & -0.0701 & -0.0706 & -0.0000 & 0.1426 & 0.0040 & 0.0363 & 0.1300 & 0.1998 & 0.0514 \\ -0.0058 & -0.0260 & -0.0821 & -0.1426 & 0.0000 & 0.0007 & -0.0103 & -0.0282 & 0.0514 & 0.4937 \end{bmatrix}$$

The column vector of eigenvalues Λ_P and the modal matrix X_P, that is, the matrix whose columns are made up of the eigenvectors, are computed as

$$\Lambda_P = \begin{bmatrix} 1.54687260865985 \\ 0.55390744777884 \\ 0.50669339270302 \\ 0.26282228387153 \\ 0.12187876942246 \\ 0.05658235034695 \\ 0.00625573952218 \\ 0.00160497361429 \\ 0.00001727972012 \\ 0.00002903815291 \end{bmatrix}$$

and

$$X_P = \begin{bmatrix} -0.2584 & -0.1685 & -0.1691 & -0.0346 & -0.4563 & 0.1911 & -0.4245 & 0.1102 & 0.6068 & 0.2642 \\ -0.4509 & -0.2648 & -0.2304 & -0.0776 & -0.3978 & 0.2037 & 0.0009 & 0.0373 & -0.6809 & -0.0524 \\ -0.5517 & -0.1522 & -0.0849 & -0.1348 & 0.0842 & -0.2420 & 0.4392 & -0.4102 & 0.3492 & -0.3135 \\ -0.5321 & 0.1348 & 0.1400 & -0.0065 & 0.3553 & -0.3934 & -0.1093 & 0.5373 & -0.0662 & 0.3050 \\ -0.3465 & 0.5405 & -0.0388 & 0.4707 & 0.2111 & 0.4732 & -0.1727 & -0.2410 & -0.0095 & -0.0904 \\ -0.0295 & 0.1043 & 0.0511 & -0.4547 & 0.1232 & 0.4945 & 0.1521 & 0.5148 & 0.1529 & -0.4557 \\ -0.0537 & 0.2292 & 0.1198 & -0.6320 & 0.1107 & 0.2326 & 0.0712 & -0.3678 & -0.0806 & 0.5636 \\ -0.0354 & 0.4144 & 0.1863 & -0.3279 & -0.2773 & -0.3810 & -0.5002 & -0.1605 & -0.0923 & -0.4223 \\ 0.0217 & 0.5263 & 0.0116 & 0.1227 & -0.5647 & -0.1168 & 0.5517 & 0.2068 & 0.0581 & 0.1558 \\ 0.1301 & 0.2364 & -0.9155 & -0.1559 & 0.1798 & -0.1653 & -0.0489 & 0.0513 & -0.0029 & 0.0083 \end{bmatrix}$$

showing that P is positive definite. Correspondingly,

$$\Lambda_{P_0-P} = \begin{bmatrix} -0.54687260865985 \\ 0.44609255222115 \\ 0.49330660729698 \\ 0.73717771612847 \\ 0.87812123057754 \\ 0.94341764965305 \\ 0.99374426047782 \\ 0.99839502638571 \\ 0.99998272027988 \\ 0.99997096184709 \end{bmatrix}$$

and

$$X_{P_0-P} = \begin{bmatrix} 0.2584 & 0.1685 & 0.1691 & 0.0346 & 0.4563 & -0.1911 & 0.4245 & 0.1102 & -0.6068 & 0.2642 \\ 0.4509 & 0.2648 & 0.2304 & 0.0776 & 0.3978 & -0.2037 & -0.0009 & 0.0373 & 0.6809 & -0.0524 \\ 0.5517 & 0.1522 & 0.0849 & 0.1348 & -0.0842 & 0.2420 & -0.4392 & -0.4102 & -0.3492 & -0.3135 \\ 0.5321 & -0.1348 & -0.1400 & 0.0065 & -0.3553 & 0.3934 & 0.1093 & 0.5373 & 0.0662 & 0.3050 \\ 0.3465 & -0.5405 & 0.0388 & -0.4707 & -0.2111 & -0.4732 & 0.1727 & -0.2410 & 0.0095 & -0.0904 \\ 0.0295 & -0.1043 & -0.0511 & 0.4547 & -0.1232 & -0.4945 & -0.1521 & 0.5148 & -0.1529 & -0.4557 \\ 0.0537 & -0.2292 & -0.1198 & 0.6320 & -0.1107 & -0.2326 & -0.0712 & -0.3678 & 0.0806 & 0.5636 \\ 0.0354 & -0.4144 & -0.1863 & 0.3279 & 0.2773 & 0.3810 & 0.5002 & -0.1605 & 0.0923 & -0.4223 \\ -0.0217 & -0.5263 & -0.0116 & -0.1227 & 0.5647 & 0.1168 & -0.5517 & 0.2068 & -0.0581 & 0.1558 \\ -0.1301 & -0.2364 & 0.9155 & 0.1559 & -0.1798 & 0.1653 & 0.0489 & 0.0513 & 0.0029 & 0.0083 \end{bmatrix}$$

showing that $P_0 - P$ is symmetric and regular (but not positive definite). Matrix $P_0 - P$ is needed to compute the curve $y = \|P_x(t) - P\|_2 = \|\Phi(t)(P_0 - P)\Phi^T(t)\|_2$.

(iii) *Computation of the curves* $y = D_+^k \|P_x(t) - P\|_2 = D_+^k \|\Phi(t)(P_0 - P)\Phi^T(t)\|_2, k = 0, 1, 2$

The computation of $y = D_+^k \|P_x(t) - P\|_2, k = 0, 1, 2$ for the given data is done according to Section 7 with $C = P_0 - P$. The pertinent curves are illustrated in Figures 2-4. By inspection, there are no kinks (like in the curve $y = \sqrt{t}$ at $t = 0$) so that $D_+^k \|P_x(t) - P\|_2 = D^k \|P_x(t) - P\|_2 = \frac{d^k}{dt^k} \|P_x(t) - P\|_2$, $k = 0, 1, 2$. For some details on the computation of $D_+^k \|P_x(t) - P\|_2$, $k = 1, 2$, see Appendix A6.

We have checked the results numerically by difference quotients. More precisely, setting

$$\Delta P_x(t) := P_x(t) - P, \; t > 0$$

and

$$g(t) := \|\Delta P_x(t)\|_2 = \|P_x(t) - P\|_2, \; t > 0$$

we have investigated the approximations

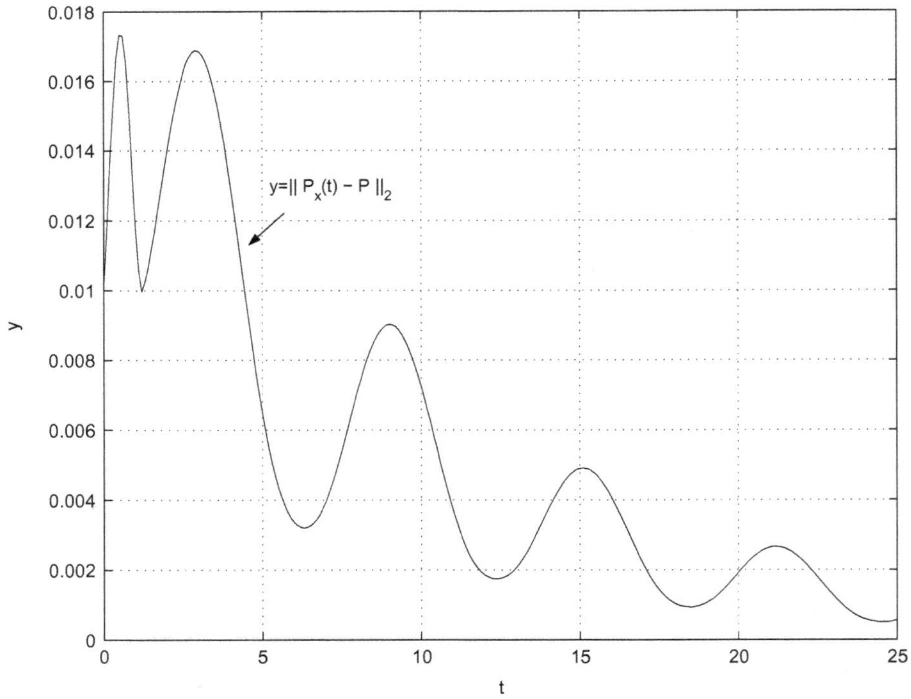

Figure 2. Curve $y = \|P_x(t) - P\|_2$, $0 \le t \le 25$, $\Delta t = 0.1$.

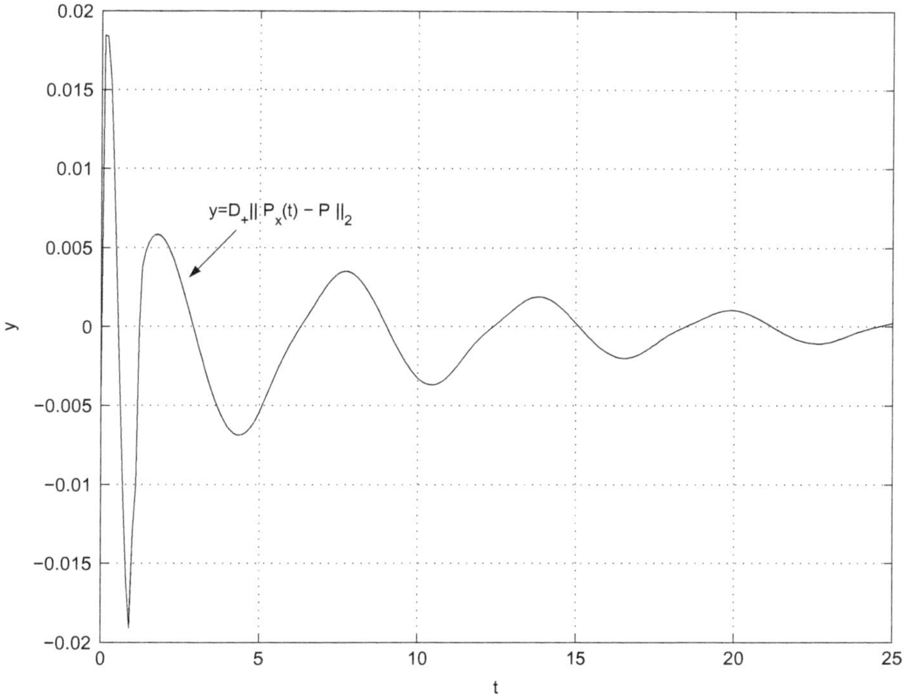

Figure 3. Right norm derivative $y = D_+ \|P_x(t) - P\|_2$, $0 \le t \le 25$, $\Delta t = 0.1$.

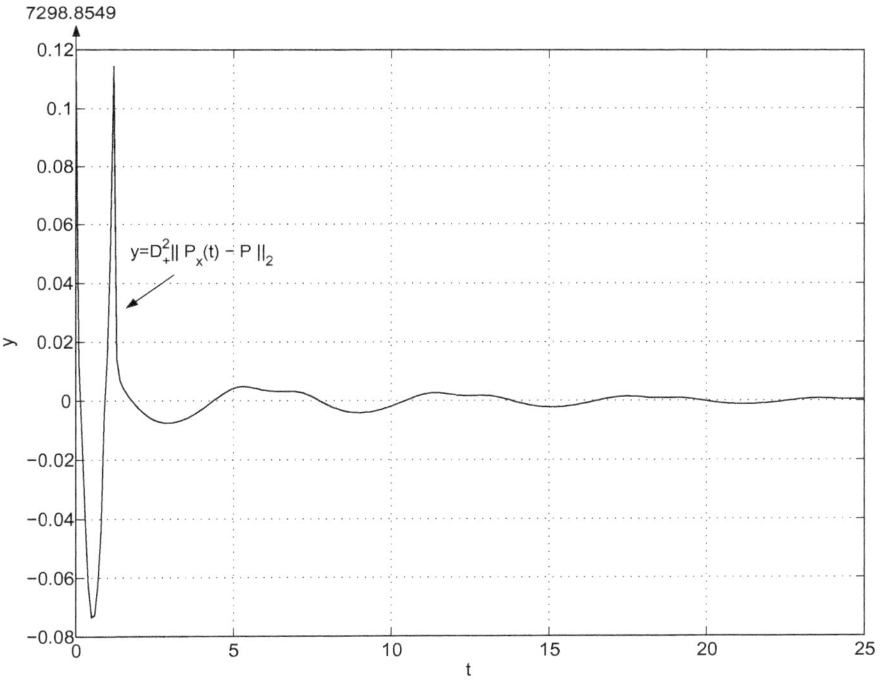

Figure 4. Second right norm derivative $y = D_+ \|P_x(t) - P\|_2$, $0 \le t \le 25$, $\Delta t = 0.1$.

$$\delta_h g(t) := \frac{g(t+h) - g(t-h)}{2h} \approx Dg(t),\ t - h \ge 0$$

and

$$\delta_{\frac{h}{2}}^2 g(t) := \delta_{\frac{h}{2}} (\delta_{\frac{h}{2}} g(t)) = \frac{g(t+h) - 2g(t) + g(t-h)}{h^2} \approx D^2 g(t),\ t - h \ge 0$$

as well as

$$\delta_h Dg(t) := \frac{Dg(t+h) - Dg(t-h)}{2h} \approx D^2 g(t),\ t - h \geq 0$$

For

$$\begin{aligned} t &= 2.5 \\ h &= 10^{-5} \end{aligned}$$

we obtain

$$\begin{aligned} D\,g(t) &= D\,\|P_x(t) - P\|_2 &&= 0.00304667381090 \\ \delta_h g(t) &= \delta_h\,\|P_x(t) - P\|_2 &&= 0.00304667381375 \end{aligned}$$

as well as

$$\begin{aligned} D^2\,g(t) &= D^2\,\|P_x(t) - P\|_2 &&= -0.00667087489483 \\ \delta^2_{\frac{h}{2}} g(t) &= \delta^2_{\frac{h}{2}}\,\|P_x(t) - P\|_2 &&= -0.00667393224019 \end{aligned}$$

and

$$\begin{aligned} D^2\,g(t) &= D^2\,\|P_x(t) - P\|_2 &&= -0.00667087489483 \\ \delta_h D\,g(t) &= \delta_h\,D\,\|P_x(t) - P\|_2 &&= -0.00667087489555 \end{aligned}$$

so that the computational results for $y = D_+^k\,\|P_x(t) - P\|_2 = D^k\,\|P_x(t) - P\|_2 = \frac{d^k}{dt^k}\,\|P_x(t) - P\|_2$, $k = 0, 1, 2$ with $t = 2.5$ are well underpinned by the difference quotients. As we see, the approximation of $D^2\,g(t) = D^2\,\|P_x(t) - P\|_2$ by $\delta_h D\,g(t)$ is much better than by $\delta^2_{\frac{h}{2}} g(t)$, which was to be expected, of course.

(iv) *Bounds on* $y = P_x(t) - P = \Phi(t)\,(P_0 - P)\,\Phi^T(t)$ *in the spectral norm* $\|\cdot\|_2$

Let $\alpha := \nu[A]$ be the spectral abscissa of the system matrix A. With the given data, we obtain

$$\alpha := \nu[A] = -0.05023936121946$$

so that the system matrix A is asymptotically stable.

The upper bound on $y = \|P_x(t) - P\|_2 = \|\Phi(t)\,(P_0 - P)\,\Phi^T(t)\|_2$ is given by $y = p_1(\varepsilon)e^{2(\alpha+\varepsilon)t}, t \geq 0$. Here, $\varepsilon = 0$ can be chosen since matrix A is diagonalizable. But, in the programs, we have chosen the machine precision $\varepsilon = eps = 2^{-52} \doteq 2.2204 \times 10^{-16}$ of MATLAB in order not to be bothered by this question.

With $\varphi_{1,\varepsilon}(t) := p_1(\varepsilon)e^{2(\alpha+\varepsilon)t}$, $t \geq 0$, the optimal constant $p_1(\varepsilon)$ in the upper bound is obtained by the two conditions

$$\begin{aligned} \|P_x(t_c) - P\|_2 &= \varphi_{1,\varepsilon}(t_c) = p_1(\varepsilon)e^{2(\alpha+\varepsilon)t_c} \\ D_+\,\|P_x(t_c) - P\|_2 &= \varphi'_{1,\varepsilon}(t_c) = 2(\alpha+\varepsilon)\,\varphi_{1,\varepsilon}(t_c) \end{aligned}$$

where t_c is the place of contact between the curves.

This is a system of two non-linear equations in the two unknowns t_c and $p_1(\varepsilon)$. By eliminating $\varphi_{1,\varepsilon}(t_c)$, this system is reduced to the determination of the zero of

$$D_+\,\|P_x(t_c) - P\|_2 - 2(\alpha+\varepsilon)\,\|P_x(t_c) - P\|_2 = 0$$

which is a single non-linear equation in the single unknown t_c. For this, MATLAB routine *fsolve* was used.

After t_c has been computed from the above equation, the best constant $p_1(\varepsilon)$ is obtained from

$$p_1(\varepsilon) = \|P_x(t_c) - P\|_2\, e^{-2(\alpha+\varepsilon)t_c}$$

From the initial guess $t_{c,0} = 3.0$, the computations deliver the values

$$\begin{aligned} t_c &= 3.14231573176783, \\ p_1(\varepsilon) &= 0.02288922631729 \end{aligned}$$

In a similar way, in the lower bound $y = p_0 e^{2\alpha t}$, we compute the best constant p_0 and the place of contact t_s. For the initial guess $t_{s,0} = 6.0$, the results are

$$\begin{aligned} t_s &= 6.20977038583445, \\ p_0 &= 0.00600530767800 \end{aligned}$$

The curve $y = \|P_x(t_c) - P\|_2$ along with the best upper and lower bounds is illustrated in Figure 5.

(v) *Applicability of the second norm derivative* $D_+^2 \|P_x(t) - P\|_2$

The first norm derivative $D_+ \|P_x(t) - P\|_2$ was employed in Point (iv). Apart from this, it can be applied to determine the *relative extrema* of the curve $y = \|P_x(t_c) - P\|_2$.

The second norm derivative $D_+^2 \|P_x(t) - P\|_2$ can be used to compute the *inflexion points*. The details are left to the reader.

8.4. Computational aspects

In this subsection, we say something about the computer equipment and the computation time for some operations.

(i) As to the *computer equipment*, the following hardware was available: an Intel Pentium D (3.20 GHz, 800 MHz Front-Side-Bus, 2x2MB DDR2-SDRAM with 533 MHz high-speed memories). As software package, we used MATLAB, Version 6.5.

(ii) The *computation time* t of an operation was determined by the command sequence $t_i = clock; operation; t = etime(clock, t_i)$; it is put out in seconds rounded to two decimal places by Matlab. For the computation of the eigenvalues of matrix A, we used the command $[XA, DA] = eig(A)$; the pertinent computation time is less than $0.01s$. To determine $\Phi(t) = e^{At}$, we employed Matlab routine $expm$. For the computation of the 251 values t, y, yu, yl in Figure 5, it took t(Table for Figure 5)$= 0.83$ s. Here, t stands for the time value running from $t_0 = 0$ to $t_e = 25$ with stepsize $\Delta t = 0.1$; y stands for the value of $\|P_x(t) - P\|_2$, yu for the value of the best upper bound $p_1(\varepsilon)e^{2(\alpha+\varepsilon)t}$ and yl for the value of the best lower bound $p_0 e^{2\alpha t}$.

9. Conclusion

In the present paper, linear stochastic vibration systems of the form $\dot{x}(t) = Ax(t) + b(t)$, $x(0) = x_0$, are investigated driven by white noise $b(t)$. If the system matrix A is asymptotically stable, then the mean vector $m_x(t)$ and the covariance matrix $P_x(t)$ both converge with $m_x(t) \to 0\ (t \to \infty)$ and $P_x(t) \to P\ (t \to \infty)$ for some symmetric positive (semi-)definite matrix P. This raises the question of the asymptotic behavior of both quantities. The pertinent investigations are made in the Euclidean norm $\|\cdot\|_2$ for $m_x(t)$ and in the spectral norm, also denoted by $\|\cdot\|_2$, for $P_x(t) - P$. The main new points are the derivation of two-sided bounds on both quantities, the derivation of the right norm derivatives $D_+^k \|P_x(t) - P\|_2$, $k = 0, 1, 2$ and, as application, the computation of the best constants in the bounds. Since we have used a new way to determine the norm derivatives $D_+^k \|P_x(t) - P\|_2$, $k = 0, 1, 2$, we have checked the obtained formulas by various difference quotients. They underpin the correctness of the numerical values for the specified data.

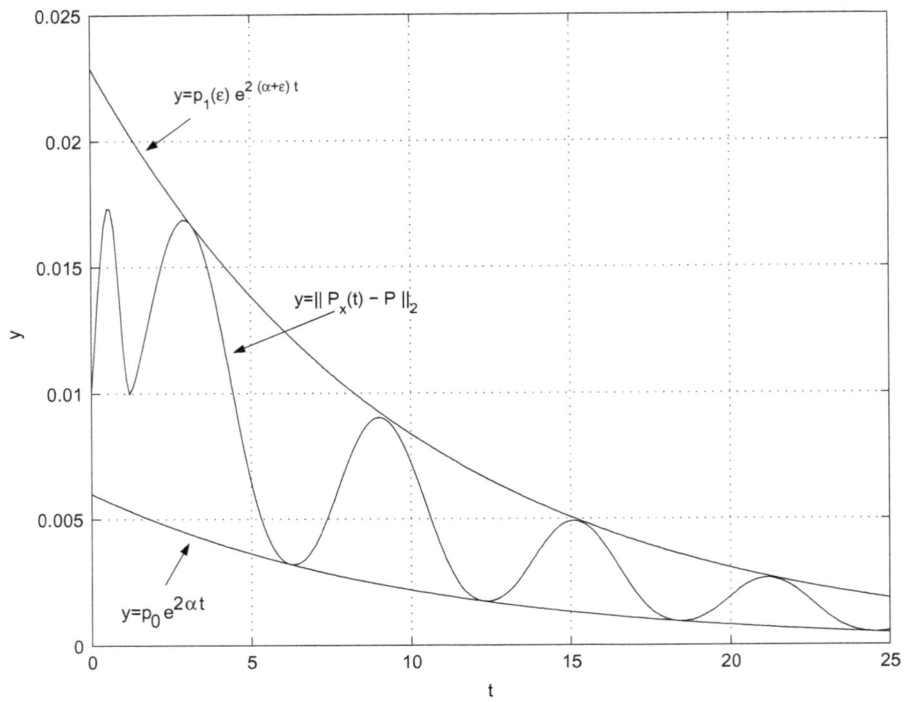

Figure 5. $y = \|P_x(t) - P\|_2$ **along with the best upper and lower bounds.**

It is reminded that the original system consists of a multi-mass vibration model with damping and white noise force excitation. By a standard method, it is cast into state-space form.

As illustration of the results, the curves $y = D_+^k \|P_x(t) - P\|_2$, $k = 0, 1, 2$ are plotted as well as the curve $y = \|P_x(t) - P\|_2$ together with the best two-sided bounds.

The computation time to generate the last figure with a 10×10 matrix A is less than a second. Of course, in engineering practice, much larger models occur. As in earlier papers, we mention that in this case engineers usually employ a method called *condensation* to reduce the size of the matrices.

We have added an Appendix to exhibit more details on some items in order to make the paper easier to comprehend.

The numerical values were given in order that the reader can check the results.

Altogether, the results of the paper should be of interest to applied mathematicians and particularly to engineers.

Funding
The authors received no direct funding for this research.

Author details
Ludwig Kohaupt[1]
E-mail: kohaupt@beuth-hochschule.de
[1] Department of Mathematics, Beuth University of Technology Berlin, Luxemburger Str. 10, D-13353, Berlin, Germany.

References

Achieser, N. I., & Glasman, I. M. (1968). *Theorie der linearen Operatoren im Hilbert-Raum* [Theory of linear operators in Hilbert space]. Berlin: Akademie-Verlag.
Heuser, H. (1975). *Funktionalanalysis* [Functional analysis]. Stuttgart: B.G. Teubner.
Kantorowitsch, L. W., & Akilow, G. P. (1964). *Funktionalanalysis in normierten Räumen* [Functional analysis in normed linear spaces]. Berlin: Akademie-Verlag. (German translation of the Russian Original).
Kato, T. (1966). *Perturbation theory for linear operators*. New York: Springer.
Kohaupt, L. (1999). Second logarithmic derivative of a complex matrix in the Chebyshev norm. *SIAM Journal on Matrix Analysis and Applications, 21*, 382–389.

Kohaupt, L. (2001). Differential calculus for some p-norms of the fundamental matrix with applications. *Journal of Computational and Applied Mathematics, 135*, 1-21.
Kohaupt, L. (2002). Differential calculus for p-norms of complex-valued vector functions with applications. *Journal of Computational and Applied Mathematics, 145*, 425–457.
Kohaupt, L. (2006). Computation of optimal two-sided bounds for the asymptotic behavior of free linear dynamical systems with application of the differential calculus of norms. *Journal of Computational Mathematics and Optimization, 2*, 127–173.
Ma, E.-Ch. (1966). A finite series solution of the matrix equation *AX - XB = C*. *SIAM Journal on Applied Mathematics, 14*, 490–495.
Müller, P. C., & Schiehlen, W. O. (1976). *Lineare Schwingungen* [Linear Vibrations]. Wiesbaden: Akademische Verlagsgesellschaft.
Niemeyer, H., & Wermuth, E. (1987). *Lineare Algebra* [Linear algebra]. Braunschweig Wiesbaden: Vieweg.
Stummel, F., & Hainer, K. (1982). *Praktische Mathematik* [Introduction to numerical analysis]. Stuttgart: B.G. Teubner.
Taylor, A. E. (1958). *Introduction to functional analysis*. New York: Wiley.

Appendix A

In this Appendix, we show more details on some items in order to make this paper more easily understandable especially for engineers and generally for a broader readership.

A1. Solution of the Lyapunov matrix equation $BX + XA = C$ by a method of Ma

In this section, we restrict ourselves to the case of diagonalizable matrices B and A since we need only this case in Section 8. The treatment of the general case can be found in Ma 1966.

Let $A \in \mathbb{C}^{n\times n}$, $B \in \mathbb{C}^{m\times m}$, and $C \in \mathbb{C}^{m\times n}$. The *problem* is to find the solution matrix $X \in \mathbb{C}^{m\times n}$ such that

$$BX + XA = C$$

We *suppose* that matrices A and B both be *diagonalizable* and that the eigenvalues $\lambda_i(A)$, $i = 1, \cdots, n$ and $\mu_j(B)$, $j = 1, \cdots, m$ *satisfy the condition*

$$\mu_j(B) + \lambda_k(A) \neq 0; k = 1, \cdots, n; j = 1, \cdots, m$$

Then, the *solution* of the equation $BX + XA = C$ can be obtained as follows.

Since A and B are diagonalizable, there exist regular matrices $U \in \mathbb{C}^{n\times n}$ and $V \in \mathbb{C}^{m\times m}$ such that

$$\tilde{A} = U^{-1} A\, U,\ \tilde{B} = V^{-1} B\, V$$

where

$$\tilde{A} = diag(\lambda_k(A))_{k=1,\cdots,n},\ \tilde{B} = diag(\mu_j(B))_{j=1,\cdots,m}$$

Define

$$\tilde{X} = V^{-1} X\, U,\ \tilde{C} = V^{-1} C\, U$$

Then, $BX + XA = C$ can be written as

$$(V^{-1} B\, V)\,(V^{-1} X\, U) + (V^{-1} X\, U)\,(U^{-1} A\, U) = V^{-1} C\, U$$

or

$$\tilde{B}\tilde{X} + \tilde{X}\tilde{B} = \tilde{C}$$

Its solution $\tilde{X} = (\tilde{X}_{jk})$ is given by

$$\tilde{X}_{jk} = \frac{\tilde{C}_{jk}}{\mu_j + \lambda_k}, j = 1, \cdots, m; k = 1, \cdots, n$$

From this, we obtain the solution of the original matrix equation $BX + XA = C$ by the relation

$$X = V\tilde{X}U^{-1}$$

Remarks (i) The solution X could depend on the transformation matrices U and V. But, it is actually independent of these matrices since the mapping $L(X) := BX + XA : \mathbb{C}^{m\times n} \mapsto \mathbb{C}^{m\times n}$ is injective. Therefore, the solution X is uniquely determined.

(ii) If $B = A^*$ and $C = C^*$, then $X = X^*$ and X is even positive semi-definite. The proof is left to the reader.

(iii) If $A \in R^{n\times n}$ and $B = A^T$ as well as $C \in R^{n\times n}$ with $C = C^T$, then $X = X^T$ and X is positive semi-definite.

A2. Proof of Lemma 2

In this section, we follow closely the line of Heuser to carry over the proof of $\sup_{\|v\|_2=1} |(Cv, v)| = \sup_{\|v\|_2=1} \|Cv\|_2 = \|C\|_2$ to derive a corresponding relation for $\inf_{\|v\|_2=1} |(Cv, v)|\|_2$ for $C \in \mathbb{C}^{n\times n}$ with $C = C^*$ (see Heuser, 1975, pp. 277–278 resp. Theorem 6.8.5).

So, let $\|\cdot\| = \|\cdot\|_2$ be the common vector norm resp. the spectral matrix norm.

Let $\lambda > 0$ be chosen arbitrarily. Then,

$$4\,\|Cx\|_2^2 = (C(\lambda x + \frac{1}{\lambda}Cx), \lambda x + \frac{1}{\lambda}Cx) - (C(\lambda x - \frac{1}{\lambda}Cx), \lambda x - \frac{1}{\lambda}Cx),\ x \in \mathbb{C}^n$$

which is proven by simplifying the right member of the equation. This entails

$$4\,\|Cx\|^2 \le |(C(\lambda x + \frac{1}{\lambda}Cx), \lambda x + \frac{1}{\lambda}Cx)| + |(C(\lambda x - \frac{1}{\lambda}Cx), \lambda x - \frac{1}{\lambda}Cx)|$$
$$= \frac{|(C(\lambda x + \frac{1}{\lambda}Cx), \lambda x + \frac{1}{\lambda}Cx)|}{\|\lambda x + \frac{1}{\lambda}Cx\|^2}\,\|\lambda x + \frac{1}{\lambda}Cx\|^2$$
$$+ \frac{|(C(\lambda x - \frac{1}{\lambda}Cx), \lambda x - \frac{1}{\lambda}Cx)|}{\|\lambda x - \frac{1}{\lambda}Cx\|^2}\,\|\lambda x - \frac{1}{\lambda}Cx\|^2$$

if $\lambda x + \frac{1}{\lambda}Cx \neq 0$ and $\lambda x - \frac{1}{\lambda}Cx \neq 0, \|x\| = 1$. Thus,

$$4\inf_{\|x\|=1} \|Cx\|^2 \le \frac{|(Cy, y)|}{(y, y)}\,\|\lambda x + \frac{1}{\lambda}Cx\|^2 + \frac{|(Cz, z)|}{(z, z)}\,\|\lambda x - \frac{1}{\lambda}Cx\|^2$$

with $y = \lambda x + \frac{1}{\lambda}Cx$ and $z = \lambda x - \frac{1}{\lambda}Cx$, if $y \neq 0$ and $z \neq 0$.

Using the parallelogram identity, we obtain

$$4\inf_{\|x\|=1} \|Cx\|^2 \le \inf_{\|y\|=1} \frac{|(Cy, y)|}{(y, y)}\,\|\lambda x + \frac{1}{\lambda}Cx\|^2 + \inf_{\|z\|=1} \frac{|(Cz, z)|}{(z, z)}\,\|\lambda x - \frac{1}{\lambda}Cx\|^2$$
$$= \inf_{\|y\|=1} \frac{|(Cy, y)|}{(y, y)} \left(\|\lambda x + \frac{1}{\lambda}Cx\|^2 + \|\lambda x - \frac{1}{\lambda}Cx\|^2\right)$$
$$= \inf_{\|y\|=1} \frac{|(Cy, y)|}{(y, y)} \left(2\,\|\lambda x\|^2 + 2\,\|\frac{1}{\lambda}Cx\|^2\right)$$
$$= \inf_{\|y\|=1} \frac{|(Cy, y)|}{(y, y)} \left(2\,\lambda^2\,\|x\|^2 + 2\,\frac{\|Cx\|^2}{\lambda^2}\right)$$

Let $\lambda^2 = \frac{\|Cx\|}{\|x\|} = \|Cx\|$ and $\|x\| = 1$. Then,

$$4 \inf_{\|x\|=1} \|Cx\|^2 \leq \inf_{\|y\|=1} \frac{|(Cy, y)|}{(y, y)} \left(2\,\|Cx\| + 2\,\|Cx\|\right)$$

or

$$\inf_{\|x\|=1} \|Cx\|^2 \leq \inf_{\|y\|=1} \frac{|(Cy, y)|}{(y, y)} \inf_{\|x\|=1} \|Cx\|$$

Now,

$$\inf_{\|x\|=1} \|Cx\|^2 = [\inf_{\|x\|=1} \|Cx\|]^2$$

This leads to

$$\boxed{\inf_{x \neq 0} \frac{|(Cx, x)|}{\|x\|^2} \geq \inf_{\|x\|=1} \|Cx\|}$$

if $\lambda x + \frac{1}{\lambda} Cx \neq 0$ and $\lambda x - \frac{1}{\lambda} Cx \neq 0, \|x\| = 1$.

Special cases (S1)–(S3):

(S1): $\boxed{\lambda x + \frac{1}{\lambda} Cx = 0, \|x\| = 1}$ This implies $-\frac{1}{\lambda}\, Cx = \lambda x$. Therefore,

$$\begin{aligned} 4\,\|Cx\|^2 &\leq \underbrace{|(C(\lambda x + \frac{1}{\lambda} Cx), \lambda x + \frac{1}{\lambda} Cx)|}_{=0} + |(C(\lambda x - \frac{1}{\lambda} Cx), \lambda x - \frac{1}{\lambda} Cx)| \\ &= |(C(\lambda x - \frac{1}{\lambda} Cx), \lambda x - \frac{1}{\lambda} Cx)| \\ &= |(C(2\,\lambda x), 2\,\lambda x)| = 4\,\lambda^2\,|(Cx, x)|,\ \|x\| = 1 \end{aligned}$$

Let

$$\lambda^2 = \frac{\|Cx\|}{\|x\|} = \|Cx\|,\ \|x\| = 1$$

Then,

$$4\,\|Cx\|^2 \leq 4\,\|Cx\|\,|(Cx, x)|$$

and therefore

$$\|Cx\| \leq |(Cx, x)|,\ \|x\| = 1$$

Thus,

$$\inf_{\|y\|=1} \|Cy\| \leq |(Cx, x)|,\ \|x\| = 1$$

so that

$$\inf_{\|y\|=1} \|Cy\| \leq \inf_{\|x\|=1} |(Cx, x)| = \inf_{x \neq 0} \frac{|(Cx, x)|}{\|x\|^2}$$

Thus, relation (22) is also proven in the special case (S1).

(S2): $\lambda x - \frac{1}{\lambda} Cx = 0$, $||x|| = 1$ This case is treated similarly as (S1).

(S3): $\lambda x + \frac{1}{\lambda} Cx = 0$ and $\lambda x - \frac{1}{\lambda} Cx = 0$, $||x|| = 1$ This leads to

$$4\, ||Cx||_2^2 = (C(\lambda x + \frac{1}{\lambda} Cx), \lambda x + \frac{1}{\lambda} Cx) - (C(\lambda x - \frac{1}{\lambda} Cx), \lambda x - \frac{1}{\lambda} Cx) = 0$$

for an x with $||x|| = 1$. Therefore, $\inf_{||x||=1} ||Cx|| = 0$ so that inequality (22) is also valid in the special case (S3).

Relation (22) with $\le$ instead of $\ge$ is trivial. On the whole, the chain of Equation (14) is proven.

Now, let $C \in \mathbb{C}^{n\times n}$ be regular, let $||\cdot||$ denote any vector norm and the associated sub matrix norm. Because for the range of C one has $R(C) = \mathbb{C}^n$, then

$$\inf_{x\neq 0} \frac{||Cx||}{||x||} = \inf_{x\neq 0} \frac{||Cx||}{||C^{-1}Cx||} = \inf_{x\neq 0} \frac{1}{\frac{||C^{-1}Cx||}{||Cx||}} = \frac{1}{\sup\limits_{x\neq 0} \frac{||C^{-1}Cx||}{||Cx||}} = \frac{1}{\sup\limits_{y\neq 0} \frac{||C^{-1}y||}{||y||}} = \frac{1}{||C^{-1}||}$$

thus showing relation (15). On the whole, the proof of Lemma 2 is completed.

Remark We have seen that the method described by Heuser (1975) to derive the relation

$$\sup_{||x||=1} |(Cx, x)| = \sup_{||x||=1} ||Cx||$$

for $C \in \mathbb{C}^{n\times n}$ with $C = C^*$ can be carried over to prove the relation

$$\inf_{||x||=1} |(Cx, x)| = \inf_{||x||=1} ||Cx||$$

As opposed to this, using Taylor's method in Taylor (1958, pp. 322–323), it seems to be impossible to prove the inf relations in a similar way as the sup relations.

Since Heuser's book is written in German, we think that it is worthwhile to make Heuser's proof idea accessible to a broad readership. Of course, the author cannot rule out that the above inf formulas have been derived before. But, he has not found such a derivation in literature.

A3. Series expansion of $\lambda_{max}(\Psi(t))$, $t_0 \le t \le t_0 + \Delta t_0$

We determine the coefficients v_j, $j = 0, 1, 2$ in the series expansion of Section 7, where we have set $v_j := v_{j,max}$, $j = 0, 1, 2$. The derivation follows a line similar to that of Kohaupt (2001, pp. 6–7). We note that the operators $\Psi(t)$ and $T^{(0)}, T^{(1)}, T^{(2)}$ are defined in a way different from that in Kohaupt (2001), however. This is the first difference.

Let $\lambda_{max}(\Psi(t))$ be the largest eigenvalue of $\Psi(t)$. Then, due to

Kato (1966, Theorem 5.11, Chapter II, pp. 115–116] and Kohaupt (1999, Lemma 2.1)

$$\lambda_{max}(\Psi(t)) = v_0 + v_1\,(t - t_0) + v_2\,(t - t_0)^2 + \cdots, \quad t_0 \le t \le t_0 + \Delta t_0$$

where the quantities v_0, v_1, and v_2 are derived now.

Let $n_{-1} := n$ and $v_k^{(0)}[T^{(0)}]$, $k = 1, \cdots, n_{-1}$ be the eigenvalues of $T^{(0)}$. Then,

$$\nu_0 = \max_{k=1,\cdots,n_{-1}} \nu_k^{(0)}[T^{(0)}]$$

Further, define

$$M_{-1}: = X: = \mathbb{C}^n$$

Let

$$V_0: = [v_1^{(\nu_0)}, \cdots, v_{n_0}^{(\nu_0)}]$$

be the matrix formed by the orthonormal set of eigenvectors $v_k^{(\nu_0)}$, $k = 1, \cdots, n_0$ associated with ν_0, and let P_{ν_0} be the orthogonal projection on the algebraic eigenspace

$$M_0: = \text{span}\{v_1^{(\nu_0)}, \cdots, v_{n_0}^{(\nu_0)}\}$$

(which is here identical with the geometric eigenspace). Then, $M_0 = P_{\nu_0} X$ with $dim\, M_0 = n_0$. Further, P_{ν_0} can be calculated by

$$P_{\nu_0} = V_0 V_0^*$$

(cf. Niemeyer & Wermuth, 1987, pp. 234–238). Let

$$\tilde{T}^{(1)}: = P_{\nu_0} T^{(1)} P_{\nu_0}$$

and

$$\nu_k^{(1)}[\tilde{T}^{(1)}],\ k = 1, \cdots, n_0$$

be the eigenvalues of $\tilde{T}^{(1)}$. Then,

$$\nu_1: = \max_{k=1,\cdots,n_0} \{\nu_k^{(1)}[\tilde{T}^{(1)}] \mid \textit{the associated eigenvector lies in } M_0\}$$

Let

$$V_1: = [v_1^{(\nu_1)}, \cdots, v_{n_1}^{(\nu_1)}]$$

be the matrix formed by the orthonormal set of eigenvectors $v_k^{(\nu_1)}$, $k = 1, \cdots, n_1$ associated with ν_1, and let P_{ν_1} be the orthogonal projection on the algebraic eigenspace

$$M_1: = \text{span}\{v_1^{(\nu_1)}, \cdots, v_{n_1}^{(\nu_1)}\}$$

Then, $M_1 = P_{\nu_1} X$ with $dim\, M_1 = n_1$. As above, P_{ν_1} can be calculated by

$$P_{\nu_1} = V_1 V_1^*$$

Let

$$\hat{T}^{(2)}: = P_{\nu_1} \tilde{T}^{(2)} P_{\nu_1}: = P_{\nu_1} (T^{(2)} - T^{(1)} S_{\nu_0} T^{(1)}) P_{\nu_1}$$

with

$$S_{\nu_0}: = \sum_{\nu_k^{(0)} \neq \nu_0} \frac{1}{\nu_k^{(0)} - \nu_0} P_{\nu_k^{(0)}}$$

(for S_{ν_0} cf. Kato, 1966, p. 40, Problem 5.10, Formula (5.32)) and for $\tilde{T}^{(2)}$ (cf. Kato, 1966, p. 116). Let

$$\nu_k^{(2)}[\hat{T}^{(2)}],\ k = 1, \cdots, n_1$$

be the eigenvalues of $\hat{T}^{(2)}$. Then,

$$\nu_2 := \max_{k=1,\cdots,n_1} \{\nu_k^{(2)}[\hat{T}^{(2)}] \mid \textit{the associated eigenvector lies in } M_1\}$$

Remark In the formula for ν_1, exactly those eigenvectors $v^{(1)}$ lie in M_0 for which $rank P_{\nu_0} = rank[P_{\nu_0}, v^{(1)}]$. (In Kohaupt (2001, p. 7, Remark), instead of P_{ν_0} it reads P_{ν_1} there, which is a typo.) Similarly one proceeds in the formula for ν_2.

The second difference to Kohaupt (2001) is that the formulas of Kohaupt (2001, Theorem 4) are not applied. Instead, one may use the relations $D_+^k \|\Psi(t_0)\|_2 = D_+^k \|s(t_0)\|_\infty$, $k = 0, 1, 2$.

A4. Formulas for $D_+^k \|s(t_0)\|_\infty$, $k = 0, 1, 2$

For these formulas, we refer to Kohaupt (2002, pp. 433–434). Let $s \in C^m(\mathbb{R}_0^+, \mathbb{R}^n)$ and $s(t) = [s_1(t), \cdots, s_n(t)]^T$, and define the following sign functionals for $i \in \{1, \cdots, n\}$:

$$s_i^{(0)} := \operatorname{sgn}[s_i(t_0)]$$

$$s_i^{(1)} := \begin{cases} \operatorname{sgn}[s_i(t_0)], & s_i(t_0) \neq 0 \\ \operatorname{sgn}[D\, s_i(t_0)], & s_i(t_0) = 0 \end{cases}$$

$$\vdots$$

$$s_i^{(m)} := \begin{cases} \operatorname{sgn}[s_i(t_0)], & s_i(t_0) \neq 0 \\ \operatorname{sgn}[D\, s_i(t_0)], & s_i(t_0) = 0,\ D\, s_i(t_0) \neq 0 \\ \vdots & \\ \operatorname{sgn}[D^m\, s_i(t_0)], & D^k\, s_i(t_0) = 0,\ k = 0, 1, \cdots, m-1 \end{cases}$$

or briefly,

$$s_i^{(k)} = \begin{cases} s_i^{(k-1)}, & s_i^{(k-1)} \neq 0 \\ \operatorname{sgn}[D^k\, s_i(t_0)], & s_i^{(k-1)} = 0 \end{cases}$$

$i = 1, \cdots, n; k = 1, \cdots, m$. With these sign functionals, define the further functionals

$$S_i^{(k)} := s_i^{(k)} \cdot D^k\, s_i(t_0),\ i = 1, \cdots, n;\ k = 0, 1, \cdots, m \tag{A1}$$

Then, the right derivatives for *real* vector functions read as follows.

Theorem 6 *($p = \infty$, real vector function) Let $s: \mathbb{R}_0^+ \to \mathbb{R}^n$ be an n-dimensional real-valued vector function that is m times continuously differentiable, and let $t_0 \in \mathbb{R}_0^+$. Suppose additionally that each two components of s are either identical or intersect each other at most finitely often near t_0. Further, let $I_{-1} = \{1, \cdots, n\}$ and I_k be the set of all indices $i_k \in I_{k-1}$ where $S_i^{(k)}$ from (23) attains its maximum, i.e.*

$$I_k := \left\{ i_k \in I_{k-1} \mid S_{i_k}^{(k)} = \max_{i \in I_{k-1}} S_i^{(k)} \right\}$$

$k = 1, \cdots, m$. Then, the right derivatives of $t \mapsto \|s(t)\|_\infty$ at $t = t_0 \geq 0$ are given by

$$D_+^k \|s(t_0)\|_\infty = \max_{i \in I_{k-1}} S_i^{(k)}, \quad k = 1, \cdots, m$$

Remark In our case, in Section 7, one has $n = 2$ and $m = 2$. Further, $s(t)$, $t_0 \leq t \leq t_0 + \Delta t_0$ is analytic so that the additional condition is automatically fulfilled.

A 5. Determination of symmetric positive semi-definite intensity matrix $Q = Q_b$ from Q_f

Since

$$b(t) = \left[\begin{array}{c} 0 \\ \hline M^{-1} f(t) \end{array}\right] = \left[\begin{array}{c|c} 0 & 0 \\ \hline 0 & M^{-1} \end{array}\right] \left[\begin{array}{c} 0 \\ \hline f(t) \end{array}\right] = B\, w(t)$$

with

$$B = \left[\begin{array}{c|c} 0 & 0 \\ \hline 0 & M^{-1} \end{array}\right], \quad w(t) = \left[\begin{array}{c} 0 \\ \hline f(t) \end{array}\right]$$

one obtains, using $M^T = M$, the relation $B^T = B$ and thus

$$b(t)\, b^T(\tau) = B w(t)\, w^T(\tau) B^T = B w(t)\, w^T(\tau) B$$

Thus, according to Müller (1976, Formula (9.7)) with $m_b(t) = 0$, one has

$$N_b(t, \tau) = E\{b(t)\, b^T(\tau)\} = B\, E\{w(t)\, w^T(\tau)\}\, B$$

Now,

$$w(t)\, w^T(\tau) = \left[\begin{array}{c|c} 0 & 0 \\ \hline 0 & f(t) f^T(t) \end{array}\right]$$

Since $f(t)$ is white noise, one has $N_f(t, \tau) = E\{f(t) f^T(\tau)\} = Q_f\, \delta(t - \tau)$, leading to

$$E\{w(t)\, w^T(\tau)\} = \left[\begin{array}{c|c} 0 & 0 \\ \hline 0 & E\{f(t) f^T(t)\} \end{array}\right] = \left[\begin{array}{c|c} 0 & 0 \\ \hline 0 & Q_f\, \delta(t - \tau) \end{array}\right] = \left[\begin{array}{c|c} 0 & 0 \\ \hline 0 & Q_f \end{array}\right] \delta(t - \tau)$$

This implies

$$N_b(t, \tau) = B \left[\begin{array}{c|c} 0 & 0 \\ \hline 0 & Q_f \end{array}\right] \delta(t - \tau)\, B = \left[\begin{array}{c|c} 0 & 0 \\ \hline 0 & M^{-1} \end{array}\right] \left[\begin{array}{c|c} 0 & 0 \\ \hline 0 & Q_f \end{array}\right] \left[\begin{array}{c|c} 0 & 0 \\ \hline 0 & M^{-1} \end{array}\right] \delta(t - \tau)$$

giving

$$N_b(t, \tau) = \left[\begin{array}{c|c} 0 & 0 \\ \hline 0 & M^{-1} Q_f M^{-1} \end{array}\right] \delta(t - \tau) = Q_b\, \delta(t - \tau)$$

with

$$Q_b = \left[\begin{array}{c|c} 0 & 0 \\ \hline 0 & M^{-1} Q_f M^{-1} \end{array}\right]$$

A6. Some details on the computation of $y = D_+^k \|P_x(t) - P\|_2, k = 1, 2$

According to Section 7 and Appendix A 3. the computation of $y = D_+^k \|P_x(t) - P\|_2 = D_+^k \|\Psi(t)\|_2 = D_+^k \|\Phi(t)\, C\, \Phi^*(t)\|_2 = D_+^k \|\Phi(t)\,(P_0 - P)\,\Phi^*(t)\|_2,\ k = 1, 2$ is based on

$$v_{1,\max}: = \max_{k=1,\cdots,n_0} \{v_k^{(1)}[\tilde{T}^{(1)}] \mid \text{the associated eigenvector lies in } M_0\}$$

and

$$v_{1,\min}: = -\max_{k=1,\cdots,n_0} \{v_k^{(1)}[-\tilde{T}^{(1)}] \mid \text{the associated eigenvector lies in } M_0\}$$

as well as

$$v_{2,\max}: = \max_{k=1,\cdots,n_1} \{v_k^{(2)}[\hat{T}^{(2)}] \mid \text{the associated eigenvector lies in } M_1\}$$

and

$$v_{2,\min}: = -\max_{k=1,\cdots,n_1} \{v_k^{(2)}[-\hat{T}^{(2)}] \mid \text{the associated eigenvector lies in } M_1\}$$

where the quantities $\tilde{T}^{(1)}$ and $\hat{T}^{(2)}$ depend on t_0.

For all $t_0 = 0(\Delta t)t_e = 0(0.1)25$, the constraint "*the associated eigenvector lies in* M_0" resp. "*the associated eigenvector lies in* M_1" was fulfilled with the only exception of that for $v_{1,\min}$ and $v_{2,\min}$ with $t_0 = 3.4$. The reason for this could not be clarified, however. In this exceptional case, we have set the quantities equal to zero. Since for $t_0 = 3.4$ we have $\|\Psi(t_0)\|_2 = \max\{|\lambda_{\max}(\Psi(t_0))|, |\lambda_{\min}(\Psi(t_0))|\} = |\lambda_{\max}(\Psi(t_0))| = \lambda_{\max}(\Psi(t_0))$, the norm derivatives are given by $D_+ \|\Psi(t_0)\|_2 = v_{1,\max}$ and $D_+^2 \|\Psi(t_0)\|_2 = 2\, v_{2,\max}$, and thus do not depend on $v_{1,\min}$ or $v_{2,\min}$ for $t_0 = 3.4$, however.

Remark It is interesting to note that for all $t_0 = 0(0.1)25$ without exception, only one of the eigenvalues $v_k^{(1)}[\tilde{T}^{(1)}]$, $v_k^{(1)}[-\tilde{T}^{(1)}]$, $k = 1, \cdots, n_0$ and $v_k^{(2)}[\hat{T}^{(2)}]$, $v_k^{(2)}[-\hat{T}^{(2)}]$, $k = 1, \cdots, n_1$ was different from zero, and further that the above-mentioned constraints can be dropped for the given data without changing the results.

Remark Finally, we want to remind the reader that, since the operator $\Psi(t) = \Phi(t)\,(P_0 - P)\,\Phi^*(t)$, $t \geq 0$, is different from zero as well as finite dimensional, it is self-adjoint and completely continuous. Therefore, according to Achieser and Glasman (1968, no. 60, p. 158) or Kantorowitsch and Akilow (1964, Chapter IX, §4.3, p. 255), $\Psi(t)$ has at least one eigenvalue different from zero for all $t \geq 0$.

Simplification of the computation of $D_+^k \|\Psi(t_0)\|_2 = D_+^k \|P_x(t_0) - P\|_2, k = 1, 2$

In the case of $|\lambda_{\max}(\Psi(t_0))| \neq |\lambda_{\min}(\Psi(t_0))|$, one can simplify the computation as follows. Let

$$\lambda(\Psi(t_0))): = \begin{cases} \lambda_{\max}(\Psi(t_0)), & |\lambda_{\max}(\Psi(t_0))| > |\lambda_{\min}(\Psi(t_0))|, \\ \lambda_{\min}(\Psi(t_0)), & |\lambda_{\max}(\Psi(t_0))| < |\lambda_{\min}(\Psi(t_0))| \end{cases}$$

Then,

$$\|\Psi(t_0)\|_2 = |\lambda(\Psi(t_0))| = :|s(t_0)|.$$

Further, according to the last *Remark* above, $\Psi(t_0) \neq 0$ and therefore $s(t_0) \neq 0$, so that in Appendix A4 with $n = 1$ and $s_1(t_0) = s(t_0)$, we obtain the signs

$$s^{(0)} = s^{(1)} = s^{(2)} = \cdots = s^{(m)} = sgn(s(t_0)) = sgn((\Psi(t_0)))$$

Thus, from Appendix A 3 , we get

$$\|\Psi(t_0)\|_2 \qquad = s^{(0)}\, v_0$$

$$D_+\|\Psi(t_0)\|_2 \qquad = s^{(0)}\, v_1$$

$$D_+^2\|\Psi(t_0)\|_2 \quad = s^{(0)}\, 2\, v_2$$

With the given data in Section 8, we had $|\lambda_{max}(\Psi(t_0))| > |\lambda_{min}(\Psi(t_0))|$ and $s^{(0)} = 1$ for all $t_0 = 0(0.1)25$. This means $\|\Psi(t)\|_2 = |\lambda_{max}(\Psi(t))| = \lambda_{max}(\Psi(t))$, $t = 0(0.1)25$.

13

Existence and properties of the Navier-Stokes equations

A.V. Zhirkin[1]*

*Corresponding author: A.V. Zhirkin, Reactor Problem Laboratory, Department of Fusion Reactors, National Research Centre "Kurchatov Institute", Kurchatov Centre of Nuclear Technologies, Academician Kurchatov Square, 1, Moscow 123182, Russia

E-mail: aleksej-zhirkin@yandex.ru

Reviewing editor: Xinguang Zhang, Curtin University, Australia

Abstract: A proof of existence, uniqueness, and smoothness of the Navier–Stokes equations is an actual problem, whose solution is important for different branches of science. The subject of this study is obtaining the smooth and unique solutions of the three-dimensional Stokes–Navier equations for the initial and boundary value problem. The analysis shows that there exist no viscous solutions of the Navier–Stokes equations in three dimensions. The reason is the insufficient capability of the divergence-free velocity field. It is necessary to modify the Navier–Stokes equations for obtaining the desirable solutions. The modified equations describe a three-dimensional flow of incompressible fluid which sticks to a body surface. The equation solutions show the resonant blowup of the laminar flow, laminar–turbulent transition, and fluid detachment that opens the way to solve the magnetic dynamo problem.

Subjects: Advanced Mathematics; Analysis - Mathematics; Applied Mathematics; Fluid Mechanics; Mathematics & Statistics; Science

Keywords: Navier–Stokes equations; incompressible fluid; three-dimensional Cauchy problem; existence and regularity of three-dimensional fluid flow; divergence-free field solution

ABOUT THE AUTHOR

Alexey V. Zhirkin is a senior researcher at the Department of Fusion Reactors of the National Research Centre "Kurchatov Institute" in Moscow, Russian Federation. He graduated from the Moscow Engineering Physics Institute (PhD) and finished the Doctor course at the Department of Quantum Engineering and Systems Science in the University of Tokyo, Japan. His research area is neutronics analysis of fusion devices. He is involved in group's research considering the design of the tokamak-based hybrid fusion–fission neutron source, including the problems of plasma confinement, magnetohydrodynamic effects, and thermal hydraulics.

PUBLIC INTEREST STATEMENT

The Navier-Stokes equations are one of the fundamentals of fluid mechanics. They describe the motion of viscous liquids, gases or plasmas, and are the basis of mathematical modeling of many natural phenomena. Their applications cover virtually all areas of science: aerodynamics and hydrodynamics, astrophysics, engineering, medicine and health, chemistry, biology, geophysics, electro and magnetic hydrodynamics, etc. However, the solutions of the Navier-Stokes equations suitable for analysis are found only for simplified problems that have little or no physical interest. There is no complete understanding of the properties of these equations. An open problem in mathematics is whether their physically reasonable (smooth) solutions always exist in three dimensions. Also it is very important to understand how the Navier-Stokes equations describe the chaotic behavior of fluids (turbulence) that is one of the most difficult problems of mathematical physics. This article is the response to the Navier-Stokes existence and smoothness challenge.

1. Introduction

A proof of existence, uniqueness, and regularity of three-dimensional fluid flows for an incompressible fluid concerns most mathematically complex unsolved problems (Constantin, 2001; Fefferman, 2000; Ladyzhenskaya, 2003). The positive result is obtained in two space dimensions (Ladyzhenskaya, 1969). In three dimensions, existence is proved for weak solutions (Leray, 1934), but their uniqueness and smoothness are not known. For strong solutions, the three-dimensional problem is not solved till now (Constantin, 2001; Fefferman, 2000; Ladyzhenskaya, 2003).

The solution of the problem can help ensure the development in different branches of science. The significance of the challenge is well described in the work (Fefferman, 2000).

"A fundamental problem in analysis is to decide whether such smooth, physically reasonable solutions exist for the Navier–Stokes equations. ... Fluids are important and hard to understand. There are many fascinating problems and conjectures about the behavior of solutions of the Euler and Navier–Stokes equations. ... Since we don't even know whether these solutions exist, our understanding is at a very primitive level. Standard methods from PDE appear inadequate to settle the problem. Instead, we probably need some deep, new ideas."

In the work of (Fefferman, 2000), the detailed statement for the Cauchy initial value problem and the problem with the initial periodical conditions of the Navier-Stokes equations is presented. We use this statement as the basic formulation of problem-solving in our research.

The regularity question considering the boundary value problem is posed in the work (Constantin, 2001):

"What are the most general conditions for smooth incompressible velocities $\vec{v}(0,\vec{r})$ that ensure, in the absence of external input of energy, that the solutions of the Navier–Stokes equations exist for all positive time and are smooth? ... A specific regularity question, still open, is for instance: Given an arbitrary infinitely differentiable, incompressible, compactly supported initial velocity field in R^3, does the solution remain smooth for all time? A version of the same question is: Given an arbitrary three-dimensional divergence-free periodic real analytic initial velocity, does the ensuing solution remain smooth for all time?"

In the work of (Ladyzhenskaya, 2003), there is a more general formulation of the problem,

> Do the Navier–Stokes equations together with the initial and boundary conditions provide the deterministic description of the dynamics of an incompressible fluid or not?

In our research, we will try to find the answers to all these questions.

2. Analysis

2.1. Statement of problem

The system of the non-stationary Navier–Stokes equations for an incompressible fluid in three dimension looks as follows:

$$\rho \cdot \partial\vec{v}(t,\vec{r})/\partial t + \rho \cdot (\vec{v}(t,\vec{r}) \cdot \nabla) \cdot \vec{v}(t,\vec{r}) = -\nabla p(t,\vec{r}) + \eta \cdot \Delta\vec{v}(t,\vec{r}) + \vec{f}_{ex}(t,\vec{r}), \tag{1}$$

$$div\ (\vec{v}(t,\vec{r})) = 0, \tag{2}$$

$$(\vec{r} \in R^3, t \geq 0)$$

where ρ is the fluid density, $\vec{v}(t,\vec{r})$ is the vector field of velocities of a fluid at time t and a point with coordinates $\vec{r} = (x,y,z)$, $p(t,\vec{r})$ is the pressure of a fluid, η is the dynamic viscosity, and $\vec{f}_{ex}(t,\vec{r})$ is the external force defined as a smooth function on $[0, \infty) \times R^3$. ∇ is the Hamilton operator and Δ is the Laplace operator. Equation (1) is the Cauchy momentum equation and Equation (2) is the mass continuity equation.

If the initial value problem is considered, the initial condition looks as follows:

$$\vec{v}(0,\vec{r}) = \vec{v}_0(\vec{r}) \quad (3)$$

where $\vec{v}_0(\vec{r})$ is an arbitrary divergence-free vector field defined in R^3.

For the boundary value problem (Constantin, 2001; Ladyzhenskaya, 2003), the boundary condition is taken for a fluid sticking to a body surface:

$$\vec{v}(t,\vec{r}_S) = 0 \quad (4)$$

where S is the surface separating the fluid and body.

The Clay Mathematics Institute suggests to prove or deny existence of solutions satisfying (1), (2), and (3) for viscous fluid ($\eta > 0$) under the following conditions (Fefferman, 2000).

For physically reasonable solutions, $\vec{v}(t,\vec{r})$ does not grow large as $\left|\vec{r}\right| \to \infty$. Hence, the forces $\vec{f}_{ex}(t,\vec{r})$ and initial conditions $\vec{v}_0(\vec{r})$ satisfy

$$\left|\partial_{\vec{r}}^{\alpha}\vec{v}_0(\vec{r})\right| \le C_{\alpha,k} \cdot (1 + \left|\vec{r}\right|)^{-k}\ on R^3, \text{ for any } \alpha \text{ and } K \quad (5)$$

and

$$\left|\partial_{\vec{r}}^{\alpha}\partial_t^{m}\vec{f}_{ex}(t,\vec{r})\right| \le C_{\alpha,m,k} \cdot (1 + \left|\vec{r}\right| + t)^{-k} \text{on } [0,\infty)\ R^3, \text{for any } \alpha\ , m \text{ and } K. \quad (6)$$

A solution of (1), (2), and (3) is accepted as physically reasonable only if it satisfies

$$\vec{v}(t,\vec{r}),\ p(t,\vec{r}),\ f_{ex}(t,\vec{r}) \in C^{\infty}([0,\infty) \times R^3), \vec{v}_0(\vec{r}) \in C^{\infty}(\vec{r} \in R^3), \quad (7)$$

and

$$\int_{R^3} \left|\vec{v}(t,\vec{r})\right|^2 \cdot d\vec{r} < const \text{ for all } t \ge 0 \ \text{(bounded energy)}. \quad (8)$$

Note. The function $\vec{f}_{ex}(t,\vec{r})$ is a smooth one on $[0, \infty) \times R^3$. So we added the condition $f_{ex}(t,\vec{r}) \in C^{\infty}([0,\infty) \times R^3)$ in (7). This condition is absent in the same expression of the work of Fefferman (2000).

Alternatively, for spatially periodic solutions of (1), (2), and (3), we assume that $\vec{v}_0(\vec{r}), \vec{f}_{ex}(t,\vec{r})$ satisfy

$$\vec{v}_0(\vec{r}) = \vec{v}_0(\vec{r} + \vec{e}_j), \vec{f}_{ex}(t,\vec{r}) = \vec{f}_{ex}(t,\vec{r} + \vec{e}_j)\ \ on\ \ [0,\infty)\ \times\ R^3, j = 1,2,3 \quad (9)$$

where $\vec{e}_j$ is jth unit vector in R^3.

In place of (5) and (6), we assume that $\vec{v}_0(\vec{r})$ is smooth and that

$$\left|\partial_{\vec{r}}^{\alpha}\partial_t^{m}\vec{f}_{ex}(t,\vec{r})\right| \le C_{\alpha,m,k} \cdot (1 + |t|)^{-k} \text{ on } [0,\infty), \times R^3, \text{for any } \alpha, m \text{ and } K. \quad (10)$$

We accept a solution of (1), (2), or (3) as physically reasonable if it satisfies

$$\vec{v}(t,\vec{r}) = \vec{v}(t,\vec{r}+\vec{e}_j), p(t,\vec{r}) = p(t,\vec{r}+\vec{e}_j) on[0,\infty)\times R^3, j = 1,2,3 \tag{11}$$

and

$$\vec{v}(t,\vec{r}),\ p(t,\vec{r}),\ f_{ex}(t,\vec{r}) \in C^{\infty}([0,\infty)\times R^3). \tag{12}$$

The problem is formulated in four statements (Fefferman, 2000).

(A) Existence and smoothness of Navier–Stokes solutions on R^3. Take $\eta > 0$ and space dimension $n = 3$. Let $\vec{v}_0(\vec{r})$ be any smooth, divergence-free vector field satisfying (5). Take $\vec{f}_{ex}(t,\vec{r})$ to be identically zero. Then, there exist smooth functions $p(t,\vec{r}), \vec{v}(t,\vec{r})$ on $[0, \infty) \times R^3$ that satisfy (1), (2), (3), (7), and (8).

(B) Existence and smoothness of Navier–Stokes solutions in R^3/Z^3. Take $\eta > 0$ and $n = 3$. Let $\vec{v}_0(\vec{r})$ be any smooth, divergence-free vector field satisfying (9); we take $\vec{f}_{ex}(t,\vec{r})$ to be identically zero. Then, there exist smooth functions $p(t,\vec{r}), \vec{v}(t,\vec{r})$ on $[0, \infty) \times R^3$ that satisfy (1), (2), (3), (11), and (12).

(C) Breakdown of Navier–Stokes solutions on R^3. Take $\eta > 0$ and $n = 3$. Then, there exist a smooth, divergence-free vector field $\vec{v}_0(\vec{r})$ on R^3 and a smooth $\vec{f}_{ex}(t,\vec{r})$ on $[0, \infty) \times R^3$, satisfying (5) and (6) on $[0, \infty) \times R^3$, for which there exist no solutions $(p(t,\vec{r}),\ \vec{v}(t,\vec{r}))$ of (1), (2), (3), (7), and (8) on $[0, \infty) \times R^3$.

(D) Breakdown of Navier–Stokes Solutions on R^3/Z^3. Take $\eta > 0$ and $n = 3$. Then, there exist a smooth, divergence-free vector field $\vec{v}_0(\vec{r})$ on R^3 and a smooth $\vec{f}_{ex}(t,\vec{r})$ on $[0, \infty) \times R^3$, satisfying (9) and (10), for which there exist no solutions $(p(t,\vec{r}),\ \vec{v}(t,\vec{r}))$ of (1), (2), (3), (11), and (12) on $[0, \infty) \times R^3$.

2.2. Idea of proof

According to the Cauchy–Kowalevski theorem, for the local existence and uniqueness of the solution of a Cauchy initial value problem for a linear partially differential equation with constant coefficients, it is enough to prove the expansibility of this solution to a uniformly converging power series in some neighborhood of every point in the solution domain. The solution can be a function of real or complex variables (Courant & Hilbert, 1962; Vladimirov, 1983).

We will extend this statement to the nonlinear Navier–Stokes equations and generalize the result obtained for a local solution to all space $R^1_+ \times R^3$ to obtain a global solution. We can do it by the following way.

At first, we carry out the statements for the boundary value problem. We define a compact simply connected domain $G \subset R^3$ with a border S. We accept that

$$\left|\partial^{\alpha}_{\vec{r}} \vec{v}_0(\vec{r})\right| \le C_{\alpha,k} \cdot (1+\left|\vec{r}\right|)^{-k} \text{ on } G \subset R^3, \text{ for any } \alpha \text{ and } K, \tag{13}$$

$$\left|\partial^{\alpha}_{\vec{r}} \partial^{m}_{t} \vec{f}_{ex}(t,\vec{r})\right| \le C_{\alpha,m,k} \cdot (1+\left|\vec{r}\right|)^{-k} \text{on}[0,T]\times G \subset R^3 (T < \infty), \text{ for any } \alpha,\ m \text{ and } K, \tag{14}$$

$$\vec{v}(t,\vec{r}),\ p(t,\vec{r}),\ \vec{f}_{ex}(t,\vec{r}) \in C^{\infty}([0,T]\times G \subset R^3), \vec{v}_0(\vec{r}) \in C^{\infty}(\vec{r} \in G \subset R^3), \tag{15}$$

$$\int_G \left|\vec{v}(t,\vec{r})\right|^2 \cdot d\vec{r} < const \quad \text{for } 0 \le t \le T (\text{bounded energy}). \tag{16}$$

The initial step of the solution existence proof is the definition of a class of generalized solutions for which a uniqueness theorem is performed (Ladyzhenskaya, 2003).

Let's designate all functions satisfying the condition (16) as $L_2(G)$. As is known from the differential equation theory (Vladimirov, 1983), any function from this space can be presented as a Fourier series converging to this function. Due to the condition (15), the Fourier decompositions to converging series can be applied both to space and time variables. So we introduce the function space $L_2(G_T)$, $G_T = [0, T] \times G \subset [0, \infty) \times R^3$ (Ladyzhenskaya, 2003), whose functions are decomposed to the Fourier series that are constructed from the complete system of functions $\{\vec{u}_n(t,\vec{r})\}, n = \overline{0,\infty}$. If this complete system is a Navier–Stokes equation solution, the functions $\vec{v}(t,\vec{r}), p(t,\vec{r}) \in L_2(G_T)$, $G_T = [0,T] \times G \subset [0,\infty) \times R^3$ presented as the expansions to this complete system are solutions of these equations.

A Fourier series is based on the use of the trigonometric functions which are presented as the exponential function with the imaginary-valued argument. So a Fourier series can be presented as a complex-valued power series (Fihtengolts, 2003a). For a continuously differentiable function on $[0, \infty) \times R^3$, the Lipschitz condition is performed for any domain G'_T, which is wider than G_T $(G_T \subset G'_T \subset [0,\infty) \times R_3)$, so the Fourier series of this function is uniformly converged on G_T (Fihtengolts, 2003a). The Fourier coefficients are uniquely determined from the boundary condition. So the local existence and uniqueness theorem can be applied to a general partial differential equation with constant coefficients (Courant & Hilbert, 1962) for the function space G_T.

We have the following reasons to suppose that this claim is also valid for the nonlinear Navier–Stokes equations. According to conditions 13–16, the functions $\vec{v}(t,\vec{r})$, $p(t,\vec{r})$, $\vec{v}_0(\vec{r})$, and $\vec{f}_{ex}(t,\vec{r})$ are k times differentiable for $k = 1, \ldots, \infty$, and continuous with all their kth-order partial derivatives on $[0, \infty) \times R^3$. The nonlinear term of the Cauchy momentum equation is k times differentiable and continuous together with its all derivatives also. Hence, $\vec{v}(t,\vec{r}), p(t,\vec{r}), \vec{v}_0(\vec{r}), \vec{f}_{ex}(t,\vec{r})$, their kth derivatives, and the nonlinear term can be presented as the uniformly converging Fourier series on G_T (Fihtengolts, 2003a). These Fourier series are differentiable and can be multiplied to other Fourier series. The result of multiplication is a Fourier series (Fihtengolts, 2003a). The Fourier coefficients of the nonlinear term are uniquely defined from the boundary condition. Assuming that the nonlinear term contains no unknowns, we obtain the unique solution of the Navier–Stokes equations as for the system of two linear equations, one of which contains the additional source presented as the Fourier series.

We can obtain the unique global solution of the initial value problem (1), (2), and (3) satisfying the statement (A) of the Clay Mathematics Institute if we take the limit of the Fourier series for $G_T = [0, T] \times G \to [0, \infty) \times R^3$. As a result, the Fourier series turns to the Fourier transforms which are presented by improper integrals. Under conditions (5)–(8), these integrals exist and converge. The similar statements can be carried out for the periodic functions satisfying (9)–(12).

The use of a trigonometric Fourier series is not the only option to obtain a solution of the Stokes–Navier equations. It is possible that for some variables (e.g. time variable), the solution is decomposed to function series based on exponential functions with a real-valued argument instead of a complex-valued one. These series are real-valued power one. Under the condition of the uniform convergence, we can reduce the nonlinear problem to a linear one and transform the series to improper integrals the same way as using the trigonometric Fourier series.

Let's verify whether the functions expandable to the uniformly converging Fourier series are really solutions of the Navier–Stokes equations. We will obtain the eigensolutions of the Navier–Stokes equations and study their relation to the Fourier series.

If $p(t,\vec{r}) \in L_2(G_T)$ for the proof of Navier–Stokes equation resolvability, it is enough to find the solutions of the mass continuity equation in $L_2(G_T)$. This is a constant coefficient homogeneous differential equation with private derivatives of the first order. It is linear and only one required unknown function $\vec{v}(t,\vec{r})$ is presented in it.

The Cauchy momentum equation is also linear with respect to the function $p(t,\vec{r})$. If we take the divergence of the right and left parts of Equation (1), we obtain that the pressure is a solution of the Poisson equation

$$\Delta p(t,\vec{r}) = div\vec{f}_{ex}(t,\vec{r}) - div(\rho \cdot (\vec{v}(t,\vec{r}) \cdot \nabla) \cdot \vec{v}(t,\vec{r})). \quad (17)$$

The solution $p(t,\vec{r}) \in L_2(G_T)$ can be easily obtained at substitution of $\vec{v}(t,\vec{r})$ in this equation.

We should take into account that, generally, the velocity vector does not coincide with a direction of pressure gradient in the mass continuity equation. Therefore, we are interested in the resolvability of the Cauchy momentum equation in the absence of the pressure gradient. Thus, it is necessary to solve the equation

$$\rho \cdot \partial\vec{v}_\perp(t,\vec{r})/\partial t + \rho \cdot (\vec{v}(t,\vec{r}) \cdot \nabla) \cdot \vec{v}_\perp(t,\vec{r}) = \eta \cdot \Delta\vec{v}_\perp(t,\vec{r}) + \vec{f}_{ex}(t,\vec{r}), \quad (18)$$

where $\vec{v}(t,\vec{r}) = \vec{v}_{//}(t,\vec{r}) + \vec{v}_\perp(t,\vec{r}), \vec{v}_{//}(t,\vec{r}) \| \nabla p(t,\vec{r}),\ \vec{v}_\perp(t,\vec{r}) \perp \nabla p(t,\vec{r})$, under the conditions (2), (3), and (5)–(8).

2.3. Solutions of mass continuity equation

Let's consider the mass continuity equation $div(\vec{v}(t,\vec{r})) = 0$. This equation defines a velocity vector field as a divergence-free one.

For a divergence-free field,

$$curl(\vec{v}_{SOL}(t,\vec{r})) \neq 0, \vec{v}_{SOL}(t,\vec{r})) = curl(\vec{H}(t,\vec{r})) \quad (19)$$

where $\Delta\vec{v}(t,\vec{r}) = 0$ is a vector potential of a vector velocity field.

Let's admit that

$$curl(\vec{v}_{SOL}(t,\vec{r})) = \vec{j}(\vec{r},t), \vec{r} \in G \subset R^3 \quad (20)$$

where $\vec{j}(t,\vec{r})$ is an arbitrary function under the condition $\vec{j}(t,\vec{r}) \in L_2(G), G \subset R^3$. In particular, $G \equiv R^3$. For this function,

$$div(\vec{j}(t,\vec{r})) = 0. \quad (21)$$

This expression is a necessary condition of resolvability of Equation (20) (Fihtengolts, 2003a).

Let's obtain the general solution of the heterogeneous system of the equations:

$$curl(\vec{v}_{SOL}(t,\vec{r})) = \vec{j}(t,\vec{r}), div_{SOL}(\vec{v}(t,\vec{r})) = 0. \quad (22)$$

We use the solution of the inverse problem of the vector analysis (Fihtengolts, 2003a).

At first, we obtain the general solution of homogeneous system of the equations:

$$div(\vec{v}_{SOL}(t,\vec{r})) = 0, curl(\vec{v}_{SOL}(t,\vec{r})) = 0. \quad (23)$$

The first equation testifies that a divergence-free field is a potential one:

$$\vec{v}_{SOL}(t,\vec{r})) = \vec{v}_{POT}(t,\vec{r})) = \nabla\Phi(\vec{r},t) \quad (24)$$

where $\Phi(t,\vec{r})$ is a scalar potential of a vector velocity field.

As a result, we obtain the Laplace equation

$$\nabla\Phi(t,\vec{r}). \tag{25}$$

Let's take a gradient of this expression. We consider that

$$\nabla\Delta(t,\vec{r}) = \Delta\nabla(t,\vec{r}) = \Delta\vec{v}_{POT}(t,\vec{r}). \tag{26}$$

We obtain the Laplace equation for the velocity:

$$\Delta\vec{v}_{POT}(t,\vec{r}) =.0 \tag{27}$$

The solutions of this equation are harmonic functions.

Let's obtain the private solution of the heterogeneous system (22). We substitute expression (19) for the vector potential in Equation (20).

Then,

$$curl(curl(\vec{H}(t,\vec{r})) = \vec{j}(t,\vec{r}) \Rightarrow$$

$$\nabla(div(\vec{H}(t,\vec{r})) - \Delta\vec{H}(t,\vec{r}) = \vec{j}(t,\vec{r}) \tag{28}$$

where $\vec{H}(t,\vec{r})$ is a vector potential of a vector velocity field.

Using the correlations $curl(\nabla(div(\vec{H}(t,\vec{r}))) = 0$ and $curl(\Delta\vec{H}(t,\vec{r})) = \Delta(curl\vec{H}(t,\vec{r})) = \Delta\vec{v}_{SOL}(t,\vec{r})$, we obtain the Poisson equation for the velocity

$$\Delta\vec{v}_{SOL}(t,\vec{r}) = -\nabla \times \vec{j}(t,\vec{r}). \tag{29}$$

The Poisson equation for the vector potential is

$$\Delta\vec{H}(t,\vec{r}) = -\vec{j}(t,\vec{r}), div(\vec{H}(t,\vec{r})) = 0. \tag{30}$$

The general solution of the heterogeneous Poisson Equation (29) includes the solution of the Laplace Equation (27) as the general solution of the homogeneous equation. Thus, the general solution of the system (22) is reduced to the general solution of Equation (29).

The vector field $\vec{v}(t,\vec{r})$ is completely defined by its scalar and vector potentials. Let's investigate the ability of potentials $\Phi(t,\vec{r})$ and $H(t,\vec{r})$ to build the complete systems of functions in three-dimensional space R^3 if these potentials are harmonic functions.

The harmonic functions can be presented as converging power series in three dimensions. For two space variables, these power series have the form of the trigonometric Fourier series. For the third variable, the power series is based on the exponential function with the real-valued argument.

The harmonic functions possess the property of completeness only for two variables. For the third variable, it is impossible to construct the complete system of functions.

For example, let's consider solutions $\Phi\ (t,\vec{r})$ of the three-dimensional Laplace equation in the rectangular Cartesian coordinates. The solutions in the form of linear functions aren't interesting to us. We investigate the solution which is looking as $\Phi\ (t,\vec{r})\tilde{}\exp\vec{k}\cdot\vec{r}$ for $\vec{k}^2 = 0$, where $\vec{k}$ is a complex-valued vector. In this solution, the value of one wave vector projection always depends on two others. The correlation between projections does not allow constructing the orthonormal basis for one of the three spatial coordinates. The complete system can be constructed only for the functions of two spatial variables.

In the cylindrical coordinates, the solutions of the Laplace equation are functions like

$$\Phi(t,\vec{r}) \sim R_n(k_p.\rho).\exp(\pm ik_\varphi \cdot \varphi).\exp(-k_\varphi.z) \tag{31}$$

where $R_n(k_\rho \cdot \rho)$ is the cylindrical Bessel function and k_ρ and k_Φ are the real eigenvalues of the problem. As well as in the previous case, we have only two independent eigenvalues for the function of three variables.

In the spherical coordinates, the solution aspiring zero at infinity looks like

$$\phi\ (t,\vec{r}) \sim (1/r^n)\ .\ Y_n(\vartheta,\varphi) \tag{32}$$

where $Y_n(\vartheta, \varphi)$ is the spherical superficial harmonics possessing the property of completeness.

The functions $1/r^n$ don't possess this property. The correlation between the eigenvalues in the Cartesian coordinate problem is kept also in the curvilinear one. The solutions depending on the curvilinear coordinates can be obtained by the expansion of the exponential function to the series on the complete system of the functions corresponding defined geometry (e.g. the cylindrical and spherical functions). Under the condition of $\vec{k}^2 = 0$, these functions lose the completeness for the three variables.

But it is much more important that the harmonic functions represent a non-viscous fluid motion. According to formula (27), there is no term with the viscosity in the Cauchy momentum equation. A viscous fluid motion is defined by not a potential (curl-free) divergence-free field but exclusively a divergence-free one which is determined by heterogeneous Equation (30).

Let's consider the case when the velocity $\vec{v}(t,\vec{r})$ is defined only by the vector potential $\vec{H}(t,\vec{r})$. We substitute the functions

$$\vec{v}(t,\vec{r}) = \vec{v}(t,\vec{k}) \cdot \exp(\vec{k} \cdot \vec{r}), \tag{33}$$

$$\vec{j}(t,\vec{r}) = \vec{j}(t,\vec{k}) \cdot \exp(\vec{k} \cdot \vec{r}) \tag{34}$$

in Equation (29). We obtain

$$\vec{v}(t,\vec{k}) = \frac{\vec{j}(t,\vec{k}) \times \vec{k}}{\vec{k}^2}, \vec{k}^2 \neq 0. \tag{35}$$

It means that

$$\vec{k} \perp \vec{v}(t,\vec{r}). \tag{36}$$

We have the same result if we substitute the function (33) in the equation $div(\vec{v}(t,\vec{r})) = 0$. We obtain the expression

$$\vec{k} \cdot \vec{v}(t,\vec{r}) = 0. \tag{37}$$

The case when

$$\vec{v}(t,\vec{r}) \sim v(t,\vec{k}) \cdot (\vec{k}/k) \tag{38}$$

leads to the expression $\vec{k}^2 = 0$. It is performed for the potential velocity $\vec{v}_{POT}(t,\vec{r})$ satisfying Equation (27).

The alternative solution is $\vec{k} \perp \vec{v}(t,\vec{r})$ where $\vec{k}$ is a complex-valued vector. It is valid for the velocity $\vec{v}_{SOL}(t,\vec{r})$ satisfying Equation (29). In this case, there exists correlation defined by the necessity to choose the vector $\vec{k}$ direction that is perpendicular to the vector $\vec{v}(t,\vec{r})$. For the problem-solving in the Cartesian or curvilinear coordinates, this correlation reduces the number of uncertain eigenvalues just as it has occurred for the Laplace equation. If one axis of the Cartesian system of coordinates is directed in parallel to the vector $\vec{v}(t,\vec{r})$, the solution becomes the function of only two spatial variables. The transformation to an arbitrary Cartesian system of coordinates and a curvilinear coordinate system leads to the result that there are only two independent spatial variables.

2.4. Solutions of Cauchy momentum equation

Under the condition $\eta > 0$, the three-dimensional Navier–Stokes equations are reduced to a two-dimensional one. Besides this, the nonlinear term is equal to 0 in the Cauchy momentum equation if there are no external sources of a fluid motion in R^3. Using expression (33), we have

$$(\vec{v}(t,\vec{r}) \cdot \nabla) \cdot \vec{v}(t,\vec{r}) = (\vec{v}_{SOL}(t,\vec{r}) \cdot \nabla) \cdot \vec{v}_{SOL}(t,\vec{r}) \sim (\vec{k}' \cdot \vec{v}_{SOL}(t,\vec{k})) \cdot \vec{v}_{SOL}(t,\vec{k}') = 0 \tag{39}$$

because the condition $\vec{k} \perp \vec{v}(t,\vec{r})$ is performed for any $\vec{k}$in $\vec{v}_{SOL}(t,\vec{k})$.

It means that there is no term with velocity in Equation (17):

$$\Delta p(t,\vec{r}) = div\vec{f}_{ex}(t,\vec{r}), \tag{40}$$

and Equation (18) is a linear one:

$$\rho \cdot \partial\vec{v}_{SOL}(t,\vec{r})/\partial t = \eta \cdot \Delta\vec{v}_{SOL}(t,\vec{r}) + \vec{f}_{ex}(t,\vec{r}). \tag{41}$$

As a result, we obtain the eigensolution of the Navier–Stokes equations for velocity in the Cartesian coordinate system with unit vectors $\vec{e}_x$, $\vec{e}_y$, and $\vec{e}_z$ under the condition $\eta > 0$ as

$$\begin{aligned}
\vec{v}_{SOL}(t,\vec{r}) &= \vec{v}_{SOL}(\vec{k}) \cdot \exp[-\eta \cdot \frac{\vec{k}^2}{\rho}] \cdot \exp(i \cdot \vec{k}\vec{r}) = \\
&\exp[-\eta \cdot \frac{(k_y')^2 + (k_z')^2}{\rho} \cdot t] \cdot v_x(k_y', k_z') \cdot \exp[i \cdot (k_y' \cdot y + k_z' \cdot z)] \cdot \vec{e}_x + \\
&\exp[-\eta \cdot \frac{(k_x'')^2 + (k_z'')^2}{\rho} \cdot t] \cdot v_y(k_x'', k_z'') \cdot \exp[i \cdot (k_x'' \cdot x + k_z'' \cdot z)] \cdot \vec{e}_y + \\
&\exp[-\eta \cdot \frac{(k_x''')^2 + (k_y''')^2}{\rho} \cdot t] \cdot v_z(k_x''', k_y''') \cdot \exp[i \cdot (k_x''' \cdot x + k_y''' \cdot y)] \cdot \vec{e}_z
\end{aligned} \tag{42}$$

where $\vec{k} = (k_x, k_y, k_z)$ and $\vec{r} = (x, y, z)$ are the real-valued vectors, $\vec{k} \perp \vec{v}(t,\vec{r})$. This formula is obtained using the expressions:

$$k_x = f_1(k_y, k_z), ky = f_2(k_x, k_z), k_z = f_3(k_x, k_y), x = f_1(y, z), y = f_2(x, z), z = f_3(x, y) \tag{43}$$

where f_1, f_2, and f_3 are linear functions.

We obtain the global solution of the problem for $\vec{f}_{ex}(t,\vec{r}) \equiv 0$ if we take the improper integral of the function (42) over the continuous spectra of k_y', k_z', k_x'', k_z'', k_x''', k_y''' and satisfy the initial condition. The solution is a smooth function at any time moment $(0 \le t < \infty)$.

For pressure, we have the general solution of the two-dimensional Poisson Equation (40).

We should say that the nonlinear term isn't equal to 0 if we use the curvilinear coordinates (e.g. spherical or cylindrical) instead of the Cartesian one. In this case, we implicitly use a source of a fluid motion at some points of space R^3 (e.g. a point or line source). But $div(\vec{v}(t,\vec{r})) \neq 0$ at the source points.

3. Alternative

The divergence-free velocity field defines viscous two-dimensional solutions of the Navier–Stokes equations. It is necessary to update the Navier–Stokes equations to consider a potential field which is not divergence-free for solving a viscous three-dimensional problem.

The necessity of updating the equations is most evident at the consideration of fluid movement near body surfaces. Sticking fluid molecules to a surface means inelastic collisions of these molecules with a body. The molecules colliding with the stuck one are also involved in the inelastic collisions since they stick to other molecules or lose kinetic energy equal to the binding energy to release the attached molecules. Therefore, there are the inelastic collisions in the whole boundary layer.

The mass continuity equation does not include the inelastic processes. It means that the field of velocities is divergence-free. It is natural to expect that the divergence-free field does not possess sufficient capabilities to describe the fluid motion in the boundary layer. We suppose that the inelastic intermolecular interactions can be taken into account in the equations by the use of the potential field where velocity divergence is not equal to zero. It distinguishes it from the velocity potential field which is already used in fluid dynamics. That field is a kind of a divergence-free one.

3.1. Modification of equations

Let's update the Navier–Stokes equations using a curl-free potential vector field. The refined mass continuity equation looks as follows

$$div(\vec{v}(t,\vec{r})) = \Delta\Phi(t,\vec{r}) \tag{44}$$

$$\vec{v}(t,\vec{r}) = \vec{v}_{SOL}(t,\vec{r}) + \vec{v}_{POT}(t,\vec{r}), \vec{v}_{POT}(t,\vec{r}) = \nabla\Phi(t,\vec{r}), \vec{v}_{SOL}(t,\vec{r})) = curl(\vec{H}(t,\vec{r})) \tag{45}$$

where $\Phi t,\vec{r})$ is the scalar potential and $\vec{H}(t,\vec{r})$ is the vector potential of the vector velocity field.

The Equation (44) always has solutions. Generally, their number is infinite (Fihtengolts, 2003a).

We use the general form of the Cauchy momentum equation

$$\rho \cdot \partial\vec{v}(t,\vec{r})/\partial t + \rho \cdot (\vec{v}(t,\vec{r}) \cdot \nabla) \cdot \vec{v}(t,\vec{r}) = -\nabla p(t,\vec{r}) + \eta \cdot \Delta\vec{v}(t,\vec{r}) + (\eta/3 + \zeta) \cdot \nabla(\nabla\vec{v}(t,\vec{r})) + \vec{f}_{ex}(t,\vec{r}) \tag{46}$$

where ζ is the second viscosity considering the intermolecular interactions. The compressibility of a fluid is not used in the derivation of this equation (Landau & Lifshitz, 1987), i.e. we can accept that $\rho = const$.

We substitute expression (44) in the term $(\eta/3 + \zeta) \cdot \nabla(\nabla\vec{v}(t,\vec{r}))$of the Cauchy momentum Equation (46). We take into account that

$$\nabla\Delta\Phi(t,\vec{r}) = \nabla\Delta\Phi\ (t,\vec{r}) = \Delta\vec{V}_{POT}(t,\vec{r}), \tag{47}$$

We obtain the Cauchy momentum equation in the form

$$\rho \cdot \partial\vec{v}(t,\vec{r})/\partial t + \rho \cdot (\vec{v}(t,\vec{r}) \cdot \nabla) \cdot \vec{v}(t,\vec{r}) = -\nabla p(t,\vec{r}) + \eta \cdot \Delta\vec{v}_{SOL}(t,\vec{r}) + (4\eta/3 + \zeta) \cdot \Delta\vec{v}_{POT}(t,\vec{r})) + \vec{f}_{ex}(t,\vec{r}). \tag{48}$$

Any vector field can be presented as a sum of a divergence-free and a curl-free vector field (Fihtengolts, 2003a). We suppose that

$$\vec{f}_{ex}(t,\vec{r}) = \vec{f}_{ex,POT}(t,\vec{r}) + \vec{f}_{ex,SOL}(t,\vec{r}), \vec{f}_{ex,POT}(t,\vec{r}) = \nabla\Phi_{ex}(t,\vec{r}),\ \vec{f}_{ex,SOL}(t,\vec{r}) = curl(\vec{H}_{ex}(t,\vec{r})) \tag{49}$$

where $\Phi_{ex}(t,\vec{r})$ is the scalar potential and $\vec{H}_{ex}(t,\vec{r})$ is the vector potential of the vector external source field.The vector field $\nabla p(t,\vec{r})$ is potential. So expression (48) can be separated into two equations

$$\rho \cdot \partial\vec{v}_{POT}(t,\vec{r})/\partial t + \rho \cdot (\vec{v}(t,\vec{r}) \cdot \nabla) \cdot \vec{v}_{POT}(t,\vec{r}) = -\nabla p(t,\vec{r}) + (4\eta/3 + \zeta) \cdot \Delta\vec{v}_{POT}(t,\vec{r})) + \vec{f}_{ex,POT}(t,\vec{r}), \tag{50}$$

$$\rho \cdot \partial\vec{v}_{SOL}(t,\vec{r})/\partial t + \rho \cdot (\vec{v}(t,\vec{r}) \cdot \nabla) \cdot \vec{v}_{SOL}(t,\vec{r}) = \eta \cdot \Delta\vec{v}_{SOL}(t,\vec{r}) + \vec{f}_{ex,SOL}(t,\vec{r}). \tag{51}$$

We consider Equation (50). We use the expressions

$$\vec{v}_{POT}(t,\vec{r}) = \nabla\Phi(t,\vec{r}),$$

$$(\vec{v}_{POT}(t,\vec{r}) \cdot \nabla) \cdot \vec{v}_{POT}(t,\vec{r}) = \frac{1}{2} \cdot \nabla\vec{v}^2_{POT}(t,\vec{r}) - \vec{v}_{POT}(t,\vec{r}) \times curl(\vec{v}_{POT}(t,\vec{r})) = \frac{1}{2} \cdot \nabla\vec{v}^2_{POT}(t,\vec{r}).$$

We obtain the solution for pressure

$$p(t,\vec{r}) = \tilde{p}(t,\vec{r}) + \rho \cdot [-\partial\Phi(t,\vec{r})/\partial t - \frac{(\nabla\Phi(t,\vec{r}))^2}{2} + \frac{1}{\rho}.(\frac{4.\eta}{3} + \zeta).\Delta\Phi(t,\vec{r})] + f(t) + \Phi_{ex}(t,\vec{r}) \tag{52}$$

where $f(t)$ is an arbitrary function of time and $\tilde{p}(t,\vec{r})$ is the solution of the equation

$$\Delta\tilde{p}(t,\vec{r}) = -\rho \cdot div[(\vec{v}_{SOL}(t,\vec{r}) \cdot \nabla) \cdot \vec{v}_{POT}(t,\vec{r})]. \tag{53}$$

In presence of the divergence-free and curl-free fields, the system of the modified Navier-Stokes equations is

$$\vec{v}(t,\vec{r}) = \vec{v}_{SOL}(t,\vec{r}) + \vec{v}_{POT}(t,\vec{r}), \vec{f}_{ex}(t,\vec{r}) = \nabla\Phi_{ex}(t,\vec{r}) + \vec{f}_{ex,SOL}(t,\vec{r}),$$

$$p(t,\vec{r}) = \tilde{p}(t,\vec{r}) + \partial.[-\partial\Phi(t,\vec{r})/\partial t - \frac{(\nabla\Phi(t,\vec{r}))^2}{2} + \frac{1}{\rho}.(\frac{4.\eta}{3} + \zeta).\Delta\Phi(t,\vec{r})] + f(t) + \Phi_{ex}(t,\vec{r}), \tag{54}$$

$$\Delta\tilde{p}(t,\vec{r}) = -\rho \cdot div[(\vec{v}_{SOL}(t,\vec{r}) \cdot \nabla) \cdot \vec{v}_{POT}(t,\vec{r})], \tag{55}$$

$$\rho \cdot \partial\vec{v}_{SOL}(t,\vec{r})/\partial t + \rho \cdot (\vec{v}(t,\vec{r}) \cdot \nabla) \cdot \vec{v}_{SOL}(t,\vec{r}) = \eta \cdot \Delta\vec{v}_{SOL}(t,\vec{r}) + \vec{f}_{ex,SOL}(t,\vec{r}), \tag{56}$$

$$div(\vec{v}(t,\vec{r})) = \Delta\Phi(t,\vec{r}). \tag{57}$$

These equations should be complemented by conditions (3) and (4).

The existence of the general solution of Equation (56) is proved in the next part. The whole system (54–57) with conditions (3) and (4) is solvable in three dimensions. The Fourier method allows obtaining the smooth solutions of these equations.

3.2. Incompressibility of fluid

If we accept that $div(\vec{v}(t,\vec{r})) \neq 0$, a crucial problem is whether the fluid is incompressible. According to the work of (Batchelor, 2000), the influence of pressure variations on the value of fluid density is negligible if

$$|\frac{1}{\rho} \cdot \frac{D\rho}{Dt}| << \frac{U}{L}|div(\vec{v}(t,\vec{r}))| << \frac{U}{L} \tag{58}$$

where L is the length scale ($\vec{v}(t,\vec{r})$ varies slightly over distances small compared with the scale), U is the value of the variations of $\left|\vec{v}(t,\vec{r})\right|$ with respect to both position and time. *It means that the velocity distribution is only approximately divergence-free.*

In our research, we suppose that assumption (58) is valid, i.e. a fluid is incompressible. We only accept the existence of a curl-free velocity field besides a divergence-free one on the condition that

$$div(\vec{v}_{SOL}(t,\vec{r})) = 0, \left|div(\vec{v}_{POT}(t,\vec{r}))\right| << \frac{U}{L}, but div(\vec{v}_{POT}(t,\vec{r})) \neq 0. \tag{59}$$

It is performed for a non-stationary flow if (Landau & Lifshitz, 1987)

$$\left|\vec{v}(t,\vec{r})\right| << c, \tag{60}$$

$$\tau >> \frac{L}{c_s}, \tag{61}$$

c_s is the speed of sound in the fluid and τ is the time scale. The last condition means that the time during which the sound passes the distance equal to the length scale has to be much less that of the time scale τ during which the motion of the fluid significantly changes, i.e. propagation of interactions in the fluid is instantaneous.

The condition (59) is necessary if we want to investigate the nonlinear phenomena of an incompressible fluid motion. Small terms in nonlinear equations can have a significant influence on behavior of a physical system in space and time as a result of a cumulative effect (Bogoliubov & Mitropolski, 1961). The equations of the viscous fluid motion become more accurate at the account of curl-free vector field.

3.3. Existence of solutions

Let's solve Equation (56).

$$\partial\vec{v}_{SOL}(t,\vec{r})/\partial t + (\vec{v}(t,\vec{r})\cdot\nabla)\vec{v}_{SOL}(t,\vec{r}) = \nu\cdot\Delta\vec{v}_{SOL}(t,\vec{r}) + \vec{f}_{ex,SOL}(t,\vec{r}), \tag{62}$$

At first, let's consider this equation without $\vec{f}_{ex,SOL}(t,\vec{r})$:

$$\partial\vec{v}_{SOL}(t,\vec{r})/\partial t + (\vec{v}(t,\vec{r})\cdot\nabla)\vec{v}_{SOL}(t,\vec{r}) = \nu\cdot\Delta\vec{v}_{SOL}(t,\vec{r}). \tag{63}$$

We define $\vec{v}_{SOL}(t,\vec{r})$ and $\vec{v}(t,\vec{r})$ as a uniformly converging series

$$\vec{v}_{SOL}(t,\vec{r}) = \vec{u}(t,\vec{r}) = \sum_{j=1}^{\infty}\vec{u}_j(t,\vec{r}), \vec{v}(t,\vec{r}) = \sum_{j=1}^{\infty}\vec{v}_j(t,\vec{r}), \tag{64}$$

$$\upsilon = \frac{\eta}{\rho}. \tag{65}$$

$$\begin{gathered}
\partial\vec{u}_1(t,\vec{r})/\partial t = \upsilon\cdot\Delta\vec{u}_1(t,\vec{r}) \\
\partial\vec{u}_2(t,\vec{r})/\partial t + (\vec{v}_1(t,\vec{r})\cdot\nabla)\cdot\vec{u}_1(t,\vec{r}) = \upsilon\cdot\Delta\vec{u}_2(t,\vec{r}), \\
\partial\vec{u}_3(t,\vec{r})/\partial t + (\vec{v}_2(t,\vec{r})\cdot\nabla)\cdot\vec{u}_1(t,\vec{r}) + (\vec{v}_1(t,\vec{r})\cdot\nabla)\cdot\vec{u}_2(t,\vec{r}) = \upsilon\cdot\Delta\vec{u}_3(t,\vec{r}), \\
\ldots \\
\partial\vec{u}_j(t,\vec{r})/\partial t + \sum_{m=1}^{j-1}[(\vec{v}_{j-m}(t,\vec{r})\cdot\nabla)\cdot\vec{u}_m(t,\vec{r}) + (\vec{v}_m(t,\vec{r})\cdot\nabla)\cdot\vec{u}_{j-m}(t,\vec{r})] = \upsilon\cdot\Delta\vec{u}_j(t,\vec{r}),
\end{gathered} \tag{66}$$

We obtained the infinite set of the inhomogeneous linear heat equations. The existence and uniqueness theorem is proved for them (Tikhonov & Samarskii, 2013). The infinite series of private solutions of the heat equations converges uniformly on $[0, \infty) \times R^3$. So we can obtain the unique solution of Equation (56) if the initial and boundary conditions are defined.

The presence $\vec{f}_{ex,SOL}(t,\vec{r})$ in the right part of the equation doesn't change the reasoning. But we will show that there exists a turbulent solution of Equation (56) besides the unique solution for a laminar flow.

At the beginning, we obtain the solution of Equation (56) for a laminar flow which can blow up as a result of a resonance effect.

3.4. Resonant blowup of laminar flow

We consider the equation

$$\partial\vec{u}(t,\vec{r})/\partial t + (\vec{v}(t,\vec{r}) \cdot \nabla)\vec{u}(t,\vec{r}) = \upsilon \cdot \Delta\vec{u}(t,\vec{r}). \tag{67}$$

Let's suppose that

$$\vec{v}(t,\vec{r}) \equiv \vec{v}_1(t,\vec{r}). \tag{68}$$

to simplify the expressions.

We obtain the system

$$\begin{aligned}
&\partial\vec{u}_1(t,\vec{r})/\partial t = \upsilon \cdot \Delta\vec{u}_1(t,\vec{r}),\\
&\partial\vec{u}_2(t,\vec{r})/\partial t + (\vec{v}(t,\vec{r}) \cdot \nabla) \cdot \vec{u}_1(t,\vec{r}) = \upsilon \cdot \Delta\vec{u}_2(t,\vec{r}),\\
&\partial\vec{u}_3(t,\vec{r})/\partial t + (\vec{v}(t,\vec{r}) \cdot \nabla) \cdot \vec{u}_2(t,\vec{r}) = \upsilon \cdot \Delta\vec{u}_3(t,\vec{r}),\\
&\ldots\\
&\partial\vec{u}_j(t,\vec{r})/\partial t + (\vec{v}(t,\vec{r}) \cdot \nabla) \cdot \vec{u}_{j-1}(t,\vec{r}) = \upsilon \cdot \Delta\vec{u}_j(t,\vec{r}),
\end{aligned} \tag{69}$$

We define the solutions in the form

$$v(t,\vec{r}) = \exp(-\omega_0^2\upsilon \cdot t) \cdot \vec{v}(\vec{r}), \tag{70}$$

$$\vec{u}_j(t,\vec{r}) = \exp(-[(j-1) \cdot \omega_0^2 + \lambda_0^2] \cdot \upsilon \cdot t) \cdot \vec{u}_j(\vec{r}), j = 1, 2, \ldots, \infty. \tag{71}$$

We obtain the linear system of the equations

$$\begin{aligned}
&\Delta\vec{u}_1(\vec{r}) + \lambda_0^2 \cdot \vec{u}_1(\vec{r}) = 0,\\
&\Delta\vec{u}_2(\vec{r}) + (\omega_0^2 + \lambda_0^2) \cdot \vec{u}_2(\vec{r}) = \frac{(\vec{v}(\vec{r}) \cdot \nabla) \cdot \vec{u}_1(\vec{r})}{\upsilon},\\
&\Delta\vec{u}_3(\vec{r}) + (2\omega_0^2 + \lambda_0^2) \cdot \vec{u}_3(\vec{r}) = \frac{(\vec{v}(\vec{r}) \cdot \nabla) \cdot \vec{u}_2(\vec{r})}{\upsilon},\\
&\ldots\\
&\Delta\vec{u}_j(t,\vec{r}) + [(j-1) \cdot \omega_0^2 + \lambda_0^2] \cdot \vec{u}_j(\vec{r}) = \frac{(\vec{v}(\vec{r}) \cdot \nabla) \cdot \vec{u}_{j-1}(\vec{r})}{\upsilon},
\end{aligned} \tag{72}$$

The solution of the first equation is

$$\vec{u}_1(\vec{r}) = \vec{u}_0 \cdot \cos(\vec{k}\vec{r}), k^2 = \lambda_0^2 \quad \cdot \tag{73}$$

Let's obtain the solutions of the second and the third equations. We suppose that

$$\vec{v}(\vec{r}) = \vec{v}_0 \cdot \cos(\vec{k}_0\vec{r}). \tag{74}$$

$$\begin{aligned}\Delta\vec{u}_2(\vec{r}) + (\omega_0^2 + \lambda_0^2) \cdot \vec{u}_2(\vec{r}) &= -\frac{(\vec{v}_0 \cdot \vec{k})}{\upsilon} \cdot \cos(\vec{k}_0\vec{r}) \cdot \sin(\vec{k}\vec{r}) \cdot \vec{u}_0 \\ &= -\frac{(\vec{v}_0 \cdot \vec{k})}{2 \cdot \upsilon} \cdot \{\sin[(\vec{k} + \vec{k}_0) \cdot \vec{r} + \sin[(\vec{k} - \vec{k}_0) \cdot \vec{r}]\} \cdot \vec{u}_0.\end{aligned} \tag{75}$$

$$\vec{u}_2(\vec{r}) = \frac{(\vec{v}_0 \cdot \vec{k}) \cdot \vec{u}_0}{2 \cdot \upsilon \cdot [(\vec{k} + \vec{k}_0)^2 - (\omega_0^2 + \lambda_0^2)]} \cdot \sin[(\vec{k} + \vec{k}_0) \cdot \vec{r}] + \frac{(\vec{v}_0 \cdot \vec{k}) \cdot \vec{u}_0}{2 \cdot \upsilon \cdot [(\vec{k} - \vec{k}_0)^2 - (\omega_0^2 + \lambda_0^2)]} \sin[(\vec{k} - \vec{k}_0) \cdot \vec{r}]. \tag{76}$$

$$\begin{aligned}\Delta\vec{u}_3(\vec{r}) + (2\omega_0^2 + \lambda_0^2) \cdot \vec{u}_3(\vec{r}) &= \frac{[(\vec{v}_0 \cdot \vec{k})^2 + (\vec{v}_0 \cdot \vec{k}) \cdot (\vec{v}_0 \cdot \vec{k}_0)] \cdot \vec{u}_0}{4 \cdot \upsilon \cdot [(\vec{k} + \vec{k}_0)^2 - (\omega_0^2 + \lambda_0^2)]} \cdot \{\cos[(\vec{k} + 2\vec{k}_0) \cdot \vec{r}] + \cos(\vec{k}\vec{r})\} \\ &+ \frac{[(\vec{v}_0 \cdot \vec{k})^2 - (\vec{v}_0 \cdot \vec{k}) \cdot (\vec{v}_0 \cdot \vec{k}_0)] \cdot \vec{u}_0}{4 \cdot \upsilon \cdot [(\vec{k} - \vec{k}_0)^2 - (\omega_0^2 + \lambda_0^2)]} \cdot \{\cos(\vec{k}\vec{r}) + \cos[(\vec{k} - 2\vec{k}_0) \cdot \vec{r}]\}.\end{aligned} \tag{77}$$

$$\begin{aligned}\vec{u}_3(\vec{r}) &= \frac{\vec{u}_0 \cdot \cos(\vec{k}\vec{r})}{4 \cdot \upsilon \cdot (2\omega_0^2 + \lambda_0^2 - \vec{k}^2)} \cdot \{\frac{(\vec{v}_0 \cdot \vec{k})^2 + (\vec{v}_0 \cdot \vec{k}) \cdot (\vec{v}_0 \cdot \vec{k}_0)}{(\vec{k} + \vec{k}_0)^2 - (\omega_0^2 + \lambda_0^2)} + \frac{(\vec{v}_0 \cdot \vec{k})^2 - (\vec{v}_0 \cdot \vec{k}) \cdot (\vec{v}_0 \cdot \vec{k}_0)}{(\vec{k} - \vec{k}_0)^2 - (\omega_0^2 + \lambda_0^2)}\} \\ &+ \frac{[(\vec{v}_0 \cdot \vec{k})^2 + (\vec{v}_0 \cdot \vec{k}) \cdot (\vec{v}_0 \cdot \vec{k}_0)] \cdot \vec{u}_0}{4 \cdot \upsilon \cdot (2\omega_0^2 + \lambda_0^2 - (\vec{k} + 2\vec{k}_0)^2) \cdot [(\vec{k} + \vec{k}_0)^2 - (\omega_0^2 + \lambda_0^2)]} \cdot \cos[(\vec{k} + 2\vec{k}_0) \cdot \vec{r}] \\ &+ \frac{[(\vec{v}_0 \cdot \vec{k})^2 - (\vec{v}_0 \cdot \vec{k}) \cdot (\vec{v}_0 \cdot \vec{k}_0)] \cdot \vec{u}_0}{4 \cdot \upsilon \cdot (2\omega_0^2 + \lambda_0^2 - (\vec{k} - 2\vec{k}_0)^2) \cdot [(\vec{k} - \vec{k}_0)^2 - (\omega_0^2 + \lambda_0^2)]} \cdot \cos[(\vec{k} - 2\vec{k}_0) \cdot \vec{r}].\end{aligned} \tag{78}$$

We study the nonlinear effect which is described as the interaction between the plane wave functions having the initial wave vectors $\vec{k}$ and $\vec{k}_0$. The third and higher order approximations of the solution obtain the term which wave vector coincident with the initial wave vector $\vec{k}$. It is the term which has the function $\cos(\vec{k}\vec{r})$. It means that there is a resonance which increases the amplitude of the solution with time (Landau & Lifshitz, 1976).

Let's study this effect more carefully.

We accept at the next step that

$$\vec{k}_0 = \vec{k}, \vec{k}^2 = \lambda_0^2, v_0 \uparrow\uparrow \vec{k}. \tag{79}$$

We obtain

$$\Delta\vec{u}_2(\vec{r}) + (\omega_0^2 + \lambda_0^2) \cdot \vec{u}_2(\vec{r}) = -\frac{v_0 \cdot k}{2 \cdot \upsilon} \cdot \sin(2\vec{k}\vec{r}) \cdot \vec{u}_0, \tag{80}$$

$$\vec{u}_2(\vec{r}) = \frac{v_0 \cdot k \cdot \vec{u}_0}{2 \cdot \upsilon \cdot (3\lambda_0^2 - \omega_0^2)} \cdot \sin(2\vec{k}\vec{r}). \tag{81}$$

$$\Delta\vec{u}_3(\vec{r}) + (2\omega_0^2 + \lambda_0^2) \cdot \vec{u}_3(\vec{r}) = \frac{v_0^2 \cdot \lambda_0^2 \cdot \vec{u}_0}{2 \cdot \upsilon^2 \cdot (3\lambda_0^2 - \omega_0^2)} \cdot [\cos(\vec{k}\vec{r}) + \cos(3\vec{k}\vec{r})], \quad (82)$$

$$\vec{u}_3(\vec{r}) = \frac{v_0^2 \cdot \lambda_0^2 \cdot \cos(\vec{k}\vec{r}) \cdot \vec{u}_0}{4 \cdot \upsilon^2 \cdot \omega_0^2 \cdot (3\lambda_0^2 - \omega_0^2)} + \frac{v_0^2 \cdot \lambda_0^2 \cdot \cos(3\vec{k}\vec{r}) \cdot \vec{u}_0}{4 \cdot \upsilon^2 \cdot (\omega_0^2 - 4\lambda_0^2) \cdot (3\lambda_0^2 - \omega_0^2)}. \quad (83)$$

Let's obtain the next order of solutions to establish the general form of the resonance amplitude. We are interested in the solutions only with $\cos(\vec{k}\vec{r})$. We define them as $\vec{u}_{2j+1}^1(\vec{r})$. We need also the solutions $\vec{u}_{2j}^1(\vec{r})$with $\sin(2\vec{k}\vec{r})$to obtain $\vec{u}_{2j+1}^1(\vec{r})$.

$$\vec{u}_3(\vec{r}) = \frac{v_0^2 \cdot \lambda_0^2 \cdot \cos(\vec{k}\vec{r}) \cdot \vec{u}_0}{2 \cdot \upsilon^2 \cdot 2\omega_0^2 \cdot (3\lambda_0^2 - \omega_0^2)} = \frac{v_0^2 \cdot \lambda_0^2 \cdot \cos(\vec{k}\vec{r}) \cdot \vec{u}_0}{2^2 \cdot \upsilon^2 \cdot \omega_0^2 \cdot (3\lambda_0^2 - \omega_0^2)}. \quad (84)$$

$$\Delta\vec{u}_4^1(\vec{r}) + (3\omega_0^2 + \lambda_0^2) \cdot \vec{u}_4^1(\vec{r}) = \frac{(\vec{v}(\vec{r}) \cdot \nabla) \cdot \vec{u}_3^1(\vec{r})}{\upsilon}, \quad (85)$$

$$\Delta\vec{u}_4^1(\vec{r}) + (3\omega_0^2 + \lambda_0^2) \cdot \vec{u}_4^1(\vec{r}) = -\frac{v_0^3 \cdot \lambda_0^3 \cdot \sin(2\vec{k}\vec{r}) \cdot \vec{u}_0}{2^2 \cdot \upsilon^3 \cdot 2\omega_0^2 \cdot (3\lambda_0^2 - \omega_0^2)} = -\frac{v_0^3 \cdot \lambda_0^3 \cdot \sin(2\vec{k}\vec{r}) \cdot \vec{u}_0}{2^3 \cdot \upsilon^3 \cdot \omega_0^2 \cdot (3\lambda_0^2 - \omega_0^2)}, \quad (86)$$

$$\vec{u}_4^1(\vec{r}) = \frac{v_0^3 \cdot \lambda_0^3 \cdot \sin(2\vec{k}\vec{r}) \cdot \vec{u}_0}{2^3 \cdot \upsilon^3 \cdot \omega_0^2 \cdot (3\lambda_0^2 - \omega_0^2) \cdot (3\lambda_0^2 - 3\omega_0^2)}. \quad (87)$$

$$\Delta\vec{u}_5^1(\vec{r}) + (4\omega_0^2 + \lambda_0^2) \cdot \vec{u}_5^1(\vec{r}) = \frac{(\vec{v}(\vec{r}) \cdot \nabla) \cdot \vec{u}_4^1(\vec{r})}{\upsilon}, \quad (88)$$

$$\Delta\vec{u}_5^1(\vec{r}) + (4\omega_0^2 + \lambda_0^2) \cdot \vec{u}_5^1(\vec{r}) = \frac{v_0^4 \cdot \lambda_0^4 \cdot \cos(\vec{k}\vec{r}) \cdot \vec{u}_0}{2^3 \cdot \upsilon^4 \cdot \omega_0^2 \cdot (3\lambda_0^2 - \omega_0^2) \cdot (3\lambda_0^2 - 3\omega_0^2)}, \quad (89)$$

$$\vec{u}_5^1(\vec{r}) = \frac{v_0^4 \cdot \lambda_0^4 \cdot \cos(\vec{k}\vec{r}) \cdot \vec{u}_0}{2^3 \cdot \upsilon^4 \cdot 4 \cdot \omega_0^4 \cdot (3\lambda_0^2 - \omega_0^2) \cdot (3\lambda_0^2 - 3\omega_0^2)} = \frac{v_0^4 \cdot \lambda_0^4 \cdot \cos(\vec{k}\vec{r}) \cdot \vec{u}_0}{2^4 \cdot \upsilon^4 \cdot 1 \cdot 2 \cdot \omega_0^4 \cdot (3\lambda_0^2 - \omega_0^2) \cdot (3\lambda_0^2 - 3\omega_0^2)}. \quad (90)$$

$$\Delta\vec{u}_6^1(\vec{r}) + (5\omega_0^2 + \lambda_0^2) \cdot \vec{u}_6^1(\vec{r}) = \frac{(\vec{v}(\vec{r}) \cdot \nabla) \cdot \vec{u}_5^1(\vec{r})}{\upsilon}. \quad (91)$$

$$\Delta\vec{u}_6^1(\vec{r}) + (5\omega_0^2 + \lambda_0^2) \cdot \vec{u}_6^1(\vec{r}) = \frac{v_0^5 \cdot \lambda_0^5 \cdot \sin(2\vec{k}\vec{r}) \cdot \vec{u}_0}{2^5 \cdot \upsilon^5 \cdot 1 \cdot 2 \cdot \omega_0^4 \cdot (3\lambda_0^2 - \omega_0^2) \cdot (3\lambda_0^2 - 3\omega_0^2)}, \quad (92)$$

$$\vec{u}_6^1(\vec{r}) = \frac{v_0^5 \cdot \lambda_0^5 \cdot \sin(2\vec{k}\vec{r}) \cdot \vec{u}_0}{2^5 \cdot \upsilon^5 \cdot 1 \cdot 2 \cdot \omega_0^4 \cdot (3\lambda_0^2 - \omega_0^2) \cdot (3\lambda_0^2 - 3\omega_0^2) \cdot (3\lambda_0^2 - 5\omega_0^2)}. \quad (93)$$

$$\Delta\vec{u}_7^1(\vec{r}) + (6\omega_0^2 + \lambda_0^2) \cdot \vec{u}_7^1(\vec{r}) = \frac{(\vec{v}(\vec{r}) \cdot \nabla) \cdot \vec{u}_5^1(\vec{r})}{\upsilon}, \quad (94)$$

$$\Delta\vec{u}_7^1(\vec{r}) + (6\omega_0^2 + \lambda_0^2) \cdot \vec{u}_7^1(\vec{r}) = \frac{v_0^6 \cdot \lambda_0^6 \cdot \cos(\vec{k}\vec{r}) \cdot \vec{u}_0}{2^5 \cdot \upsilon^6 \cdot 1 \cdot 2 \cdot \omega_0^4 \cdot (3\lambda_0^2 - \omega_0^2) \cdot (3\lambda_0^2 - 3\omega_0^2) \cdot (3\lambda_0^2 - 5\omega_0^2)}, \quad (95)$$

$$\vec{u}_7^1(\vec{r}) = \frac{v_0^6 \cdot \lambda_0^6 \cdot \cos(\vec{k}\vec{r}) \cdot \vec{u}_0}{2^5 \cdot \upsilon^6 \cdot 1 \cdot 2 \cdot 6 \cdot \omega_0^6 \cdot (3\lambda_0^2 - \omega_0^2) \cdot (3\lambda_0^2 - 3\omega_0^2) \cdot (3\lambda_0^2 - 5\omega_0^2)}$$
$$= \frac{v_0^6 \cdot \lambda_0^6 \cdot \cos(\vec{k}\vec{r}) \cdot \vec{u}_0}{2^6 \cdot \upsilon^6 \cdot 1 \cdot 2 \cdot 3 \cdot \omega_0^6 \cdot (3\lambda_0^2 - \omega_0^2) \cdot (3\lambda_0^2 - 3\omega_0^2) \cdot (3\lambda_0^2 - 5\omega_0^2)}. \quad (96)$$

As a result, we suppose that the resonance solution is

$$\vec{u}^{res}(t, \vec{r}) = \cos(\vec{k}\vec{r}) \cdot \vec{u}_0 \cdot \sum_{n=0}^{\infty} \frac{v_0^{2n} \cdot \lambda_0^{2n} \cdot \exp[-(2n \cdot \omega_0^2 + \lambda_0^2) \cdot \upsilon \cdot t]}{2^{2n} \cdot \upsilon^{2n} \cdot \omega_0^{2n} \cdot n! \cdot \prod_{m=1}^{n} (3\lambda_0^2 - (2m-1) \cdot \omega_0^2)}, \quad \frac{3\lambda_0^2}{\omega_0^2} \neq 2m - 1. \quad (97)$$

We define

$$A_n(t, v_0, \omega_0, \lambda_0) = \frac{v_0^{2n} \cdot \lambda_0^{2n} \cdot \exp[-(2n \cdot \omega_0^2 + \lambda_0^2) \cdot \upsilon \cdot t]}{2^{2n} \cdot \upsilon^{2n} \cdot \omega_0^{2n} \cdot n! \cdot \prod_{m=1}^{n} (3\lambda_0^2 - (2m-1) \cdot \omega_0^2)}, \quad (98)$$

$$A_n(v_0, \omega_0, \lambda_0) = \frac{v_0^{2n} \cdot \lambda_0^{2n}}{2^{2n} \cdot \upsilon^{2n} \cdot \omega_0^{2n} \cdot n! \cdot \prod_{m=1}^{n} (3\lambda_0^2 - (2m-1) \cdot \omega_0^2)}, \quad (99)$$

$$\vec{u}^{res}(t, \vec{r}) = \cos(\vec{k}\vec{r}) \cdot \vec{u}_0 \cdot \sum_{n=0}^{\infty} A_n(t, v_0, \omega_0, \lambda_0) = \cos(\vec{k}\vec{r}) \cdot \vec{u}_0 \cdot \sum_{n=0}^{\infty} A_n(v_0, \omega_0, \lambda_0) \cdot \exp[-(2n \cdot \omega_0^2 + \lambda_0^2) \cdot \upsilon \cdot t]. \quad (100)$$

$$\left| \frac{A_{n+1}(\vec{v}_0, \vec{k}, \omega_0, \lambda_0)}{A_n(\vec{v}_0, \vec{k}, \omega_0, \lambda_0)} \right| = \left| \frac{v_0^2 \cdot \lambda_0^2 \cdot \exp(-2 \cdot \omega_0^2 \cdot \upsilon \cdot t)}{2^2 \cdot \upsilon^2 \cdot \omega_0^2 \cdot (n+1) \cdot (3\lambda_0^2 - (2n+1) \cdot \omega_0^2)} \right| \to 0 \text{ for } n \to \infty, \forall t \in [0, \infty). \quad (101)$$

According to d'Alembert's ratio test (Fihtengolts, 2003b), the obtained series converges absolutely for any $t \in [0, \infty)$ if

$$\frac{3\lambda_0^2}{\omega_0^2} \neq 2m - 1. \quad (102)$$

Let's investigate the behavior of the solution if

$$\frac{3\lambda_0^2}{\omega_0^2} = 2m - 1. \quad (103)$$

For example, let's solve the system of Equation (72) if

$$\vec{k}_0 = \vec{k}, \lambda_0^2 = \vec{k}^2, \omega_0^2 = 3 \cdot \lambda_0^2, v_0 \uparrow\uparrow \vec{k}. \quad (104)$$

We have the second equation as

$$\Delta \vec{u}_2(t, \vec{r}) - \frac{1}{\upsilon} \cdot \partial \vec{u}_2(t, \vec{r}) / \partial t = \frac{v_0 \cdot k}{2 \cdot \upsilon} \cdot \exp(-4\lambda_0^2 \cdot t) \cdot \sin(2\vec{k}\vec{r}) \cdot \vec{u}_0. \quad (105)$$

We find the solution as

$$\vec{u}_2(\vec{r}) = u_2 \cdot t \cdot \exp(-4\lambda_0^2 \cdot \upsilon \cdot t) \cdot \sin(2\vec{k}\vec{r}) \cdot \vec{u}_0. \quad (106)$$

We obtain

$$-4 \cdot k^2 \cdot u_0 \cdot t - \frac{u_2}{\upsilon} + 4\lambda_0^2 \cdot u_0 \cdot t = \frac{v_0 \cdot k}{2 \cdot \upsilon}, \tag{107}$$

$$u_2 = -\frac{v_0 \cdot k}{2}. \tag{108}$$

$$\vec{u}_2(\vec{r}) = -\frac{v_0 \cdot \lambda_0 \cdot \vec{u}_0}{2} \cdot t \cdot \exp(-4\lambda_0^2 \cdot \upsilon \cdot t) \cdot \sin(2\vec{k}\vec{r}). \tag{109}$$

The next equation is

$$\Delta\vec{u}_3(\vec{r}) - \frac{1}{\upsilon}\partial\vec{u}_3(t,\vec{r})/\partial t = -\frac{v_0^2 \cdot \lambda_0^2 \cdot \vec{u}_0}{2 \cdot \upsilon} \cdot t \cdot \exp(-7\lambda_0^2 \cdot \upsilon \cdot t) \cdot [\cos(\vec{k}\vec{r}) + \cos(3\vec{k}\vec{r})]. \tag{110}$$

We separate it into two equations:

$$\Delta\vec{u}_3(\vec{r}) - \frac{1}{\upsilon}\partial\vec{u}_3(t,\vec{r})/\partial t = -\frac{v_0^2 \cdot \lambda_0^2 \cdot \vec{u}_0}{2 \cdot \upsilon} \cdot t \cdot \exp(-7\lambda_0^2 \cdot \upsilon \cdot t) \cdot \cos(\vec{k}\vec{r}) \tag{111}$$

and

$$\Delta\vec{u}_3(\vec{r}) - \frac{1}{\upsilon}\partial\vec{u}_3(t,\vec{r})/\partial t = -\frac{v_0^2 \cdot \lambda_0^2 \cdot \vec{u}_0}{2 \cdot \upsilon} \cdot t \cdot \exp(-7\lambda_0^2 \cdot \upsilon \cdot t) \cdot \cos(3\vec{k}\vec{r}). \tag{112}$$

We find the solution of the first equation as

$$\vec{u}_3(t,\vec{r}) = u_3 \cdot (t + a) \cdot \exp(-7\lambda_0^2 \cdot \upsilon \cdot t) \cdot \cos(\vec{k}\vec{r}) \cdot \vec{u}_0 \tag{113}$$

where a is an unknown constant.

$$(-k^2 + 7\lambda_0^2) \cdot u_3 \cdot t + (-k^2 + 7\lambda_0^2) \cdot u_3 \cdot a - \frac{u_3}{\upsilon} = -\frac{v_0^2 \cdot \lambda_0^2}{2 \cdot \upsilon} \cdot t, \tag{114}$$

$$u_3 = -\frac{v_0^2}{12 \cdot \upsilon}, a = \frac{1}{6 \cdot \upsilon \cdot \lambda_0^2}. \tag{115}$$

We find the solution of the second equation as

$$\vec{u}_3(t,\vec{r}) = u_3 \cdot (t + a) \cdot \exp(-7\lambda_0^2 \cdot \upsilon \cdot t) \cdot \cos(3\vec{k}\vec{r}) \cdot \vec{u}_0 \tag{116}$$

where a is an unknown constant.

$$(-9k^2 + 7\lambda_0^2) \cdot u_3 \cdot t + (-9k^2 + 7\lambda_0^2) \cdot u_3 \cdot a - \frac{u_3}{\upsilon} = -\frac{v_0^2 \cdot \lambda_0^2}{2 \cdot \upsilon} \cdot t, \tag{117}$$

$$u_3 = \frac{v_0^2}{4 \cdot \upsilon}, a = -\frac{1}{2 \cdot \upsilon \cdot \lambda_0^2}. \tag{118}$$

We obtain the solution of Equation (110) as the sum of the solutions of Equations (111) and (112):

$$\vec{u}_3(t,\vec{r}) = -\frac{v_0^2}{12 \cdot \upsilon} \cdot (t + \frac{1}{6 \cdot \upsilon \cdot \lambda_0^2}) \cdot \exp(-7\lambda_0^2 \cdot \upsilon \cdot t) \cdot \cos(\vec{k}\vec{r}) \cdot \vec{u}_0$$
$$+\frac{v_0^2}{4 \cdot \upsilon} \cdot (t - \frac{1}{2 \cdot \upsilon \cdot \lambda_0^2}) \cdot \exp(-7\lambda_0^2 \cdot \upsilon \cdot t) \cdot \cos(3\vec{k}\vec{r}) \cdot \vec{u}_0. \tag{119}$$

We have obtained the secular terms increasing the solution for some period of time. The time of growth (the blowup time) is not infinite due to the presence of the exponential function. This time is proportional to the value $\frac{1}{\upsilon\cdot\lambda_0^2}$. It can be very large if $\upsilon\cdot\lambda_0^2 << 1$ and becomes infinitesimal if $\lambda_0^2 \to 0$. The maximum of the time distribution of the solution approaches infinity.

3.5. Laminar flow without resonance

The increase of the solution as the result of the resonance of two fields is possible if a physical nature of the values $\vec{v}(t,\vec{r})$ and $\vec{u}(t,\vec{r})$is different. For example, one is a fluid velocity, another is a magnetic field. If $\vec{v}(t,\vec{r})$ and $\vec{u}(t,\vec{r})$ describe a divergence-free and curl-free fluid velocity field, the sum of them is a velocity of a fluid. It is considered that there exists no spontaneous growth of a velocity (kinetic energy) for a closed physical system in the absence of an external source (Landau & Lifshitz, 1976). Actually, there is a change of basic values of solution parameters to be compared with their unperturbed values in higher approximations of nonlinear equations. This leads to the suppression of the resonance effect. We can remove the resonance terms using the method of Poincare (Landau & Lifshitz, 1976).

We suppose that

$$\vec{k} = \vec{k}^{(0)} + \vec{k}^{(1)} + \vec{k}^{(2)} + \ldots, (k^{(0)})^2 = \lambda_0^2, v_0 \uparrow\uparrow \vec{k}, \vec{k}^{(0)} \uparrow\uparrow \vec{k}^{(1)} \uparrow\uparrow \vec{k}^{(2)} \uparrow\uparrow \ldots. \tag{120}$$

Our purpose is to delete the resonance term with $\cos(\vec{k}\vec{r})$.

We modify the first equation of the system (72):

$$\Delta\vec{u}_1(\vec{r}) + \lambda_0^2\cdot\vec{u}_1(\vec{r}) = 0,$$

$$\frac{\lambda_0^2}{k^2}\cdot\Delta\vec{u}_1(\vec{r}) + \lambda_0^2\cdot\vec{u}_1(\vec{r}) = -(1-\frac{\lambda_0^2}{k^2})\cdot\Delta\vec{u}_1(\vec{r}),$$

$$\begin{aligned}\frac{\lambda_0^2}{k^2}\cdot\Delta\vec{u}_1(\vec{r}) + \lambda_0^2\cdot\vec{u}_1(\vec{r}) &= -\frac{(\vec{k}^{(0)} + \vec{k}^{(1)} + \vec{k}^{(2)} + \ldots)^2 - \lambda_0^2}{k^2}\cdot\Delta\vec{u}_1(\vec{r})\\ &= -\frac{(k^{(1)})^2 + (k^{(2)})^2 + \ldots + 2\cdot\lambda_0\cdot k^{(1)} + 2\cdot\lambda_0\cdot k^{(2)} + 2\cdot k^{(1)}\cdot k^{(2)} + \ldots}{k^2}\cdot\Delta\vec{u}_1(\vec{r}).\end{aligned} \tag{121}$$

We construct the system using Equation (72) (Landau & Lifshitz, 1976):

$$\frac{\lambda_0^2}{k^2}\cdot\Delta\vec{u}_1(\vec{r}) + \lambda_0^2\cdot\vec{u}_1(\vec{r}) = 0,$$

$$\Delta\vec{u}_2(\vec{r}) + (\omega_0^2 + \lambda_0^2)\cdot\vec{u}_2(\vec{r}) = \frac{(\vec{v}(\vec{r})\cdot\nabla)\cdot\vec{u}_1(\vec{r})}{\upsilon} - 2\cdot\lambda_0\cdot k^{(1)}\cdot\Delta\vec{u}_1(\vec{r}),$$

$$\Delta\vec{u}_3(\vec{r}) + (2\omega_0^2 + \lambda_0^2)\cdot\vec{u}_3(\vec{r}) = \frac{(\vec{v}(\vec{r})\cdot\nabla)\cdot\vec{u}_2(\vec{r})}{\upsilon} - \frac{(k^{(1)})^2 + 2\cdot\lambda_0\cdot k^{(2)}}{k^2}\cdot\Delta\vec{u}_3(\vec{r}), \tag{122}$$

$$\ldots$$

$$\frac{\lambda_0^2}{k^2}\cdot\Delta\vec{u}_j(t,\vec{r}) + [(j-1)\cdot\omega_0^2 + \lambda_0^2]\cdot\vec{u}_j(\vec{r}) = \frac{(\vec{v}(\vec{r})\cdot\nabla)\cdot\vec{u}_{j-1}(\vec{r})}{\upsilon} - \frac{(k^{(1)})^{j-1} + 2\cdot\lambda_0\cdot k^{(j-1)} + \ldots}{k^2}\cdot\Delta\vec{u}_1(t,\vec{r}),$$

The solution of the first equation is

$$\vec{u}_1(\vec{r}) = \vec{u}_0 \cdot \cos(\vec{k}\vec{r}). \tag{123}$$

The second equation is

$$\Delta\vec{u}_2(\vec{r}) + (\omega_0^2 + \lambda_0^2) \cdot \vec{u}_2(\vec{r}) = -\frac{V_0 \cdot \lambda_0}{2 \cdot \upsilon} \cdot \sin(2\vec{k}\vec{r}) \cdot \vec{u}_0 - \frac{2 \cdot \lambda_0 \cdot k^{(1)}}{k^2} \cdot \Delta\vec{u}_1(\vec{r}). \tag{124}$$

There is no the term with $\cos(\vec{k}\vec{r})$in the solution if $\vec{k}^{(1)} = 0$. We have the solution of the second equation

$$\vec{u}_2(\vec{r}) = \frac{V_0 \cdot \lambda_0 \cdot \vec{u}_0}{2 \cdot \upsilon \cdot (3\lambda_0^2 - \omega_0^2)} \cdot \sin(2\vec{k}\vec{r}). \tag{125}$$

The third equation is

$$\Delta\vec{u}_3(\vec{r}) + (2\omega_0^2 + \lambda_0^2) \cdot \vec{u}_3(\vec{r}) = \frac{V_0^2 \cdot \lambda_0^2 \cdot \vec{u}_0}{2 \cdot \upsilon^2 \cdot (3\lambda_0^2 - \omega_0^2)} \cdot [\cos(\vec{k}\vec{r}) + \cos(3\vec{k}\vec{r})] - \frac{2 \cdot \lambda_0 \cdot k^{(2)}}{k^2} \cdot \Delta\vec{u}_1(\vec{r}). \tag{126}$$

$$\vec{u}_3(\vec{r}) = \frac{V_0^2 \cdot \lambda_0^2 \cdot \cos(3\vec{k}\vec{r}) \cdot \vec{u}_0}{4 \cdot \upsilon^2 \cdot (\omega_0^2 - 4\lambda_0^2) \cdot (3\lambda_0^2 - \omega_0^2)} \quad if, k^{(2)} = -\frac{V_0^2 \cdot \lambda_0}{4 \cdot \upsilon^2 \cdot (3\lambda_0^2 - \omega_0^2)} \tag{127}$$

If $\omega_0^2 = \lambda_0^2$, we obtain

$$\vec{u}_1(\vec{r}) = \vec{u}_0 \cdot \cos(\vec{k}\vec{r}), \tag{128}$$

$$\vec{u}_2(\vec{r}) = \frac{V_0 \cdot \vec{u}_0}{4 \cdot \upsilon \cdot \lambda_0} \cdot \sin(2\vec{k}\vec{r}), \vec{k}^{(1)} = 0, \tag{129}$$

$$\vec{u}_3(\vec{r}) = -\frac{V_0^2 \cdot \cos(3\vec{k}\vec{r}) \cdot \vec{u}_0}{24 \cdot \upsilon^2 \cdot \lambda_0^2} \quad , \quad k^{(2)} = -\frac{V_0^2}{8 \cdot \upsilon^2 \cdot \lambda_0}. \tag{130}$$

3.6. Laminar–turbulent transition

Let's consider the inhomogeneous equation

$$\partial\vec{u}(t,\vec{r})/\partial t + (\vec{v}(t,\vec{r}) \cdot \nabla)\vec{u}(t,\vec{r}) = \upsilon \cdot \Delta\vec{u}(t,\vec{r}) + \vec{f} \cdot \exp(-\lambda_0^2 \cdot \upsilon \cdot t) \cdot \cos(\vec{\kappa} \cdot \vec{r}), \tag{131}$$

$\vec{v}(t,\vec{r}) = \vec{v}_{POT}(t,\vec{r}) + \vec{v}_{SOL}(t,\vec{r})$ is the flow velocity,

$\vec{u}(t,\vec{r}) = \vec{v}_{SOL}(t,\vec{r})$ is the divergence-free flow velocity, and

$\vec{v}_{POT}(t,\vec{r})$ is the curl-free flow velocity.

We suppose that

$$\vec{v}(t,\vec{r}) = 2\upsilon \cdot \vec{\lambda} + \vec{V}(t,\vec{r}), \tag{132}$$

$\vec{\lambda}$ is the constant vector and $\vec{v}(t,\vec{r}) = \vec{\lambda} + \vec{V}(t,\vec{r})$ is the flow velocity vector.

We consider the linearized equation

$$\Delta\vec{u}_1(t,\vec{r}) - \frac{1}{\upsilon}\partial\vec{u}_1(t,\vec{r})/\partial t = 2\cdot(\vec{\lambda}\cdot\nabla)\vec{u}_1(t,\vec{r}) + \vec{f}\cdot\exp(-\lambda_0^2\cdot\upsilon\cdot t)\cdot\cos(\vec{\kappa}\cdot\vec{r}). \tag{133}$$

We perform the qualitative analysis of the resonance for this nonlinear equation (Landau & Lifshitz, 1976).

We find the solution of this equation as

$$\vec{u}_1(t,\vec{r}) = \vec{u}_0\cdot\exp(-\lambda_0^2\cdot\upsilon\cdot t)\cdot\cos(\vec{\kappa}\cdot\vec{r}). \tag{134}$$

We define

$$\vec{f}(t,\vec{r}) = \mathrm{Re}\{\vec{f}\cdot\exp[-\lambda_0^2\cdot\upsilon\cdot t + i\cdot(\vec{\kappa}\cdot\vec{r})]\}, \tag{135}$$

$$\vec{u}_1(t,\vec{r}) = \mathrm{Re}\{\vec{u}_0\cdot\exp[-\lambda_0^2\cdot\upsilon\cdot t + i\cdot(\vec{\kappa}\cdot\vec{r})]\}. \tag{136}$$

We obtain

$$\vec{u}_0 = \frac{\vec{f}}{(\lambda_0^2-\kappa^2) - 2i\cdot(\vec{\lambda}\cdot\vec{\kappa})} = \frac{\vec{f}\cdot\exp(i\cdot\delta)}{\sqrt{(\lambda_0^2-\kappa^2)^2 + 4\cdot(\vec{\lambda}\cdot\vec{\kappa})^2}}, tg\delta = \frac{2(\vec{\lambda}\cdot\vec{\kappa})}{\kappa^2-\lambda_0^2}, \tag{137}$$

$$\vec{u}_1(t,\vec{r}) = \frac{\vec{f}}{\sqrt{(\lambda_0^2-\kappa^2)^2 + 4\cdot(\vec{\lambda}\cdot\vec{\kappa})^2}}\cdot\exp(-\lambda_0^2\cdot\upsilon\cdot t)\cdot\cos(\vec{\kappa}\cdot\vec{r}+\delta). \tag{138}$$

We define the absolute value of the amplitude

$$u_0 = \frac{f}{\sqrt{(\lambda_0^2-\kappa^2)^2 + 4\cdot(\vec{\lambda}\cdot\vec{\kappa})^2}}. \tag{139}$$

We use the approximate expressions

$$\kappa^2 - \lambda_0^2 = (\kappa+\lambda_0)\cdot(\kappa-\lambda_0) \approx 2\lambda_0\cdot\varepsilon \tag{140}$$

where $\varepsilon = \kappa - \lambda_0$,

$$2\cdot(\vec{\lambda}\cdot\vec{\kappa}) \approx 2\cdot(\lambda\cdot\lambda_0). \tag{141}$$

We obtain

$$u_0 = \frac{f}{2\lambda_0\cdot\sqrt{\varepsilon^2+\lambda^2}}. \tag{142}$$

$$u_0^2 = \frac{f^2}{4\lambda_0^2\cdot(\varepsilon^2+\lambda^2)}. \tag{143}$$

According to formulas (127) and (130), we accept that the absolute value of the wave vector $\vec{\kappa}$ is dependent on the amplitude v_0 due to the nonlinear effect. We define this dependence as (Landau & Lifshitz, 1976)

$$\lambda_0 + \chi\cdot v_0^2. \tag{144}$$

where χ is the function dependent on the anharmonic coefficients (see (127) and (130)).

It means that we replace the value λ_0 to $\lambda_0 + \chi \cdot v_0^2$ in $\kappa - \lambda_0$ of expression (143) (Landau & Lifshitz, 1976). As a result, we use the expression $\varepsilon - \chi \cdot v_0^2$ instead of ε in (143). We take into account that

$$\vec{v}(t,\vec{r}) = (\vec{u}_0 + \vec{v}_{POT}) \cdot \exp(-\lambda_0^2 \cdot \upsilon \cdot t) \cdot \cos(\vec{\kappa} \cdot \vec{r}), \vec{v}_0 = \vec{u}_0 + \vec{v}_{POT} \tag{145}$$

where $\vec{v}_{POT}$ is the amplitude of the curl-free velocity, $\vec{u}_0 \perp \vec{v}_{POT}$, $\vec{v}_{POT} || \vec{\kappa}$.

If the directions of $\vec{v}(t,\vec{r})$ and $\vec{\kappa}$ are defined, we obtain that

$$u_0 = v_0 \cdot \sin\theta, \theta = const, 0 \le \theta \le \pi, \tag{146}$$

where θ is the constant angle between $\vec{v}(t,\vec{r})$ and $\vec{\kappa}$.

We obtain

$$u_0^2 \cdot [(\varepsilon - \chi_0 \cdot u_0^2)^2 + \lambda^2] = \frac{f^2}{4\lambda_0^2}, \chi_0 = \frac{\chi}{\sin^2\theta} \text{ or} \tag{147}$$

$$\varepsilon = \chi_0 \cdot u_0^2 \pm \sqrt{\left(\frac{f}{2\lambda_0 \cdot u_0}\right)^2 - \lambda^2}. \tag{148}$$

There is no difference between this expression and the formula presented in the work of (Landau & Lifshitz, 1976). Let's use the results of the analysis in this formula.

The greatest amplitude is

$$u_{0,max} = \frac{f}{2\lambda_0 \cdot \lambda}. \tag{149}$$

If

$$f_0 > f_k = \sqrt{\frac{32\lambda_0^2 \cdot \lambda^3}{3\sqrt{3}|\chi_0|}} \tag{150}$$

we obtain the bifurcation solutions for the amplitude u_0 besides the unique one. So it is possible to describe laminar–turbulent transition and an opposite process. A flow separation is also considered. In addition, there are resonances at external forces whose wave vectors and frequencies are greatly different from their own ones (Landau & Lifshitz, 1976).

$$tg\delta = \frac{2(\vec{\lambda} \cdot \vec{\kappa})}{\kappa^2 - \lambda_0^2} \approx \frac{\lambda}{\varepsilon - \chi \cdot u_0^2}. \tag{151}$$

The phase shift between the external force and the velocity is dependent on the velocity amplitude and varies stochastically at the turbulence.

3.7. Equations for potential field and curl-free velocity

We use the equation

$$div(\vec{v}(t,\vec{r})) = \nabla\Phi(t,r)).$$

Its solution as a plane wave satisfies the Helmholtz equation

$$\Delta\Phi(t,\vec{r}) + \xi\,.\Phi(t,\vec{r}) = 0 \tag{152}$$

$\Phi(t,\vec{r})$ is the scalar potential of the velocity vector field and ξ is the value connected with the inter-molecular interactions. In general, ξ is dependent on space coordinates and can be complex-valued.

The equivalent system of the equations for the velocity is

$$\Delta\vec{v}_{POT}(t,\vec{r}) + \xi\cdot\vec{v}_{POT}(t,\vec{r}) = 0, \tag{153}$$

$$div(\vec{v}_{SOL}(t,\vec{r})) = 0. \tag{154}$$

These equations allow constructing the complete system of functions in space R^3.

We suppose that ξ is a real value and

$$\xi > 0. \tag{155}$$

The solution of Equation (153) is

$$\vec{v}_{POT}(t,\vec{r}) = \vec{v}_{POT}(t)\cdot\cos(\vec{k}\vec{r}+\delta),\ k^2 = \xi, \delta \text{ is the phase shift.} \tag{156}$$

Let's verify the incompressibility condition

$$\left|div(\vec{v}(t,\vec{r}))\right| << \frac{U}{L} \tag{157}$$

for this solution.

$$\left|div(\vec{v}(t,\vec{r}))\right| = \left|(\vec{v}(t)\cdot\vec{k})\cdot\sin(\vec{k}\vec{r})\right| = \left|(\vec{v}_{POT}(t)\cdot k)\cdot\sin(\vec{k}\vec{r})\right|. \tag{158}$$

$$\left|\vec{v}_{POT}(t)\right| \sim U. \tag{159}$$

We obtain from the boundary conditions that

$$\cos(\vec{k}\cdot\vec{L}) = 0 \tag{160}$$

where $\vec{L} = (\pm L_x, \pm L_y, \pm L_z)$ is the boundary position.

We have that

$$\left|\vec{k}\right| \sim \frac{1}{L}. \tag{161}$$

We obtain from the scale definition that variations of $\left|\cos(\vec{k}\vec{r})\right|$ are small if $\left|\vec{r}\right| << L$. Therefore,

$$\left|\vec{k}\vec{r}\right| << 1. \tag{162}$$

$$\left|\sin(\vec{k}\vec{r})\right| \approx \left|\vec{k}\vec{r}\right| \sim \frac{\left|\vec{r}\right|}{L} << 1. \tag{163}$$

It means that condition (157) is performed. The fluid is incompressible.

We have also that

$$\xi \sim \frac{1}{L^2}. \tag{164}$$

It reflects the fact that the energy of the intermolecular interactions is extremely small.

It was shown in the previous part that laminar–turbulent transition is possible for a divergence-free velocity. We can construct equations which allow this for a potential of a curl-free field or a curl-free velocity.

Let's transform expression (152) to the equation similar to the one for oscillations of a mechanical system with an own frequency ω under the influence of a small nonlinear perturbation $\varepsilon.F(t,\vec{r},\Phi(t,\vec{r}),\partial\Phi(t,\vec{r})/\partial x,\partial\Phi(t,\vec{r})/\partial y,\partial\Phi(t,\vec{r}/\partial z)$ (Bogoliubov & Mitropolski, 1961):

$$\nabla(\Phi(t,\vec{r})+\omega^2.\Phi(t,\vec{r})=\epsilon.F(t,\vec{r},\Phi(t,\vec{r}),\partial\Phi(t,\vec{r})/\partial x,\partial\Phi(t,\vec{r})/\partial y,\partial\Phi(t,\vec{)}/\partial z), \tag{165}$$

where ε is the small positive parameter and $F(t,\vec{r},\Phi(t,\vec{r}),\partial\Phi(t,\vec{r})/\partial x,\partial\Phi(t,\vec{r})/\partial y,\partial\Phi(t,\vec{)}/\partial z)$is the function which can be expanded to a Fourier series for the variable $\vec{r}$. The Fourier coefficients are polynomials in relation to $\Phi(t,\vec{r}),\partial\Phi(t,\vec{r})/\partial x,\partial\Phi(t,\vec{r})/\partial y and \partial\Phi(t,\vec{)}/\partial z)$. For instance, the equation is similar to the one for driven small nonlinear oscillations (Landau & Lifshitz, 1976) is

$$\nabla(\Phi(t,\vec{r})+(\vec{\gamma}.\nabla)\Phi(t,\vec{r})+\vec{\omega}^2.\Phi(t,\vec{r})=f(t).\cos(\omega'.\vec{r})-\alpha.\Phi(t,\vec{r})^2-\beta.\Phi(t,\vec{r})^3 \tag{166}$$

where $\vec{\gamma}=(\gamma_x,\gamma_y,\gamma_z)$ is the attenuation parameter, $\vec{\omega}=(\omega_x,\omega_y,\omega_z)$ are the own frequencies of system oscillations, *f(t)* is the amplitude of the source of fluid motion which is presented by the periodic function with the frequency $\vec{\omega}'=(\omega_x',\omega_y',\omega_z')$, and α, β are the positive constants.

The similar equations can be derived for the curl-free velocity using the expression (153).

The solutions of these equations describe the development of turbulence at the resonance as the equations for the divergence-free velocity (Landau & Lifshitz, 1976).

3.8. Magnetic dynamo

We consider a mechanism of self-generating a magnetic field through movement of an electrically conducting fluid.

We take the equations of magnetohydrodynamics (MHD) for an incompressible fluid (Landau & Lifshitz, 1976) and modify them. The modified MHD equations look as follows

$$\rho\cdot\partial\vec{v}(t,\vec{r})/\partial t+\rho\cdot(\vec{v}(t,\vec{r})\cdot\nabla)\cdot\vec{v}(t,\vec{r})=-\nabla\left(p(t,\vec{r})+\frac{H^2}{8\pi}\right)+\eta\cdot\Delta\vec{v}(t,\vec{r})$$
$$+\frac{1}{4\pi}(\vec{H}(t,\vec{r})\cdot\nabla)\cdot\vec{H}(t,\vec{r})+(4\eta/3+\zeta)\cdot div(\vec{v}(t,\vec{r})), \tag{167}$$

$$\partial\vec{H}(t,\vec{r})/\partial t=\frac{c^2}{4\pi\sigma}\Delta\vec{H}(t,\vec{r})+(\vec{H}(t,\vec{r})\cdot\nabla)\vec{v}(t,\vec{r})-(\vec{v}(t,\vec{r})\cdot\nabla)\vec{H}(t,\vec{r})-\vec{H}\cdot div(\vec{v}(t,\vec{r})), \tag{168}$$

$$div(\vec{H}(t,\vec{r}))=0, \tag{169}$$

$$div(\vec{v}(t,\vec{r}))=\Delta\Phi(t,\vec{r}), \tag{170}$$

$\vec{H}(t,\vec{r})$ is the magnetic field, c is the speed of light, and σ is electrical conductivity.

Equation (167) can be separated into parts the same way as Equation (48). As a result, we obtain the equations for pressure as (54) and (55), and the equation for velocity which is (see (56))

$$\rho \cdot \partial \vec{v}_{SOL}(t,\vec{r})/\partial t + \rho \cdot (\vec{v}(t,\vec{r}) \cdot \nabla) \cdot \vec{v}_{SOL}(t,\vec{r}) = \eta \cdot \Delta \vec{v}_{SOL}(t,\vec{r}) + \vec{f}_{ex,SOL}(t,\vec{r}), \tag{171}$$

$$\vec{f}_{ex,SOL}(t,\vec{r}) = \frac{1}{4\pi}(\vec{H}(t,\vec{r}) \cdot \nabla) \cdot \vec{H}(t,\vec{r}). \tag{172}$$

All these equations can be solved the way described in the previous part. The solution of Equation (168) obtains the resonan.t terms which are dependent on the velocity amplitude, v_0^n, $n = 1, 2, 3, \ldots$. The time distribution grows to the maximum value and then vanishes exponentially. It means that the amplitude of the magnetic field obtains very high values for a short period of time, even if the initial amplitude u_0 is comparatively small. According to Equation (168), the increased magnetic field produces a high value of the amplitude of the fluid velocity as a result of resonance. It causes the next growth of the amplitude of the magnetic field. There exists positive feedback between the moving current conducting fluid and the magnetic field. The growth of the velocity and magnetic field is accelerated, i.e. there is a blowup. It means that the amplitude of the magnetic field obtains very high values for a short period of time, even if the initial amplitude u0 is comparatively small. As a result of strengthening, the amplitude of the magnetic field can exceed the threshold (150) and turbulence is developed. The solutions of the equations have lost their uniqueness. The bifurcation occurs for the system parameters. The system is unstable and any fluctuation of the velocity parameters may cause flow detachment. The amplitude of the magnetic field falls down together with the amplitude of the fluid velocity. A new growth of the fluid velocity and magnetic field is originated from small stochastic perturbations in the conducting moving fluid (Landau & Lifshitz, 1976). The initial conditions of this growth define the generation of a magnetic excursion or reversal as a result.

4. The conclusion

It is possible to give a certain answer to the Clay Mathematics Institute's question. There exist no smooth functions $p(t,\vec{r}), \vec{v}(t,\vec{r})$ on $[0, \infty) \times R^3$ that satisfy the claims (A) or (B). For $\eta > 0$, the smooth solutions of the Navier–Stokes equations exist only in space of two variables.

We obtained the general solution of the mass continuity equation and studied it for the condition $\eta > 0$. The result is that there exists no smooth, divergence-free vector field $\vec{v}_0(\vec{r})$ on R^3 satisfying (C) or (D).

The answers to the questions formulated in the works of (Constantin, 2001; Ladyzhenskaya, 2003) are the following.

There are no conditions for smooth incompressible viscous velocities $\vec{v}(0,\vec{r})$ that ensure, in the absence of external input of energy, that the solutions of the Navier–Stokes equations exist for all positive time in three dimensions and are smooth.

The Navier–Stokes equations together with the initial and boundary conditions don't provide the deterministic description of the dynamics of an incompressible viscous fluid in space of three dimensions.

Under the conditions (A) and (B), the divergence-free solutions can be obtained only if the Stokes–Navier equation system takes the two-dimensional and linear form.

For obtaining the viscous solutions of the nonlinear problem, it is necessary to break the strict divergence-free condition by introducing an implicit or explicit source in the right part of the mass continuity equation. Using this source, we apply a potential field whose Laplacian isn't equal to 0 at all points of R^3.

The exact solutions of the Navier–Stokes equations presented in the works of (Bertozzi & Majda, 2002; Landau & Lifshitz, 1987) are obtained using the linearization of the solutions or the infringement of the strict divergence-free condition and/or (2) by the introduction of a fictitious source in the right part of the mass continuity equation. In the last case, the solutions are obtained in curvilinear coordinates. The examples are a viscous fluid rotating by a disk (an infinite disk source), a fluid flow in a diffuser and confuser (an infinite linear source), and the Landau problem for a flooded jet (a point source) (Landau & Lifshitz, 1987).

However, we suppose that only a source geometrical factor is not enough for the description of nonlinear processes considering a viscous fluid flow. It is necessary to consider the molecular inelastic interactions in the mass continuity equation. These interactions change the intermolecular binding energy on the expense of the kinetic energy of a fluid. Their presence in the equation is carried out by the use of the curl-free potential velocity field. Due to this field, the mass continuity equation can be reduced to the Helmholtz equation, whose solutions give a complete system of functions in R^3.

The modified equations describe an incompressible fluid flow. Due to the inelastic interactions, an incompressible fluid sticks to a body surface. The analysis of the equation solutions shows the blowup of the laminar flow at the resonance, transition of the laminar flow to the turbulent one, and fluid detachment. This opens the way to solve the magnetic dynamo problem.

Funding

The author received no direct funding for this research.

Author details

A.V. Zhirkin[1]

E-mail: aleksej-zhirkin@yandex.ru

[1] Reactor Problem Laboratory, Department of Fusion Reactors, National Research Centre "Kurchatov Institute", Kurchatov Centre of Nuclear Technologies, Academician Kurchatov Square, 1, Moscow 123182, Russia.

References

Batchelor, G. K. (2000). *An introduction to fluid dynamics*. Cambridge: Cambridge University Press. ISBN 0-521-66396-2

Bertozzi, A., & Majda, A. (2002). *Vorticity and incompressible flows*. Cambridge: Cambridge University Press.

Bogoliubov, N. N., & Mitropolski, Y. A. (1961). *Asymptotic methods in the theory of nonlinear oscillations*. New York, NY: Gordon and Breach.

Constantin, P. (2001). *Some open problems and research directions in the mathematical study of fluid dynamics, in mathematics unlimited-2001 and beyond* (pp. 353–360). Berlin: Springer Verlag.

Courant, R., & Hilbert, D. (1962). *Methods of mathematical physics*. (Vol. 2 Paperback). ISBN 0-471-50439-4

Fefferman, C. H. (2000). *Existence and smoothness of the Navier-Stokes equation* (pp. 1–5). Cambridge, MA. Retrieved from http://claymath.org/Millenium_Prize_Problems/Navier-Stokes_Equations,ClayMathematics Institute

Fihtengolts, G. M. (2003a). *Course of differential and integral calculus*. (Vol. 3). Fizmatlit ISBN: 5-9221-0155-2. Retrieved from http://www.ebook300.com

Fihtengolts, G. M. (2003b). *Course of differential and integral calculus* (Vol. 2) Fizmatlit ISBN: 5-9221-0155-2. Retrieved from http://www.ebook300.com

Ladyzhenskaya, O. A. (1969). *The mathematical theory of viscous incompressible flows* (2nd ed.). New York, NY: Gordon and Breach.

Ladyzhenskaya, O. A. (2003). Sixth problem of the millennium: Navier–Stokes equations, existence and smoothness. *Russian Mathematical Surveys, 58*, 251. http://dx.doi.org/10.1070/RM2003v058n02ABEH000610

Landau, L. D., & Lifshitz, E. M. (1976). *Mechanics*. (Vol. 1, 3rd ed.). Oxford: Butterworth Heinemann. ISBN 978-0-750-62896-9

Landau, L. D., & Lifshitz, E. M. (1976). *Electrodynamics of continuous media* (Vol. 8, 2nd ed.). Oxford: Butterworth Heinemann. ISBN 978-0-7506-2634-7

Landau, L. D., & Lifshitz, E. M. (1987). *Fluid mechanics* (Vol. 6, 2nd ed.). Butterworth Heinemann. ISBN 978-0-080-33933-7.

Leray, J. (1934). Essai sur le mouvement d'un liquide visqueux emplissant l'espace [An essay on the motion of a viscous fluid filling the space]. *Acta Mathematica, 63*, 193–248.

Tikhonov, A. N., & Samarskii, A. A. (2013). *Equations of mathematical physics*. New York, NY: Dover Publications.

Vladimirov, V. S. (1983). *The equations of mathematical physics* (2nd English ed.). Moscow: Mir. MR 0764399, Zbl 0207.09101

14

δ-Dynamic chromatic number of Helm graph families

N. Mohanapriya[1,2], J. Vernold Vivin[3]* and M. Venkatachalam[1]

*Corresponding author: J. Vernold Vivin, Department of Mathematics, University College of Engineering Nagercoil (Anna University Constituent College), Konam, Nagercoil 629 004, Tamil Nadu, India

E-mail: vernoldvivin@yahoo.in

Reviewing editor: Feng Qi, Tianjin Polytechnic University, China

Abstract: An r-dynamic coloring of a graph G is a proper coloring c of the vertices such that $|c(N(v))| \geq min\{r, d(v)\}$, for each $v \in V(G)$, where $N(v)$ and $d(v)$ denote the neighborhood and the degree of v, respectively. The r-dynamic chromatic number of a graph G is the minimum k such that G has an r-dynamic coloring with k colors. In this paper, we obtain the δ-dynamic chromatic number of middle, total, and central of helm graph, where $\delta = \min_{v \in V(G)} \{d(v)\}$.

Subjects: Advanced Mathematics; Combinatorics; Discrete Mathematics; Mathematics & Statistics; Science

Keywords: *r*-dynamic coloring; middle graph; total graph; central graph

2000 Mathematics subject classifications: 05C15

1. Introduction

Throughout this paper all graphs are finite and simple. The r-dynamic chromatic number was first introduced by Montgomery (2001). An r-dynamic coloring of a graph G is a map c from $V(G)$ to a set of colors such that (i) if $uv \in E(G)$, then $c(u) \neq c(v)$, and (ii) for each vertex

ABOUT THE AUTHORS

N. Mohanapriya is working as an assistant professor in the Department of Mathematics, RVS Technical Campus—Coimbatore, Tamil Nadu, India. She is pursuing her Part-time (Category B) PhD in Bharathiar University, Coimbatore in the field of graph theory, particularly in graph colorings. She has published five research papers in International and National Journals and Conferences.

J. Vernold Vivin is working as an assistant professor in the Department of Mathematics, University College of Engineering Nagercoil, (Anna University, Constituent College), Konam, Nagercoil. His research areas of interests are graph theory and fractional differential equations. He has published more than 50 research papers in International and National Journals and Conferences.

M. Venkatachalam is working as an assistant professor in the Department of Mathematics, RVS Technical Campus—Coimbatore, Tamil Nadu, India. His research areas of interests are graph theory and inventory models. He has published more than 20 research papers in International and National Journals and Conferences.

PUBLIC INTEREST STATEMENT

Graph coloring is one of the research areas that shaped the graph theory as we know it today; the attempts to prove many more theorems in graph colorings have inspired many notions that became important on their own. Also motivated the study of colorings of various families of graphs, including the graphs embedded in the surfaces of bounded genus. Even though a computer-assisted proof of the theorems was eventually found, many natural problems motivated by it remain unsolved and the study of colorings of planar graphs and of graphs on surfaces is one of the most active areas of research in modern graph theory. This research paper outlines some of recently developed coloring, i.e. dynamic coloring. The new results in this paper are demonstrated by giving detailed proof based on our recent papers and to the best of our knowledge.

$v \in V(G)$, $|c(N(v))| \geq min\{r, d(v)\}$, where $N(v)$ denotes the set of vertices adjacent to v and $d(v)$ is its degree. The r-dynamic chromatic number of a graph G, written $\chi_r(G)$, is the minimum k such that G has an r-dynamic proper k-coloring. The 1-dynamic chromatic number of a graph G is equal to its chromatic number. The 2-dynamic chromatic number of a graph has been studied under the name dynamic chromatic number in Ahadi, Akbari, Dehghana, and Ghanbari (2012), Akbari, Ghanbari, and-Jahanbakam (2009, 2010), Alishahi (2012), Lai, Montgomery, and Poon (2003). There are many upper bounds and lower bounds for $\chi_d(G)$ in terms of graph parameters. For example,

For a graph G with $\Delta(G) \geq 3$, Lai et al. (2003) proved that $\chi_d(G) \leq \Delta(G) + 1$. An upper bound for the dynamic chromatic number of a d-regular graph G in terms of $\chi(G)$ and the independence number of G, $\alpha(G)$, was introduced in Dehghan and Ahadi (2012). In fact, it was proved that $\chi_d(G) \leq \chi(G) + 2\log_2 \alpha(G) + 3$. Taherkhani gave in (2016) an upper bound for $\chi_2(G)$ in terms of the chromatic number, the maximum degree Δ and the minimum degree δ, i.e. $\chi_2(G) - \chi(G) \leq \left\lceil (\Delta e)/\delta \log\left(2e\left(\Delta^2 + 1\right)\right)\right\rceil$.

Li, Yao, Zhou, and Broersma proved in (2009) that the computational complexity of $\chi_d(G)$ for a 3-regular graph is an NP-complete problem. Furthermore, Liu and Zhou (2008) showed that to determine whether there exists a 3-dynamic coloring, for a claw free graph with maximum degree 3, is NP-complete.

In this paper, we study $\chi_r(G)$ when r is δ, the minimum degree of the graph. We find the δ-dynamic chromatic number for middle, total, and central graph of helm graph.

2. Results

Let G be a graph with vertex set $V(G)$ and edge set $E(G)$. The middle graph (Michalak, 1981) of G, denoted by $M(G)$ is defined as follows. The vertex set of $M(G)$ is $V(G) \cup E(G)$. Two vertices x, y of $M(G)$ are adjacent in $M(G)$ in case one of the following holds: (i) x, y are in $E(G)$ and x, y are adjacent in G. (ii) x is in $V(G)$, y is in $E(G)$, and x, y are incident in G.

Let G be a graph with vertex set $V(G)$ and edge set $E(G)$. The total graph (Michalak, 1981) of G, denoted by $T(G)$ is defined in the following way. The vertex set of $T(G)$ is $V(G) \cup E(G)$. Two vertices x, y of $T(G)$ are adjacent in $T(G)$ in case one of the following holds: (i) x, y are in $V(G)$ and x is adjacent to y in G. (ii) x, y are in $E(G)$ and x, y are adjacent in G. (iii) x is in $V(G)$, y is in $E(G)$ and x, y are incident in G.

The central graph (Vernold Vivin, 2007) $C(G)$ of a graph G is obtained from G by subdividing each edge of G exactly once and then joining each pair of vertices of the original graph which were previously non-adjacent.

The helm graph H_n is the graph obtained from an n-wheel graph by adjoining a pendent edge at each node of the cycle. Where $V(H_n) = \{v\} \cup \{v_1, v_2, \ldots, v_{n-1}\} \cup \{u_1, u_2, \ldots, u_{n-1}\}$ and $E(H_n) = \{e_i{:}1 \leq i \leq n-1\} \cup \{e_i'{:}1 \leq i \leq n-1\} \cup \{s_i{:}1 \leq i \leq n-1\}$, where e_i is the edge vv_i $(1 \leq i \leq n-1)$, e_i' is the edge $v_iv_{i+1}(1 \leq i \leq n-2)$ and s_i is the edge $v_iu_i(1 \leq i \leq n-1)$. This notation is valid through the entire paper.

THEOREM 2.1 *Let $n \geq 8$. The δ-dynamic chromatic number of the middle graph of a helm of order $2n-1$ is $\chi_\delta(M(H_n)) = n$.*

Proof By the definition of middle graph, $V(M(H_n)) = V(H_n) \cup E(H_n) = \{v\} \cup \{v_i{:}1 \leq i \leq n-1\} \cup \{u_i{:}1 \leq i \leq n-1\} \cup \{e_i{:}1 \leq i \leq n-1\} \cup \{e_i'{:}1 \leq i \leq n-1\} \cup \{s_i{:}1 \leq i \leq n-1\}$.

The vertices v and $\{e_i{:}1 \leq i \leq n-1\}$ induce a clique of order K_n in $M(H_n)$. Thus, $\chi_\delta(M(H_n)) \geq n$.

Consider the following n-coloring of $M(H_n)$:

For $1 \leq i \leq n-1$, assign the color c_i to e_i and assign the color c_n to v. For $1 \leq i \leq n-1$, assign the color c_n to u_i, $\deg(u_i) = \delta(M(H_n)) = 1$. For $1 \leq i \leq n-1$, assign to e_i' one of the allowed colors—such color exists, because $\deg(u_i) = 8$. For $1 \leq i \leq n-1$, if any, assign to vertex v_i one of the allowed colors—such color exists, because $\deg(v_i) = 4$. For $1 \leq i \leq n-1$, if any, assign to vertex s_i one of the allowed colors—such color exists, because $\deg(s_i) = 3$. An easy check shows that $N(v)$ contains an induced clique of order 5, for every $v \in V(M(H_n))$. Thus, this coloring is a δ-dynamic coloring. Hence, $\chi_\delta(M(H_n)) \leq n$. Therefore, $\chi_\delta(M(H_n)) = n, \forall\, n \geq 8$. □

THEOREM 2.2 *Let $n \geq 9$. The δ-dynamic chromatic number of the total graph of a helm of order $2n-1$ is $\chi_\delta(T(H_n)) = n$.*

Proof By the definition of total graph $V(T(H_n)) = V(H_n) \cup E(H_n) = \{v\} \cup \{v_i : 1 \leq i \leq n-1\} \cup \{u_i : 1 \leq i \leq n-1\} \cup \{e_i : 1 \leq i \leq n-1\} \cup \{e_i' : 1 \leq i \leq n-1\} \cup \{s_i : 1 \leq i \leq n-1\}$. The vertices v and $\{e_i : 1 \leq i \leq n-1\}$ induce a clique of order K_n in $T(H_n)$. Thus, $\chi_\delta(T(H_n)) \geq n$.

Consider the following n-coloring of $T(H_n)$:

For $1 \leq i \leq n-1$, assign the color c_i to e_i and assign the color c_n to v. For $1 \leq i \leq n-1$, assign to u_i one of the allowed colors—such color exists, because $\delta(u_i) = \deg(u_i) = 2$. For $1 \leq i \leq n-1$, assign to e_i' one of the allowed colors—such color exists, because $\deg(u_i) = 8$. For $1 \leq i \leq n-1$, if any, assign to vertex v_i one of the allowed colors—such color exists, because $\deg(v_i) = 8$. For $1 \leq i \leq n-1$, if any, assign to vertex s_i one of the allowed colors—such color exists, because $\deg(s_i) = 5$. An easy check shows that $N(v)$ contains an induced clique of order 5, for every $v \in V(M(H_n))$. Thus, this coloring is a δ-dynamic coloring. Hence, $\chi_\delta(T(H_n)) \leq n$. Therefore, $\chi_\delta(T(H_n)) = n, \forall\, n \geq 9$. □

THEOREM 2.3 *Let $n \geq 4$. The δ-dynamic chromatic number of the central graph of a helm of order $2n-1$ is $\chi_\delta(C(H_n)) = 2n-1$.*

Proof By the definition of central graph, subdividing each edge of H_n exactly once and then joining each pair of vertices of H_n which were non-adjacent. Let $V(C(H_n)) = V(H_n) \cup E(H_n) = \{v\} \cup \{v_i : 1 \leq i \leq n-1\} \cup \{u_i : 1 \leq i \leq n-1\} \cup \{e_i : 1 \leq i \leq n-1\} \cup \{e_i' : 1 \leq i \leq n-1\} \cup \{s_i : 1 \leq i \leq n-1\}$. Clearly, the graph induced by $\{v_{2i} : i = 1, 2, \ldots, \lfloor (n-1)/2 \rfloor\}$ is a complete graph. Thus, a proper coloring assigns at least $\lfloor (n-1)/2 \rfloor$ colors to them. The same happens with the subgraph induced by $\{v_{2i-1} : i = 1, 2, \ldots, \lfloor (n-1)/2 \rfloor\}$. Moreover, if we are considering a δ-dynamic coloring when $n-1$ is odd v_{n-1} should have a different color from v_{2i-1}, $i = 1, 2, \ldots, (n-1)/2$, because v_{n-1} and v_1 are the only neighbors of e_{n-1}', and v_{n-1} is adjacent to v_{2i-1}, $i = 2, \ldots, (n-1)/2$. A similar reasoning also shows that in a δ-dynamic coloring, the colors assigned to odd vertices should be different to the colors assigned to even vertices and that all of them should be different from the color assigned to v. It is also shown that in a δ-dynamic coloring, the colors assigned to vertices $\{v\} \cup \{v_i : 1 \leq i \leq n-1\}$ should be different to the colors assigned to the vertices of $\{u_i : 1 \leq i \leq n-1\}$. Since, the vertices $\{u_i : 1 \leq i \leq n-1\}$ and v induces a clique of order K_n in $C(H_n)$ and the vertices $\{v_i : 1 \leq i \leq n-1\}$ adjacent to $\{u_j : 1 \leq j \leq n-1\}\ \forall\, i \neq j$. But, any three consecutive vertices of the path must be colored differently in any dynamic coloring. Since, the first and third vertices are the only neighbors of the second vertex and must be colored differently (by the condition of dynamic coloring) and also differently from the second vertex. So, the same color to $\{v_i : 1 \leq i \leq n-1\}$ and $\{u_j : 1 \leq j \leq n-1\}\ \forall\, i = j$ is impossible. Thus, $\chi_\delta(C(H_n)) \geq 2n-1$.

Consider the following $2n-1$-coloring of $C(H_n)$:

For $1 \leq i \leq n-1$, assign the color c_i to v_i and assign the color c_n to v. For $1 \leq i \leq n-1$, assign the color c_{n+i} to u_i. For $1 \leq i \leq n-1$, assign to vertices s_i, e_i' and e_i one of the allowed colors—such color exists, because $\deg(s_i) = \deg(e_i') = \deg(e_i) = \delta(C(H_n)) = 2$. An easy check shows that this coloring is a δ-dynamic coloring. Hence, $\chi_\delta(C(H_n)) \leq 2n-1$. Therefore, $\chi_\delta(C(H_n)) = 2n-1$. □

Acknowledgements
With due Respect, the authors sincerely thank the referee for his careful reading, excellent comments and fruitful suggestions that have resulted in the improvement of the quality of this manuscript.

Funding
The authors received no direct funding for this research.

Author details
N. Mohanapriya[1,2]
E-mail: n.mohanamaths@gmail.com
J. Vernold Vivin[3]
E-mail: vernoldvivin@yahoo.in
ORCID ID: http://orcid.org/0000-0002-3027-2010
M. Venkatachalam[1]
E-mail: venkatmaths@gmail.com
[1] RVS Faculty of Engineering, Department of Mathematics, RVS Technical Campus - Coimbatore, Coimbatore 641 402, Tamil Nadu, India.
[2] Bharathiar University, Coimbatore 641 046, Tamil Nadu, India.
[3] Department of Mathematics, University College of Engineering Nagercoil (Anna University Constituent College), Konam, Nagercoil 629 004, Tamil Nadu, India.

References

Ahadi, A., Akbari, S., Dehghana, A., & Ghanbari, M. (2012). On the difference between chromatic number and dynamic chromatic number of graphs. *Discrete Mathematics, 312*, 2579–2583.

Akbari, S., Ghanbari, M., & Jahanbekam, S. (2009). On the list dynamic coloring of graphs. *Discrete Applied Mathematics, 157*, 3005–3007.

Akbari, S., Ghanbari, M., & Jahanbekam, S. (2010). On the dynamic chromatic number of graphs. *Combinatorics and Graphs, in: Contemporary Mathematics - American Mathematical Society, 531*, 11–18.

Alishahi, M. (2012). Dynamic chromatic number of regular graphs. *Discrete Applied Mathematics, 160*, 2098–2103.

Dehghan, A., & Ahadi, A. (2012). Upper bounds for the 2-hued chromatic number of graphs in terms of the independence number. *Discrete Applied Mathematics, 160*, 2142–2146.

Lai, H. J., Montgomery, B., & Poon, H. (2003). Upper bounds of dynamic chromatic number. *Ars Combinatoria, 68*, 193–201.

Li, X., Yao, X., Zhou, W., & Broersma, H. (2009). Complexity of conditional colorability of graphs. *Applied Mathematics Letters, 22*, 320–324.

Li, X., & Zhou, W. (2008). The 2nd-order conditional 3-coloring of claw-free graphs. *Theoretical Computer Science, 396*, 151–157.

Michalak, D. (1981). On middle and total graphs with coarseness number equal 1, Springer Verlag graph theory. In *Lagow proceedings* (pp. 139–150). New York: Springer Verlag Berlin Heidelberg.

Montgomery, B. (2001). *Dynamic coloring of graphs* (PhD thesis). West Virginia University. Ann Arbor, MI: ProQuest LLC.

Taherkhani, A. (2016). On r-Dynamic chromatic number of graphs. *Discrete Applied Mathematics, 201*, 222–227.

Vernold Vivin, J. (2007). *Harmonious coloring of total graphs, n-leaf, central graphs and circumdetic graphs* (PhD thesis). Coimbatore: Bharathiar University.

15

On quadratic Gauss sums and variations thereof

M.L. Glasser[1,2] and Michael Milgram[3]*

*Corresponding author: Michael Milgram, Consulting Physicist, Geometrics Unlimited Ltd., Box 1484, Deep River, Ontario, Canada, K0J 1P0

E-mail: mike@geometrics-unlimited.com

Reviewing editor: Regina Burachik, University of South Australia, Australia

Abstract: A number of new terminating series involving $\sin(n^2/k)$ and $\cos(n^2/k)$ are presented and connected to Gauss quadratic sums. Several new closed forms of generic Gauss quadratic sums are obtained and previously known results are generalized.

Subjects: Science; Mathematics & Statistics; Advanced Mathematics; Analysis - Mathematics; Complex Variables; Mathematical Analysis; Real Functions; Special Functions; Number Theory

Keywords: Gauss quadratic sum; *cosine* sum; *sine* sum; finite trigonometric sums

MSC Subject classifications: 11T23; 11T24; 11L03; 30B50; 42A32; 65B10

1. Introduction

In a recent work (Glasser & Milgram, 2014), a number of new integrals were evaluated analytically, and in the process, we noted that the limiting case of some of those integrals reduced to simpler known forms that involved trigonometric series with quadratic dummy indices of summation. A reasonably thorough search of the literature indicates that such series are not very well tabulated—in fact only two were found in tables (Hansen, 1975)—those two corresponding to the real and imaginary part of Gauss quadratic sums, the series associated with classical number theory (Apostol, 1998), and, recently applied to quantum mechanics (Armitage & Rogers, 2000; Gheorghiu & Looi, 2010). Hardy and Littlewood (1914), in their original work considered integrals that are similar to those considered here (see (2.1) below) but not in closed form; a comprehensive historical summary can be found in Berndt and Evans (1981, Section 2.3).

In Section 2, we summarize the integrals upon which our results are based; all eight variations of alternating quadratic sums analogous to classical quadratic Gauss sums are developed in Section 3.1. These new sums are then connected to the classical sums (Section 3.2), and that relationship is used to find new closed forms for odd-indexed variants of the classical sums (Section 3.3), all of which are believed to be new. The upper summation limit of an interesting subset of these results is extended by the addition of a new parameter p (Section 4.1), and a template proof is provided in the Appendix. Miscellaneous results arising during our investigation are also listed (Section 4.2). The trigonometric

ABOUT THE AUTHORS

Michael Milgram and Larry Glasser's research interests generally coincide—both are motivated by a desire to investigate various aspects of classical analysis as they might someday be applied to problems in physics. In pursuing this objective, their curiosity sometimes takes them far afield, as this work demonstrates.

PUBLIC INTEREST STATEMENT

Scientists frequently need to add numbers up, and this paper presents a new way of determining the sum of a general collection of complicated things without doing any addition. Mathematicians always like to know that such things will work for any case that they may try, so when a new result is presented, it must be proven valid in a relatively rigorous manner. That is why this paper is somewhat longer than it needs to be—the new results must be proven, rather than simply stated.

forms of the results presented in the first four sections are finally rewritten in the canonical form of a generalized Gauss quadratic sum involving the roots of unity, again leading to new expressions, best described as alternating and extended Gauss quadratic sums (Section 5). It is recognized that many of the results derived here can be obtained by clever—and lengthy—manipulation (Chapman, 2014) of known results from number theory starting from the classical Gauss sum. This work provides a simple method of obtaining sums that are not listed in any of the tables or reference works usually consulted (e.g. Hansen, 1975), particularly motivated by the fact that number theoretic methods are not well known in other fields where such sums arise (Armitage & Rogers, 2000; Gheorghiu & Looi, 2010).

2. The basic integrals

In the previous work (Glasser & Milgram, 2014, Equation 3.33), with $k \in \mathbb{N}$, $\{a, b, s\} \in \Re$, we evaluated a family of related integrals of the form

$$\int_0^\infty \frac{v^s\,(v^2+1)^{s/2}\left(\sin(a\,v^2)\cos\left(s\arctan(1/v)\right)\cosh(a\,v)-\cos(a\,v^2)\sin\left(s\arctan(1/v)\right)\sinh(a\,v)\right)}{\cosh\left(\pi\,(2\,k-1)\,v\right)}dv$$
$$= \frac{(-1)^{k+1}}{2^{2\,s}\,(2\,k-1)^{(2\,s+1)}}\sum_{n=1}^{k-1}(-1)^n\left((2\,k-1)^2-(2\,n)^2\right)^s\sin\left(\frac{a\left((2\,k-1)^2-(2\,n)^2\right)}{4\,(2\,k-1)^2}\right)$$
$$-\frac{1}{2}\,\frac{(-1)^k\sin(a/4)}{2\,k-1} \tag{2.1}$$

where $4a \leq (2k-1)\pi$, and, for particular choices of s, this family of integrals (Glasser & Milgram, 2014, Equations 3.28, 3.34, 3.46 with s=0 and $b \to i\,a$ and 3.37 and 3.41 combined) gives rise (respectively) to the following particular results:

$$\int_0^\infty \frac{\cos\left(a\,v^2\right)\sinh\left(2\,a\,v\right)}{\sinh\left(k\,\pi\,v\right)}dv = \frac{(-1)^{k+1}}{k}\sum_{n=1}^{k-1}(-1)^n\sin\left(\frac{a\left(k^2-n^2\right)}{k^2}\right)+\frac{(-1)^{k+1}\sin(a)}{2\,k} \tag{2.2}$$

$$\int_0^\infty \frac{e^{i\,a\,v^2}\cosh(av)}{\cosh((2k-1)\pi v)}dv = \frac{(-1)^{(k+1)}\exp(i\,a/4)}{2\,k-1}\left(\frac{1}{2}+\sum_{n=1}^{k-1}(-1)^n\exp\left(-\frac{ian^2}{(2\,k-1)^2}\right)\right) \tag{2.3}$$

$$\int_0^\infty \frac{e^{i\,a\,v^2}\sinh(a\,v)}{\sinh\left(2\,\pi\,k\,v\right)}dv = \frac{i}{4\,k}(-1)^k\exp(i\,a/4)\sum_{n=1-k}^{k}(-1)^n\exp\left(\frac{-i\,a\,n^2}{4\,k^2}\right) \tag{2.4}$$

$$\int_0^\infty \frac{e^{i\,a\,v^2}\cosh(\pi\,(2\,k-1)\,v)\cosh(a\,v)}{\cosh(2\,\pi\,b)+\cosh(2\,\pi\,v\,(2\,k-1))}dv = -\frac{1}{4\,(2\,k-1)\cosh(\pi\,b)}\sum_{n=1}^{2\,k-1}(-1)^n\cosh(X_n)\,e^{i\,A_n} \tag{2.5}$$

where

$$A_n = \frac{a\,(4\,b^2-4\,n^2+1+8\,k\,n-4\,k)}{4\,(2\,k-1)^2} \tag{2.6}$$

and

$$X_n = \frac{2\,a\,b\,(-n+k)}{(2\,k-1)^2} \tag{2.7}$$

3. Basic derivations

3.1. Alternating Gauss quadratic sums

All the main results in this section depend upon the following well-known result (Gradshteyn & Ryzhik, 1980, Equation 3.691) (see cover image)

$$\int_0^\infty \sin\left(av^2\right) dv = \int_0^\infty \cos\left(av^2\right) dv = 1/4\sqrt{\frac{2\pi}{a}}, a > 0 \tag{3.1}$$

In (2.2), we evaluate the limit $a = k\pi/2$, and, after setting $k \to 2k$, we eventually arrive at

$$\sum_{n=1}^{2k} (-1)^n \sin\left(\frac{\pi n^2}{4k}\right) = (-1)^k \sqrt{\frac{k}{2}} \tag{3.2}$$

after setting $k \to 2k - 1$, we find

$$\sum_{n=1}^{2k-2} (-1)^n \cos\left(\frac{\pi n^2}{2\,(2k-1)}\right) = -\frac{(-1)^k \sqrt{2k-1}}{2} - \frac{1}{2} \tag{3.3}$$

Both of these results may be characterized as *Alternating Gauss Quadratic Sums* for which we can find no references in the literature (e.g. Apostol, 1998). In Equations 2.3 and Equations 2.4 let $a = (2k-1)\pi$ and $a = 2k\pi$, respectively, and, by comparing real and imaginary parts, eventually arrive at

$$\sum_{n=1}^{2k} (-1)^n \cos\left(\frac{\pi n^2}{4k}\right) = \frac{-1 + (-1)^k + \sqrt{2}\sqrt{k}(-1)^k}{2} \tag{3.4}$$

and

$$\sum_{n=1}^{2k-1} (-1)^n \sin\left(\frac{\pi n^2}{2\,(2k-1)}\right) = \frac{(-1)^k \left(1 + \sqrt{2k-1}\right)}{2} \tag{3.5}$$

All of these are believed to be new.

By setting $s = 0, a = (2k-1)\pi/4$ in Equation 2.1 we obtain companions to the above:

$$\sum_{n=1}^{2k-1} (-1)^n \left(\cos\left(\frac{\pi n^2}{4k-1}\right) + \sin\left(\frac{\pi n^2}{4k-1}\right)\right) = -\frac{1}{2} + \frac{1}{2}(-1)^k \sqrt{4k-1} \tag{3.6}$$

after setting $k \to 2k$, and

$$\sum_{n=1}^{2k-2} (-1)^n \left(\cos\left(\frac{\pi n^2}{4k-3}\right) - \sin\left(\frac{\pi n^2}{4k-3}\right)\right) = -\frac{1}{2} - \frac{1}{2}(-1)^k \sqrt{4k-3} \tag{3.7}$$

after setting $k \to 2k - 1$.

If $a = (2k-1)\pi$ in Equation 2.3, after setting $k \to 2k$ we find

$$\sum_{n=1}^{2k-1} (-1)^n \left(\cos\left(\frac{\pi n^2}{4k-1}\right) - \sin\left(\frac{\pi n^2}{4k-1}\right)\right) = -\frac{1}{2} - \frac{(-1)^k \sqrt{4k-1}}{2} \tag{3.8}$$

and, after setting $k \to 2k - 1$ we obtain

$$\sum_{n=1}^{2k-2} (-1)^n \left(\cos\left(\frac{\pi n^2}{4k-3}\right) + \sin\left(\frac{\pi n^2}{4k-3}\right)\right) = -\frac{1}{2} - \frac{(-1)^k \sqrt{4k-3}}{2} \tag{3.9}$$

Adding and subtracting Equations 3.6 and 3.8 gives

$$\sum_{n=1}^{2k-1} (-1)^n \cos\left(\frac{\pi n^2}{4k-1}\right) = -1/2 \tag{3.10}$$

and

$$\sum_{n=1}^{2k-1} (-1)^n \sin\left(\frac{\pi n^2}{4k-1}\right) = \frac{1}{2}(-1)^k \sqrt{4k-1} \tag{3.11}$$

while performing the same operations on Equations 3.7 and 3.9 yields

$$\sum_{n=1}^{2k-2} (-1)^n \cos\left(\frac{\pi n^2}{4k-3}\right) = -\frac{1}{2} - \frac{(-1)^k \sqrt{4k-3}}{2} \tag{3.12}$$

and

$$\sum_{n=1}^{2k-2} (-1)^n \sin\left(\frac{\pi n^2}{4k-3}\right) = 0 \tag{3.13}$$

all of which, together with Equations 3.2–3.5 are believed to be new; note the fourfold modularity of these results, and the fact that Equations 3.10–3.13 can be also written, respectively, (and more conveniently)

$$\sum_{n=1}^{4k-1} (-1)^n \cos\left(\frac{\pi n^2}{4k-1}\right) = 0 \tag{3.14}$$

$$\sum_{n=1}^{4k-1} (-1)^n \sin\left(\frac{\pi n^2}{4k-1}\right) = (-1)^k \sqrt{4k-1} \tag{3.15}$$

$$\sum_{n=1}^{4k-3} (-1)^n \cos\left(\frac{\pi n^2}{4k-3}\right) = -(-1)^k \sqrt{4k-3} \tag{3.16}$$

$$\sum_{n=1}^{4k-3} (-1)^n \sin\left(\frac{\pi n^2}{4k-3}\right) = 0 \tag{3.17}$$

3.2. Variations

An interesting finite sum identity/equality arises from Equation 2.5. Set $b = 0$, $a = (2k-1)\pi$ in the real and imaginary parts to discover

$$\sum_{n=1}^{2k-1} (-1)^n \sin\left(\frac{\pi\,(2n-1)\,(-2n-1+4k)}{4(2k-1)}\right) = \sum_{n=1}^{2k-1} (-1)^n \cos\left(\frac{\pi(2n-1)(-2n-1+4k)}{4(2k-1)}\right) = -\sqrt{k-\frac{1}{2}} \tag{3.18}$$

By removing odd multiples of $\pi/2$ from the arguments of each of the above trigonometric functions, Equation 3.18 can alternatively be written in the form of *odd-indexed quadratic Gauss sums*

$$\sum_{n=1}^{2k-1} \sin\left(\frac{1}{4}\frac{\pi(2n-1)^2}{2k-1}\right) = \sum_{n=1}^{2k-1} \cos\left(\frac{1}{4}\frac{\pi(2n-1)^2}{2k-1}\right) = \sqrt{k-\frac{1}{2}} \tag{3.19}$$

The fact that the arguments of both the sine and cosine terms are fractional multiples of $\pi/4$ may explain the intriguing equality of these sine and cosine sums of equal argument, which also hints at the existence of some more general property. The only previously tabulated sums of this genre, corresponding to Gauss' classical result for quadratic sums (Apostol, 1998, Equation 9.10(30); Berndt & Evans' 1981) are to be found in the following tabular listing (Hansen, 1975, Equations 16.1.1 and 19.1.1):

$$\sum_{n=1}^{k} \sin\left(\frac{2\pi n^2}{k}\right) = \frac{\sqrt{k}}{2}\left(1+\cos\left(k\pi/2\right)-\sin\left(k\pi/2\right)\right) \tag{3.20}$$

and

$$\sum_{n=1}^{k} \cos\left(\frac{2\pi n^2}{k}\right) = \frac{\sqrt{k}}{2}\left(1+\cos\left(k\pi/2\right)+\sin\left(k\pi/2\right)\right) \tag{3.21}$$

This will be discussed further in Section 5.

By adding and subtracting the above, and applying simple trigonometric identities, we also find

$$\sum_{n=1}^{2k-1} (-1)^n \cos\left(\frac{\pi\left(-2n^2+4kn-k\right)}{2\left(2k-1\right)}\right) = 0 \tag{3.22}$$

which can alternatively be rewritten as

$$\sum_{n=-2k}^{2k} (-1)^n \sin\left(\frac{\pi n^2}{4k+1}\right) = \sum_{n=1-2k}^{2k-1} (-1)^n \cos\left(\frac{\pi n^2}{4k-1}\right) = 0 \tag{3.23}$$

and

$$\sum_{n=1}^{2k-1} (-1)^n \sin\left(\frac{\pi(-2n^2+4kn-k)}{2(2k-1)}\right) = -\sqrt{2k-1} \tag{3.24}$$

which is equivalent to

$$\sum_{n=-2k}^{2k} (-1)^n \cos\left(\frac{\pi n^2}{4k+1}\right) = (-1)^k \sqrt{4k+1} \tag{3.25}$$

or, in a different form,

$$\sum_{n=1-2k}^{2k-1} (-1)^n \sin\left(\frac{\pi n^2}{4k-1}\right) = (-1)^k \sqrt{4k-1} \tag{3.26}$$

Similarly, Equation 3.19 can be rewritten as

$$\sum_{n=-2k}^{2k} \sin\left(\frac{\pi(2n+1)^2}{4(4k+1)}\right) = \sum_{n=-2k}^{2k} \cos\left(\frac{\pi(2n+1)^2}{4(4k+1)}\right) = \frac{\sqrt{8k+2}}{2} \tag{3.27}$$

All of these are reminiscent of special values of a truncated Jacobi theta function

$$\theta_L(z,\tau) \equiv \sum_{n=-L}^{L} e^{(i\pi(n^2\tau+2nz))} = \sum_{n=-L}^{L} \left(\cos\left(\pi\left(n^2\tau+2nz\right)\right)+i\sin\left(\pi\left(n^2\tau+2nz\right)\right)\right) \tag{3.28}$$

also recognized as a generalized Gauss quadratic sum (Apostol, 1998, Theorem 9.16).

3.3. Odd indexed quadratic sums

Break Equations 3.2 and 3.3 into their even and odd parts to find

$$\sum_{n=0}^{k} \sin\left(\frac{\pi n^2}{k}\right) - \sum_{n=1}^{k} \sin\left(\frac{\pi (2n-1)^2}{4k}\right) = \frac{(-1)^k \sqrt{2k}}{2} \tag{3.29}$$

and

$$\sum_{n=1}^{k} \cos\left(\frac{\pi n^2}{k}\right) - \sum_{n=1}^{k} \cos\left(\frac{\pi (2n-1)^2}{4k}\right) = -\frac{1}{2} + \frac{(-1)^k (1+\sqrt{2k})}{2} \tag{3.30}$$

Employing the method outlined in the Appendix, and using Equation 4.13 (see next section below) we find

$$\sum_{n=0}^{k} \sin\left(\frac{\pi n^2}{k}\right) = \frac{\sqrt{2k}\left(1+(-1)^k\right)}{4} \tag{3.31}$$

and

$$\sum_{n=1}^{k} \cos\left(\frac{\pi n^2}{k}\right) = \begin{cases} \sqrt{2k}/2 & \text{if } k \text{ is even} \\ -1 & \text{if } k \text{ is odd} \end{cases} \tag{3.32}$$

From Equations 3.31 and 3.32, the odd indexed quadratic sums appearing in Equations 3.29 and 3.30 eventually give

$$\sum_{n=1}^{k} \sin\left(\frac{\pi (2n-1)^2}{4k}\right) = \sum_{n=1}^{k} \cos\left(\frac{\pi (2n-1)^2}{4k}\right) = -\frac{\sqrt{2k}\left((-1)^k - 1\right)}{4} \tag{3.33}$$

a generalization of Equation 3.19. This is discussed further in Section 5.

4. Other results

4.1. Generalizations

Along the foregoing lines, we also note that some of the above may be generalized by extending the upper limit of summation. If p is a positive integer, it may be shown (by induction and removing all integral multiples of π from the arguments of the corresponding trigonometric functions—see the Appendix for a sample proof) that Equation 3.2 generalizes to

$$\sum_{n=1}^{4kp} (-1)^n \sin\left(\frac{\pi n^2}{4k}\right) = p(-1)^k \sqrt{2k} \tag{4.1}$$

Equation 3.5 generalizes to

$$\sum_{n=1}^{2(2k-1)p} (-1)^n \sin\left(\frac{\pi n^2}{2(2k-1)}\right) = (-1)^k p \sqrt{2k-1} \tag{4.2}$$

Equation 3.15 generalizes to

$$\sum_{n=1}^{(4k-1)p} (-1)^n \sin\left(\frac{\pi n^2}{4k-1}\right) = (-1)^k p \sqrt{4k-1} \tag{4.3}$$

and Equation 3.17 generalizes to

$$\sum_{n=1}^{(4k-3)p} (-1)^n \sin\left(\frac{\pi n^2}{4k-3}\right) = 0 \tag{4.4}$$

Similar properties also hold for the alternating cosine sums. Equation 3.3 generalizes to

$$\sum_{n=1}^{2(2k-1)p} (-1)^n \cos\left(\frac{\pi n^2}{2(2k-1)}\right) = -(-1)^k p\sqrt{2k-1} \tag{4.5}$$

Equation 3.4 generalizes to

$$\sum_{n=1}^{4kp} (-1)^n \cos\left(\frac{\pi n^2}{4k}\right) = p(-1)^k\sqrt{2k} \tag{4.6}$$

the variation Equation 3.14 generalizes to

$$\sum_{n=1}^{(4k-1)p} (-1)^n \cos\left(\frac{\pi n^2}{4k-1}\right) = 0 \tag{4.7}$$

and Equation 3.16 generalizes to

$$\sum_{n=1}^{(4k-3)p} (-1)^n \cos\left(\frac{\pi n^2}{4k-3}\right) = -(-1)^k p\sqrt{4k-3} \tag{4.8}$$

4.2. Other miscellaneous results

Ancillary to proofs of the above, we obtained a number of miscellaneous results, all of which can be proven by using the methods outlined in the previous section with the help of the Appendix. To prove Equation 4.17 for p odd, the following companion to Equation 3.3 is needed

$$\sum_{n=1}^{2k-1} \cos\left(\frac{\pi n^2}{2(2k-1)}\right) = \frac{\sqrt{2k-1}}{2} - \frac{1}{2} \tag{4.9}$$

and the following, needed for the proof of Equation 4.16 is a companion to Equation 3.4

$$\sum_{n=1}^{2k} \cos\left(\frac{\pi n^2}{4k}\right) = \frac{\sqrt{2k}+(-1)^k-1}{2} \tag{4.10}$$

Further, based on Equations 3.3 and 3.20 with $k \to 4(2k-1)$ followed by a lengthy series of reductions of the upper limit, we have

$$\sum_{n=1}^{2k-1} \sin\left(\frac{\pi n^2}{2(2k-1)}\right) = \frac{\sqrt{2k-1}}{2} - \frac{(-1)^k}{2} \tag{4.11}$$

Further, Equation 3.19 generalizes to

$$\sum_{n=1}^{p(2k-1)} \sin\left(\frac{\pi(2n-1)^2}{4(2k-1)}\right) = \frac{p\sqrt{4k-2}}{2} \tag{4.12}$$

and Equation 3.29 generalizes to

$$\sum_{n=1}^{pk} \sin\left(\frac{2\pi n^2}{k}\right) = \frac{1}{2}p\sqrt{k}\left(1+\cos\left(\frac{k\pi}{2}\right) - \sin\left(\frac{k\pi}{2}\right)\right) \tag{4.13}$$

Equation 3.2 generalizes to

$$\sum_{n=1}^{2pk}(-1)^n \sin\left(\frac{\pi n^2}{4k}\right) = p(-1)^k\sqrt{k/2} \tag{4.14}$$

Equation 3.3 generalizes to

$$\sum_{n=1}^{p(2k-1)}(-1)^n \cos\left(\frac{\pi n^2}{2\,(2k-1)}\right) = -\frac{1-(-1)^p}{4} - \frac{(-1)^k p\sqrt{2k-1}}{2} \tag{4.15}$$

Equation 3.4 generalizes to

$$\sum_{n=1}^{2pk}(-1)^n \cos\left(\frac{\pi n^2}{4k}\right) = \frac{p(-1)^k\sqrt{2k}}{2} - \frac{1-(-1)^{pk}}{2} \tag{4.16}$$

Equation 3.5 generalizes to

$$\sum_{n=1}^{p(2k-1)}(-1)^n \sin\left(\frac{\pi n^2}{2\,(2k-1)}\right) = \frac{(-1)^k(1+2p\sqrt{2k-1}-(-1)^p)}{4} \tag{4.17}$$

Along with the above, we have also found

$$\sum_{n=1}^{p(4k-1)-2k}(-1)^n \cos\left(\frac{\pi n^2}{4k-1}\right) = -\frac{1}{2} \tag{4.18}$$

$$\sum_{n=1}^{p(4k-3)-2k+1}(-1)^n \cos\left(\frac{\pi n^2}{4k-3}\right) = -\frac{1}{2} - (-1)^k\sqrt{4k-3}\,(p-\frac{1}{2}) \tag{4.19}$$

and

$$\sum_{n=1}^{p(4k-1)-2k}(-1)^n \sin\left(\frac{\pi n^2}{4k-1}\right) = (-1)^k\left(p-\frac{1}{2}\right)\sqrt{4k-1} \tag{4.20}$$

All of the above may be proven using the Appendix as a template.

5. Connection with Gauss quadratic sums

As usual, denote the "Extended Gauss Quadratic Sum" for $j, k, m, p \in \mathbb{N}$ by

$$G_p(j;k;\theta) = \sum_{n=1}^{kp}\exp\left(\frac{2i\pi j n^2}{k} + 2\pi i\theta n\right) \tag{5.1}$$

and write the (classical) "Gauss Quadratic Sums" as

$$G_1(j;k;0) = \sum_{n=1}^{k}\cos\left(\frac{2\pi j n^2}{k}\right) + i\sum_{n=1}^{k}\sin\left(\frac{2\pi j n^2}{k}\right) \tag{5.2}$$

Consistent with this notation, we define the "Extended Alternating Gauss Quadratic Sum" by

$$G_p^A(j;k) \equiv \sum_{n=1}^{pk}(-1)^n\exp\left(\frac{i\pi j n^2}{k}\right) = G_{p/2}(j;2k;1/2) \tag{5.3}$$

Along with Gauss' classical result, equivalent to Equations 3.20 and 3.21, (one of whose proofs involves contour integrals similar to the method used to obtain our results quoted in Section 2)

$$G_1\left(1;k;0\right) = \frac{1}{2}\sqrt{k}(1+i)\left(1+\exp(-i\pi k/2)\right) \tag{5.4}$$

the entire content of Section 4.1 can be newly summarized as a special case of Equation 5.3—in terms of four roots of unity—by

$$G_p^A\left(1;k\right) = \sum_{n=1}^{pk}(-1)^n \exp\left(\frac{i\pi n^2}{k}\right) = p\sqrt{k}\exp\left(i\pi\left(1/4+\lfloor k/4\rfloor - \bmod(k,4)/4\right)\right) \tag{5.5}$$

where $\lfloor\ldots\rfloor$ is the "floor" function. As an extension to Equation 5.4, with reference to Equations 3.19, 3.33 and 4.12, we also obtain the following Generalized Odd-indexed Quadratic Gauss sum

$$\sum_{n=1}^{pk}\exp\left(\frac{i\pi(2n-1)^2}{4k}\right) = (1+i)p\sqrt{2k}(1-(-1)^k)/4 \tag{5.6}$$

which could be compared to the known result (Armitage & Rogers, 2000, Equation 4.9).

$$\sum_{n=0}^{2k-1}\exp\left(\frac{i\pi(n-m)^2}{2k}\right) = \sqrt{2ki} \tag{5.7}$$

write Equation 5.7 in the form of Equation 5.6 by setting $k \to k/2, m = 1/2$ and noting that the result is formally only valid (if it is technically valid at all) for k even suggests that

$$\sum_{n=0}^{k-1}\exp\left(\frac{i\pi(n-1/2)^2}{k}\right) = \sqrt{ik} \tag{5.8}$$

which surprisingly reduces to Equation 5.6 for *odd* values of k when $p = 1$, after correcting for the difference in the summation limits between the two results. *Although interesting as a special limiting case of Equations 5.6, 5.8 is only correct when k is odd.*

According to our notational definitions, an equivalent form of Equation 5.6 is

$$G_{p/2}\left(1;2k;-\frac{1}{2k}\right) = \sum_{n=1}^{pk}\exp\left(\frac{i\pi(n^2-n)}{k}\right) = \frac{1}{4}(1+i)\exp\left(-i\pi/(4k)\right)p\sqrt{2k}\left(1-(-1)^k\right) \tag{5.9}$$

the asymptotics of which have recently been revisited (Paris, 2014) without reference to the closed form Equation 5.9. An alternate (canonical) form of Equation 5.6 is

$$\begin{aligned}\sum_{n=0}^{p(k-1)}\exp\left(\frac{i\pi(n^2-n)}{k}\right) &= \frac{p}{4}(1+i)\exp(-i\pi/4k)\sqrt{2k}(1-(-1)^k)\\ &\quad + 1-(-1)^{(p^2k+p)}\sum_{n=1}^{p}\exp\left(\frac{i\pi n(n-1)}{k}\right)\end{aligned} \tag{5.10}$$

We emphasize that Equation 5.10 is valid for all positive integers p, k and thereby generalizes the (ought-to-be) known result

$$\sum_{n=0}^{k-1}\exp\left(\frac{i\pi\left(n^2-n\right)}{k}\right) = \frac{1}{2}(1+i)\exp\left(\frac{-i\pi}{4k}\right)\sqrt{2k} \tag{5.11}$$

which is only valid for odd values of k. Since Equation 5.11 appears not to be well known, we note that it can be simply obtained by applying the generalized Landsberg–Schaar identity (Berndt & Evans, 1981, Equation 2.8).

$$\sqrt{j}\sum_{n=0}^{k-1}\exp\left(\frac{i\pi\left(jn^2+nm\right)}{k}\right)=\sqrt{k}\exp\left(\frac{i\pi\left(jk-m^2\right)}{4jk}\right)\sum_{n=0}^{j-1}\exp\left(-\frac{i\pi(n^2k+nm)}{j}\right) \quad (5.12)$$

to the left-hand side of Equation 5.11, with the caveat that Equation 5.12 is only valid when $jk+m$ is even. Utilizing the freedom to choose p leads to further interesting results. For example, let $p\rightarrow pk$ in Equation 5.10 and, with reference to Equation 5.9, obtain

$$\sum_{n=0}^{pk(k-1)}\exp\left(\frac{i\pi(n^2-n)}{k}\right)=1+\frac{1}{4}p(1+i)\exp\left(-\frac{i\pi}{4k}\right)\sqrt{2k}(k-1)(1-(-1)^k) \quad (5.13)$$

Alternatively, let $p\rightarrow p(k-1)$ to obtain

$$\sum_{n=0}^{p(k-1)^2}\exp\left(\frac{i\pi n(n-1)}{k}\right)=1+\sum_{n=1}^{p}\exp\left(\frac{i\pi n(n-1)}{k}\right)+\frac{1}{4}(1+i)(k-2)p\sqrt{2k}\exp\left(-\frac{i\pi}{4k}\right)\left(1-(-1)^k\right) \quad (5.14)$$

Acknowledgements
We thank N.E. Frankel who drew our attention to Hardy and Littlewood.

Funding
This work was partially performed while Prof Glasser was a visitor at the Physics Department, Universidad de Valladolid. Dr Milgram's unfunded contribution to this work is tolerated by Geometrics Unlimited, Ltd. There are no grants involved in either case.

Author details
M.L. Glasser[1,2]
E-mail: laryg@clarkson.edu
Michael Milgram[3]
E-mail: mike@geometrics-unlimited.com
[1] Department of Physics, Clarkson University, Potsdam 13699-5820, New York, USA.
[2] Dpto. de Fisica Teórica, Universidad de Valladolid, Valladolid 47011, Spain.
[3] Consulting Physicist, Geometrics Unlimited Ltd., Box 1484, Deep River, Ontario, Canada, K0J 1P0.

References

Apostol, T. M. (1998). *Introduction to analytic number theory.* New York, NY: Springer.

Armitage, V., & Rogers, A. (2000). Gauss sums and quantum mechanics. *Journal of Physics A: Mathematical and General, 33*, 5993. Retrieved from http://arxiv.org/abs/quant-ph/0003107, doi:10.1088/0305-4470/33/34/305

Berndt, B. C., & Evans, R. J. (1981). The determination of Gauss sums. *Bulletin (New Series) of the American Mathematical Society, 5*, 107–129.

Chapman, R., (2014). *Some classical Gauss sums* (private communication). Requiring several pages, Chapman obtains Equation 3.19 starting from Equation 5.4 using only classical number theoretic formulae.

Gheorghiu, V., & Looi, S.Y. (2010). Construction of equally entangled bases in arbitrary dimensions via quadratic Gauss sums and graph states. *Physical Review A, 81*, 62341–62346. Retrieved from http://link.aps.org/doi/10.1103/PhysRevA.81.062341, http://arxiv.org/abs/1004.1633, doi:10.1103/PhysRevA.81.062341

Glasser, M. L., & Milgram, M. (2014). *Master theorems for a family of integrals*. Retrieved from http://arxiv.org/pdf/1403.2281v2.pdf; Integral transforms and special functions. (2014, June 13), electronic eprint available from doi:10.1080/10652469.2014.924114

Gradshteyn, I. S., & Ryzhik, I. M. (1980). *Table of integrals, series and products.* New York, NY: Academic Press.

Hansen, E. R. (1975). *A table of series and products.* Englewood Cliffs, NJ: Prentice Hall.

Hardy, G. H., & Littlewood, J. E. (1914). Some problems of diophantine approximation. *Acta Mathematica, 37*, 193–238; also available in Collected papers of G. H. Hardy. Vol. 1, Oxford, 1966. Retrieved from https://archive.org/details/CollectedPapersOfG.H.Hardy-Volume1

Paris, R. B. (2014). An asymptotic expansion for the generalized quadratic Gauss sum revisited. Retrived from http://arxiv.org/pdf/1403.7973v1.pdf

Appendix

The following proof of Equation 4.3 is a template for the proof of all results presented in Sections 3.2 and 4. We proceed by induction with the claim

$$\sum_{n=1}^{p(4k-1)} (-1)^n \sin\left(\frac{\pi n^2}{4k-1}\right) = (-1)^k p\sqrt{4k-1} \tag{A.1}$$

The case $p = 1$ is known to be true from (3.15). Set $p = 2$ in (A1) giving

$$\sum_{n=1}^{8k-2} (-1)^n \sin\left(\frac{\pi n^2}{4k-1}\right) = 2(-1)^k \sqrt{4k-1} \tag{A.2}$$

Splitting the sum into two parts gives

$$\sum_{n=1}^{4k-1} (-1)^n \sin\left(\frac{\pi n^2}{4k-1}\right) + \sum_{n=4k}^{8k-2} (-1)^n \sin\left(\frac{\pi n^2}{4k-1}\right) = 2(-1)^k \sqrt{4k-1} \tag{A.3}$$

and shifting the summation indices of the second sum, yields

$$\sum_{n=1}^{4k-1} (-1)^n \sin\left(\frac{\pi n^2}{4k-1}\right) + \sum_{n=1}^{4k-1} (-1)^{(n+4k-1)} \sin\left(\frac{\pi (n+4k-1)^2}{4k-1}\right) = 2(-1)^k \sqrt{4k-1} \tag{A.4}$$

Identify the first sum via (3.15), square the argument of the sine term and remove common factors in the second sum to obtain

$$(-1)^k \sqrt{4k-1} + \sum_{n=1}^{4k-1} (-1)^{(n-1)} \sin\left(\pi\left(\frac{n^2}{4k-1} + 2n + 4k - 1\right)\right) = 2(-1)^k \sqrt{4k-1} \tag{A.5}$$

and after removing integral multiples of π find

$$(-1)^k \sqrt{4k-1} + \sum_{n=1}^{4k-1} (-1)^n \sin\left(\frac{\pi n^2}{4k-1}\right) = 2\,(-1)^k \sqrt{4k-1} \tag{A.6}$$

yielding the requisite identity because of (3.15). Therefore, (A1) is true for $p = 1$ and $p = 2$.

Assuming (A1) for any p, let $p \to p + 1$, giving

$$\sum_{n=1}^{(p+1)(4k-1)} (-1)^n \sin\left(\frac{\pi n^2}{4k-1}\right) = (-1)^k\,(p+1)\,\sqrt{4k-1} \tag{A.7}$$

split the summation into two parts as follows

$$\sum_{n=1}^{p(4k-1)} (-1)^n \sin\left(\frac{\pi n^2}{4k-1}\right) + \sum_{n=p(4k-1)}^{(p+1)(4k-1)} (-1)^n \sin\left(\frac{\pi n^2}{4k-1}\right) = (-1)^k\,(p+1)\,\sqrt{4k-1} \tag{A.8}$$

apply (A1) to the first term, shift the summation index of the second, expand and simplify as in the transition from (A4) to (A5) to obtain

$$(-1)^k p\sqrt{4k-1} - (-1)^p \sum_{n=0}^{4k-1} (-1)^{(n+1)} \sin\left(\pi\left(\frac{n^2}{4k-1} + 2np + p^2\right)\right)$$
$$= (-1)^k (p+1)\sqrt{4k-1} \tag{A.9}$$

Now, remove integral multiples of π from the argument of the sine function, leaving

$$(-1)^k p\sqrt{4k-1} + (-1)^{(p+p^2)} \sum_{n=1}^{4k-1} (-1)^n \sin\left(\frac{\pi n^2}{4k-1}\right) = (-1)^k (p+1)\sqrt{4k-1} \tag{A.10}$$

which reduces to an identity because of (A1) and the fact that

$$(-1)^{p+p^2} = 1 \tag{A.11}$$

for all integers p. Hence by the usual logic of induction (A1) is true. QED. Note: In some proofs, it is expedient to reverse the order of summation.

Permissions

All chapters in this book were first published in CM, by Cogent OA; hereby published with permission under the Creative Commons Attribution License or equivalent. Every chapter published in this book has been scrutinized by our experts. Their significance has been extensively debated. The topics covered herein carry significant findings which will fuel the growth of the discipline. They may even be implemented as practical applications or may be referred to as a beginning point for another development.

The contributors of this book come from diverse backgrounds, making this book a truly international effort. This book will bring forth new frontiers with its revolutionizing research information and detailed analysis of the nascent developments around the world.

We would like to thank all the contributing authors for lending their expertise to make the book truly unique. They have played a crucial role in the development of this book. Without their invaluable contributions this book wouldn't have been possible. They have made vital efforts to compile up to date information on the varied aspects of this subject to make this book a valuable addition to the collection of many professionals and students.

This book was conceptualized with the vision of imparting up-to-date information and advanced data in this field. To ensure the same, a matchless editorial board was set up. Every individual on the board went through rigorous rounds of assessment to prove their worth. After which they invested a large part of their time researching and compiling the most relevant data for our readers.

The editorial board has been involved in producing this book since its inception. They have spent rigorous hours researching and exploring the diverse topics which have resulted in the successful publishing of this book. They have passed on their knowledge of decades through this book. To expedite this challenging task, the publisher supported the team at every step. A small team of assistant editors was also appointed to further simplify the editing procedure and attain best results for the readers.

Apart from the editorial board, the designing team has also invested a significant amount of their time in understanding the subject and creating the most relevant covers. They scrutinized every image to scout for the most suitable representation of the subject and create an appropriate cover for the book.

The publishing team has been an ardent support to the editorial, designing and production team. Their endless efforts to recruit the best for this project, has resulted in the accomplishment of this book. They are a veteran in the field of academics and their pool of knowledge is as vast as their experience in printing. Their expertise and guidance has proved useful at every step. Their uncompromising quality standards have made this book an exceptional effort. Their encouragement from time to time has been an inspiration for everyone.

The publisher and the editorial board hope that this book will prove to be a valuable piece of knowledge for researchers, students, practitioners and scholars across the globe.

List of Contributors

Bai-Ni Guo
School of Mathematics and Informatics, Henan Polytechnic
University, Jiaozuo City, Henan Province, 454010, China.

Feng Qi
College of Mathematics, Inner Mongolia University for Nationalities, Tongliao City, Inner Mongolia Autonomous Region, 028043, China
Department of Mathematics, College of Science, Tianjin Polytechnic University, Tianjin City, 300387, China

Piyush Kumar Bhandari
Department of Mathematics, Shrinathji Institute of Technology & Engineering, Nathdwara, Rajasthan 313301, India

S.K. Bissu
Department of Mathematics, Government College of Ajmer, Ajmer, Rajasthan 305001, India

M.R. Krishnamurthy
Department of Studies and Research in Mathematics, Kuvempu University, Shankaraghatta, Shimoga 577 451, Karnataka, India

B.C. Prasannakumara
Government First Grade College, Koppa, Chikkamagaluru 577126, Karnataka, India

B.J. Gireesha
Department of Studies and Research in Mathematics, Kuvempu University, Shankaraghatta, Shimoga 577 451, Karnataka, India
Department of Mechanical Engineering, Cleveland State University, Cleveland, OH 44115, USA

Rama S.R. Gorla
Department of Mechanical Engineering, Cleveland State University, Cleveland, OH 44115, USA

A. Guezane-Lakoud
Faculty of Sciences, Department of Mathematics, Laboratory of Advanced Materials, Badji Mokhtar-Annaba University, P.O. Box 12, 23000 Annaba, Algeria

William A. Sethares and James A. Bucklew
Department of Electrical and Computer Engineering, University of Wisconsin, Madison, WI, USA

Arsalan Rahmani
Department of Mathematics, University of Kurdistan, Pasdaran Boulevard, P. O. Box 416, Sanandaj, Iran

Kishor R. Gaikwad
Department of Mathematics Nanded Education Society's, Science College, Nanded, 431605 Maharashtra, India

Muhammad Aslam Noor, Khalida Inayat Noor and Muhammad Uzair Awan
Department of Mathematics, COMSATS Institute of Information Technology, Islamabad, Pakistan.

Feng Qi
Department of Mathematics, College of Science, Tianjin Polytechnic University, Tianjin City 300387, China
Institute of Mathematics, Henan Polytechnic University, Jiaozuo City, Henan Province 454010, China

Praveen Ailawalia
Department of Applied Sciences and Humanities, M.M. University, Sadopur, Ambala City, Haryana, India.

Sunil Kumar Sachdeva
Department of Applied Sciences, D.A.V. Institute of Engineering and Technology, Kabir Nagar, Jalandhar, Punjab, India Punjab Technical University, Jalandhar, Punjab, India

Devinder Pathania
Department of Applied Sciences, Guru Nanak Engineering College, Ludhiana, Punjab, India

Rajneesh Kumar
Department of Mathematics, Kurukshetra University, Kurukshetra, Haryana, India

Nidhi Sharma
Department of Mathematics, MM University, Mullana, Ambala, Haryana, India

Parveen Lata
Department of Basic and Applied Sciences, Punjabi University, Patiala, Punjab, India

Divya Ahluwalia
Department of Mathematics, University of Petroleum and Energy Studies, Dehradun, India

N. Sukavanam and Urvashi Arora
Department of Mathematics, Indian Institute of Technology Roorkee, Roorkee 247667, India

Ludwig Kohaupt
Department of Mathematics, Beuth University of Technology Berlin, Luxemburger Str. 10, D-13353 Berlin, Germany

A.V. Zhirkin
Department of Fusion Reactors, National Research Centre "Kurchatov Institute", Kurchatov Centre of Nuclear Technologies, Academician Kurchatov Square, 1, Moscow 123182, Russia

N. Mohanapriya
RVS Faculty of Engineering, Department of Mathematics, RVS Technical Campus - Coimbatore, Coimbatore 641 402, Tamil Nadu, India
Bharathiar University, Coimbatore 641 046, Tamil Nadu, India

J. Vernold Vivin
Department of Mathematics, University College of Engineering Nagercoil (Anna University Constituent College), Konam, Nagercoil 629 004, Tamil Nadu, India

M. Venkatachalam
RVS Faculty of Engineering, Department of Mathematics, RVS Technical Campus - Coimbatore, Coimbatore 641 402, Tamil Nadu, India

M.L. Glasser
Department of Physics, Clarkson University, Potsdam 13699-5820, New York, USA
Dpto. de Física Teórica, Universidad de Valladolid, Valladolid 47011, Spain

Michael Milgram
Consulting Physicist, Geometrics Unlimited Ltd., Box 1484, Deep River, Ontario, Canada, K0J 1P0

Index

www.ingramcontent.com/pod-product-compliance
Lightning Source LLC
Chambersburg PA
CBHW082013190326
41458CB00010B/3169